Dear Professor Ca

With best wishes,

Yasuhiro Iwasawa

TAILORED METAL CATALYSTS

CATALYSIS BY METAL COMPLEXES

TAILORED METAL CATALYSTS

Edited by

YASUHIRO IWASAWA
Department of Chemistry, Faculty of Science,
The University of Tokyo, Japan

D. REIDEL PUBLISHING COMPANY

A MEMBER OF THE KLUWER ACADEMIC PUBLISHERS GROUP

DORDRECHT / BOSTON / LANCASTER / TOKYO

Library of Congress Cataloging-in-Publication Data

CIP

Main entry under title:

Tailored metal catalysts.

(Catalysis by metal complexes)
Includes bibliographies and index.
1. Metal catalysts. I. Iwasawa, Yasuhiro, 1946 – II. Series.
QD505.T33 1986 541.3′95 85-28324
ISBN 90-277-1866-0

Published by D. Reidel Publishing Company,
P.O. Box 17, 3300 AA Dordrecht, Holland.

Sold and distributed in the U.S.A. and Canada
by Kluwer Academic Publishers
190 Old Derby Street, Hingham, MA 02043, U.S.A.

In all other countries, sold and distributed
by Kluwer Academic Publishers Group,
P.O. Box 322, 3300 AH Dordrecht, Holland.

Printed in The Netherlands

TABLE OF CONTENTS

PREFACE

Well tailored metal catalysts are catalysts of the new generation resulting from scientific development at the boundary between homogeneous and heterogeneous chemistry. The main factors involved in making tailored metal catalysts are not those of traditional impregnation in which the chemistry is in general unknown and ill-defined, or of simple ion exchange which involves long-range forces with little control on the local structure through definite and special bond direction.

Tailored Metal Catalysts thus has a rather different emphasis from normal review publications in the field of catalysis. Here we concentrate more on the distinct surface chemistry and catalytic properties of important established materials with well-characterized active structures or precursors, although at the same time providing a systematic presentation of relevant data.

Many pioneering works have been undertaken in the field of tailored metal catalysts since the early research on polymer-attached homogeneous metal complexes by the British Petroleum Company Ltd. and the Mobil Oil Corporation around 1969; transition metal complexes attached on polymers by Grubbs (1971), Heinemann (1971), Manassen (1971), Pittman (1971), Bursian *et al.* (1972), Kagan (1973), Bailar (1974); transition metal complexes attached on inorganic oxides by Allum *et al.* (1972), Ballard (1973), Candlin and Thomas (1974), Murrell (1974), Yermakov (1974); metal carbonyls/polymers by Moffat (1970); metal carbonyls/inorganic oxides by Parkyns (1965), Davie *et al.* (1969), Banks *et al.* (1969), Howe (1973), Burwell (1975); metal carbonyl clusters/polymers by Collman (1972); metal carbonyl clusters/inorganic oxides by Robertson and Webb (1974), Anderson (1974), Smith *et al.* (1975).

In the middle and late 1970s the concept of attached catalysts was developed and various types of examples were described in the literature. An important subject in the 1980's and the future may be the tailoring of novel surface compounds which have no direct homogeneous analogs. Surface tailoring has strategic advantages in the synthesis of catalysts with surface structures and compositions which are unusual in homogeneous systems and acutely difficult to make in traditional heterogeneous systems. Tailored metal catalysts may contribute to a complete understanding of the origin of heterogeneous catalysis and to elegant improvements in solid catalysts. I believe that *Tailored Metal Catalysts* can provide a valuable contribution toward present and future fundamental and practical catalyst research.

We appreciate the assistance of Dr. Howe, one of the authors, for the correction of the English in this book. We also grateful to the various authors and editors who kindly gave us permission to reproduce diagrams from their papers.

Yasuhiro Iwasawa

The University of Tokyo,
Tokyo, Japan

April 1984

Y. IWASAWA

INORGANIC OXIDE-ATTACHED METAL CATALYSTS

1. Introduction

The heterogenization of homogeneous catalysts has been a very active and attractive area of research. From an industrial viewpoint, this could provide an ideal method for combining the advantages, and simultaneously avoiding the disadvantages, of homogeneous and heterogeneous catalysts. Furthermore, not only does this type of research provide potential preparations of new types of catalyst systems with specific properties quite different from homogeneous catalysts, but it also allows the development of chemistry in the boundary between homogeneous and heterogeneous systems. The use of such chemically and structurally controllable surface systems rather than the more conventional supported catalysts with heterogeneous metal centers, can lead to novel information on the origin of catalysis as well as providing methods for the future design of intriguing catalysts when studied in combination with modern physical techniques for surface characterization.

In this chapter the properties and catalytic characteristics of inorganic oxide-attached metal catalysts are described. The term 'attached' refers to the situation where structurally well-defined molecules are connected by specific covalent bonds or stable definite bonds to supporting surfaces. This description can involve both the anchoring of a metal complex via a methylene chain to a surface and the direct immobilizing of a metal complex without the involvement of any spacer groups.

The formation of salts by ion-exchange absorption on Brönsted acid sites, the hydrolytic adsorption of multivalent metal ions or poorly defined physically supported species will not be discussed in this chapter. Rather, it is concerned with well-characterized transition metal complexes attached by chemical bonding to inorganic oxide surfaces such as silica, alumina, etc. Particular emphasis will be given to the tailoring of the stereochemical and electronic environments of metal centers on surfaces through such attachments.

Y. Iwasawa (ed.), Tailored Metal Catalysts, 1–85.

2. General Features of Homogeneous and Heterogeneous Catalysts [1–4b]

Homogeneous catalysts possess several advantages:

1. Homogeneity— a uniform active structure.
2. Efficiency— theoretically all the atoms or molecules of the catalyst are available for reactants.
3. Reproducibility— as a result of well-controlled active sites.
4. High selectivity— often more specific than a heterogeneous catalyst, even for asymmetric induction.
5. Mild reaction conditions— high activity under mild operating conditions.
6. Controllability— ready chemical modification allowing the control of the electronic and steric properties of a metal center.
7. Readily studied reaction mechanisms— ease of investigation of mechanism at the molecular level.

The disadvantages of homogeneous catalysts are:

1. Separation— a major disadvantage is the problem of separating an expensive catalyst from the reaction products with the separation requiring special treatment which usually destroys the catalyst.
2. Solubility— the range of solvents suitable for a homogeneous catalyst is often limited by their solubility; a compatible solvent is not necessarily the most suitable for a high reaction rate.

The advantages of heterogeneous catalysts are considered to be as follows:

1. Separation and recovery— ready removal of catalysts together with their recovery and reactivation for re-use.
2. Durability— catalysts are capable of use in fixed and fluidized beds.
3. High activity— over a wide range of reactions, e.g., hydrogenolysis, olefin rearrangement, reduction of aromatics and esters, etc., homogeneous catalysts are much less active or inactive.
4. Thermal stability— for a wide variety of reaction conditions.
5. Ease of handling— relatively easy handling in large-scale operation.

Heterogeneous catalysts have the following disadvantages:

1. Heterogeneity— catalysts usually contain more than one active site.
2. Ill-defined surfaces— leading to difficulty in design and improvement as a result of ill-defined active sites.
3. Less selectivity— mainly as a result of the heterogeneous nature of the reaction sites and also frequently from severe reaction conditions.
4. Low efficiency— only metal atoms at surface sites may be accessible and effective in catalysis.

3. General Aspects of the Development of Attached Metal Catalysts

3.1. FIRST GENERATION ATTACHED CATALYSTS

The first generation of so-called hybrid catalysts resulted from the corresponding growth in interest in the 1970s which stimulated studies of the attachment of homogeneous catalysts onto inorganic oxide supports through reaction sequences which preserved the structure of the homogeneous metal complex and presumably its mode of reaction. During this period not only was catalyst recovery studied but the application of the resulting attached catalysts under new boundary conditions was developed.

Inorganic oxide-attached catalysts may be stabilized at elevated temperatures and under severe reaction conditions where their homogeneous counterparts are unstable. Thus the diphosphine Rh complex **1**, obtained by attaching *trans*-$[Cl_3Si(CH_2)_2PPh_2]_2RhCl(CO)$ onto a silica gel, was stable under quite severe reaction conditions and also after exposure to corrosive solvents [5].

$$[(\nmid Si-O)_2SiCl(CH_2)_2PPh_2]_2RhCl(CO)$$
(1)

A chiral complex, $[Rh(COD)PNNP(+\text{ or }-)]^+ ClO_4^-$ (COD = 1,5-cyclooctadiene; PNNP(+) = *N,N'*-bis((*R*)(+)- or (*S*)(−)-methylbenzyl)-*N,N'*-bis(diphenylphosphine)ethylenediamine), supported on smectite clays such as hectorite, bentonite, nontronite and halloysite catalyzed the asymmetric hydrogenation of α-acetamidoacrylic acid with good efficiency [6]. Activation was performed before hand using hydrogen at 20 atm pressure and led to a stable catalyst. In contrast, prehydrogenation of the homogeneous rhodium complex leads to its decomposition even at atmospheric pressure with the formation of metallic rhodium.

From an engineering standpoint, another outstanding advantage of attached catalysts relative to their homogeneous counterparts is the ability to support virtually any desired catalyst concentration, a situation which may be impossible under homogeneous conditions because of solubility limitations. Smaller reactors can be used for higher catalyst concentrations, thus reducing the cost of a chemical plant. Whereas the range of suitable solvents for a homogeneous catalyst is often limited by the solubility characteristics of the catalyst, this clearly presents no problem for an attached catalyst where the active elements are all available for interaction with reactant molecules. Attached heterogeneous catalysts may also be used for gas-phase reactions.

In large-scale production processes, difficulties may sometimes arise from strong corrosive attack on reactors. Such problems may be overcome through the use of attached catalysts. Since inorganic oxides are insoluble and nonvolatile, reactions involving toxic compounds attached to oxide supports provide a means of making systems of this kind more acceptable environmentally.

In principle, attached catalysts could be constructed to retain the advantages of both homogeneous and heterogeneous catalysts. In general, however, the first generation of supported metal complex catalysts have proved to be less active than their homogeneous counterparts in solution. This result is not unexpected because the mobility of a complex bonded to a surface is lower and steric factors are apt to be more important in influencing its reactivity.

3.2. SECOND GENERATION ATTACHED CATALYSTS

Homogeneous catalysts are usually coordinatively unsaturated; they thus have a tendency to aggregate, blocking the coordination sites necessary for catalysis. The catalytic activity of a homogeneous metal complex which is only active in its monomeric form may be enhanced by attaching it to a rigid inorganic oxide. Thus, the preparation of titanocene monomers with sustained catalytic activity has been achieved by site isolation in a crosslinked polystyrene matrix [7]. More recently, titanocene monomers have been site-isolated on silica using the following reaction [8];

$$\text{⫽Si—OH} + [(OEt)SiMe_2CpTiCpCl_2] \longrightarrow \text{⫽Si—OSiMe}_2\text{CpTiCpCl}_2 \tag{2}$$

Following reduction with BuLi or isopropyl magnesium bromide, the silica-attached complex exhibited sustained activity as an olefin hydrogenation catalyst. The reason for the enhancement of the catalytic activity appears to be an increase in the concentration of active coordinatively unsaturated species, rather than an increase in the activity of each active site (turnover frequency). However, the nature of the active species has not been established unequivocally since $CpTiCl_3$ is itself capable of binding irreversibly to silica and this reduced form has also been observed to be active [9]. The contribution arising from the formation of new ill-defined surface species may also be important.

An increase in activity through attachment of a transition metal complex has also been observed in butadiene polymerization and propene oligomerization with the catalyst obtained through the interaction of $Ni(\eta^3\text{-}C_3H_5)_2$ with silica or alumina [10, 11];

$$\text{⫽M—OH} + Ni(\eta^3\text{-}C_3H_5)_2 \longrightarrow \text{⫽M—ONi}(\eta^3\text{-}C_3H_5) + C_3H_6 \tag{3}$$

where ⫽M = Si atom on SiO_2 or Al atom on Al_2O_3. For compounds of the type $[Ni(C_3H_5)X]_2$, the reactivity of the active sites (propagation centers) in butadiene polymerization varies as the inorganic ligand is changed in the series

Cl < Br < I << [}Si—O] [10]. Such an increase is accompanied by a change in the stereospecificity of the active sites.

Major progress in the application of second generation attached metal catalysts has been achieved in ethene polymerization, where attached benzyl, allyl, cyclopentadienyl and hydroborate complexes of various transition metals have been studied [12–14]:

Ti	:	$Ti(CH_2C_6H_5)_4$
V	:	$V(\eta^3\text{-}C_3H_5)_3$
Cr	:	$Cr(\eta^3\text{-}C_3H_5)_3$, $Cr(\eta^5\text{-}C_5H_5)_2$
Zr	:	$Zr(\eta^3\text{-}C_3H_5)_4$, $Zr(CH_2C_6H_5)_4$, $Zr(BH_4)_4$
Nb	:	$Nb(\eta^3\text{-}C_3H_5)_3$
Hf	:	$Hf(\eta^3\text{-}C_3H_5)_3$, $Hf(CH_2C_6H_5)_4$

As soluble homogeneous catalysts these metal complexes are practically inactive for ethene polymerization. Industrial production of high-density polyethylene may also be achieved through the use of catalysts prepared by attaching $Cr(C_5H_5)_2$ onto SiO_2 [14, 15], polymerization taking place in the gas phase at 368–378 K and 25 atm pressure to give a polymer yield as great as ca. 4.5×10^5 mol per g atom Cr.

Species which are not stable in homogeneous systems may be formed on a surface through site isolation. Thus $Co_2(CO)_8$ and $Fe_2(CO)_9$ can be attached to a modified silica surface to form mononuclear structures [16, 17] as follows [18]:

$$\}Si\text{—}OSiMe_2Cp + Co_2(CO)_8$$
$$\longrightarrow \underset{\substack{(\mathbf{4}) \\ (\nu_{CO}\ 2020,\ 1965\ cm^{-1})}}{\}Si\text{—}OSiMe_2CpCo(CO)_2} \overset{h\nu}{\nrightarrow} \text{dinuclear species}$$

$$\}Si\text{—}OSiMe_2Cp + Fe_2(CO)_9$$
$$\longrightarrow \underset{\substack{(\mathbf{5}) \\ (\nu_{CO}\ 2030,\ 1960\ cm^{-1})}}{\}Si\text{—}OSiMe_2CpFeH(CO)_2} \overset{\Delta}{\nrightarrow} \text{dinuclear species}$$

In contrast to their unsupported analogs, no dinuclear or polynuclear species are formed from the SiO_2-attached species **4** and **5** even at temperatures up to 473 K. The cobalt species thus obtained catalyzed the hydroformylation of olefins in toluene solution between 373 and 423 K.

Dimerization may take place at the surface, however, depending upon the length of the spacer groups separating the metal centers from the surface. The attached Rh catalysts **6** used in the liquid-phase hydrogenation of alkenes at 1.1 atm pressure H_2 and 310–340 K are deactivated at a chain length $n > 4$

[19]. Active site isolation and coordination unsaturation become difficult when longer spacer groups are used.

$$\wr Si-OSi(CH_2)_nPPh_2 + Rh_2Cl_2(C_2H_4)_4$$

$$\longrightarrow [\wr Si-OSi(CH_2)_nPPh_2]_xRhCl(C_2H_4)_y \quad (6)$$

$$(n = 1-6)$$

Supported Ni–allyl oligomerization catalysts have been prepared either by simultaneously reacting $Ni(CO)_4$ and C_3H_5Cl with γ-Al_2O_3 [20] or by treating previously absorbed carbonyl with allene [21]. Such catalysts possessed a monomeric structure in contrast to the dimer formed in solution.

3.3. THIRD GENERATION ATTACHED CATALYSTS

Materials of this kind may not possess direct analogs among homogeneous catalysts. They may also be new surface compounds which are impossible to make by traditional methods of catalyst preparation. Third generation catalysts can be prepared by tailoring a first-stage attached structure through the use of well-defined chemical treatments; in this way it is possible to develop novel types of chemistry on surfaces which are difficult to achieve in solution. Third generation attached catalysts may thus provide new catalytic and physicochemical properties as well as being useful for the study of theoretical problems in heterogeneous catalysis such as the determination of active structures, compositions, electronic configurations, the oxidation states of metals, reaction mechanisms, etc.

The reaction of a dinuclear complex of chromium, $Cr_2(\eta^3\text{-}C_3H_5)_4$, which possesses quadruple $Cr \equiv Cr$ bond (0.1975 nm in length), with the surface hydroxyl groups of γ-Al_2O_3 (pore diameter $\geqslant$ 3.5 nm) results in the formation of a complete monolayer of the Cr^{II} ion (**7**). This species was active in the hydrogenation of propene at 240 K. Structure **7** may be transformed quantitatively into a complete monolayer of the dinuclear Cr^{III} species **8** and then to the monochromate(VI) structure **9** as the diffuse reflectance UV–vis spectra illustrated in Figure 1 show. All these structures are capable of mutual transformation [22]. The generation of Cr^{II} ions on γ-Al_2O_3 is difficult by traditional methods of preparation while the production of a two-dimensional phase of chromium (2+, 3+, 6+) on an inorganic oxide surface with a uniform distribution of chromium ions is impossible by the usual methods of support.

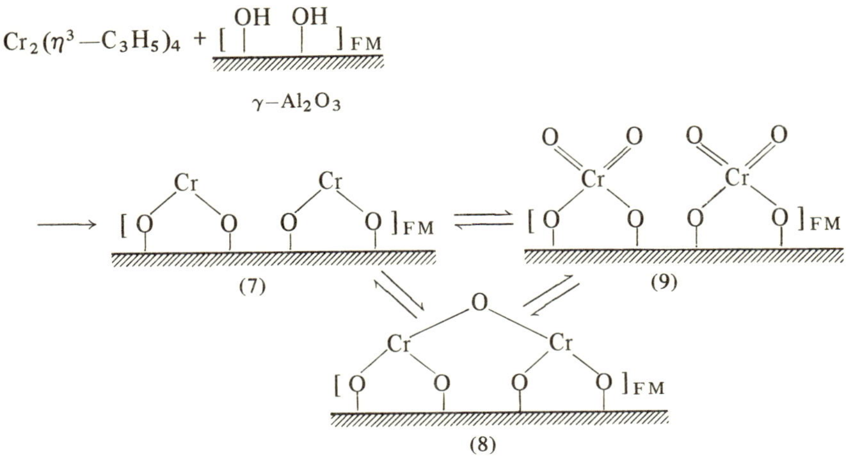

(FM = full monolayer)

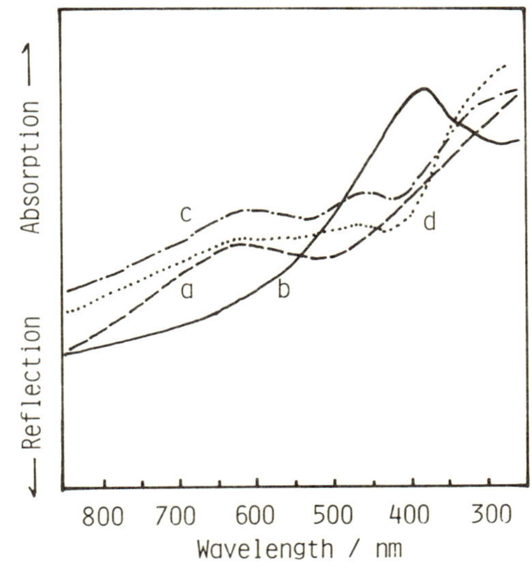

Fig. 1. UV–vis diffuse reflectance spectra of γ-Al_2O_3-attached chromium monolayer (a, b, c) and dimer (d). Curve a: species (7), evacuated at 343 K; curve b: species (9), oxidized with O_2 for 1 h at 773 K; curve c: species (8), reduced with H_2 for 1 h at 573 K or oxidized with O_2 for 30 min at 263 K; and curve d: $[(\mathbin{\rlap{/}{\,}}Al-O)_2Cr-allyl]_2$.

The two-dimensional Cr^{III} layer attached to SiO_2 by Iwasawa, the Mo monomers attached to SiO_2 by Yermakov and Iwasawa, and to Al_2O_3 by Iwasawa, and the Mo dimers attached to SiO_2 and Al_2O_3 by Iwasawa may also be included among third generation catalysts. These are described in greater detail later.

Another interesting example is the reaction of $\}Si-OSi(CH_2)_2PPh_2$ with $IrCl(CO)_2$ (*p*-toluidine) in the presence of zinc and CO (at 370 kPa pressure) which resulted in the formation of a surface-bond Ir_4 cluster (**10**) [23, 24]. Although such bond clusters may not themselves be classified as third generation catalysts, this experimental result does suggest the possible synthesis of unusual surface-bound species.

$$\}Si-OSi(CH_2)_2PPh_2[Ir_4(CO)_{11}]$$

(10)

4. Physical Techniques for the Characterization of Surface-attached Species

Progress in tailoring attached metal catalysts has been hampered by a lack of detailed structural information regarding surface species. The characterization of surface compounds has often been poor and plagued by many of the problems which arise in attempting to characterize traditional heterogeneous catalysts. This situation is particularly acute when the overall loadings involved are small in comparison with the large amounts of support materials employed. Again, with insoluble inorganic oxide supports many techniques can no longer be applied. It may be said that well-tailored metal catalysts are only possible when confirmatory evidence regarding molecular structures, compositions, local environments, and the electronic character and populations of the attached metal species is available.

Table I lists the various physical techniques available for surface characterization of attached metal catalysts and suggests the necessity for combining some of these techniques if a satisfactory analysis of a surface is to be achieved.

The following techniques may also provide valuable information on these problems (see also [24b, c]).

Infrared (IR) spectroscopy [25–31]. This is a vital probe for following the course of attachment of transition metal complexes to surfaces as well as the transformation of the surface species so obtained. However, most inorganic oxides exhibit strong bulk absorption within the low wavenumber region (roughly < 1000 cm^{-1}) which masks much of the useful absorption range. Thus, when account is taken of the particle size condition ($\lambda >> d$) which determines the wavelength range over which suitable low scattering losses occur, the accessible wavenumber range for transmission IR spectroscopy is generally limited to the mid-infrared region (4000–1000 cm^{-1}) for surface studies.

Physical techniques for the surface characterization of supported metal/metal oxide catalysts

Information requied	Method available[a]	Nature of incident beam	Type of particle detected	Name of technique
	AES	electron	electron	
chemical analysis	XPS	photon (X-ray)	electron	
	UPS	photon (UV)	electron	
	SIMS	ion	ion	electron spectroscopy
composition	ISS	ion	ion	
	EPMA	electron	photon (X-ray)	
structure/ arrangement	XRF	photon (X-ray)	photon (X-ray)	ion spectroscopy
	EXAFS	photon (X-ray)	photon (X-ray)	
bond distance/ number	XANES	photon (X-ray)	photon (X-ray)	
	EELS	electron	electron	photon spectroscopy
chemical state/ bond character	IETS	electric field	electron	
	IR/FT-IR	photon (IR)	photon (IR)	
electron density of states	Raman	photon (vis)	photon (Raman)	other spectroscopy
	DRS	photon (UV–vis)	photon (UV–vis)	
oxidation states	PAS	photon (UV–vis)	phonon	
	Photoluminescence	photon (UV–vis)	photon (UV–vis)	mass spectroscopy
distribution	NMR (MASNMR)	photon (meterwave)	photon (meterwave)	
topography/shape	Mössbauer	photon (γ-ray)	photon (γ-ray/electron)	
	ESR	photon (microwave)	photon (microwave)	others
	TEM/SEM/STEM	electron	electron	
	TDS/TPDE/TPR	thermal	chemical species	

[a] AES: Auger Electron Spectroscopy, XPS: X-ray Photoelectron Spectroscopy, UPS: Ultraviolet Photoelectron Spectroscopy, SIMS: Secondary Ion Mass Spectroscopy, ISS: Ion Scattering Spectroscopy, EPMA: Electron Probe Micro-Analysis, XRF: X-ray Fluorescence, EXAFS: Extended X-ray Absorption Fine Structure, XANES: X-ray Absorption Near-Edge Structure, EELS: Electron Energy Loss Spectrscopy, IETS: Inelastic Electron Tunneling Spectroscopy, IR: Infrared spectroscopy, Raman: Raman spectroscopy, DRS: VU–vis Diffuse Reflectance Spectroscopy, PAS: Photo-Acoustic Spectroscopy, Photoluminescence: Photoluminescence spectroscopy, NMR: Nuclear Magnetic Resonance Spectroscopy, MASNMR: Magic Angle Spinning NMR, TEM: Transmission Electron Microscopy, SEM: Scanning Electron Microscopy, STEM: Scanning Transmission Electron Microscopy, TDS: Thermal Desorption Spectroscopy, TPDE: Temperature Programmed DEcomposition, TPR: Temperature Programmed Reduction.

Assuming typical values of 100 m^2g^{-1} and 20 mg cm^{-2} for the specific surface area and weight of the geometric area irradiated respectively, the lower limit for the observation of, for example, a C—H stretching vibration will be a few per cent of a monolayer coverage if background interference is to be avoided.

A very useful modern method in IR spectroscopy is Fourier transform infrared spectroscopy (FT–IR). This technique enables measurements to be made an samples exhibiting high background interference with the result that the sensitivity can be increased by more than an order of magnitude. Typical designs of cells for *in situ* work at low and high temperatures have been discussed in the literature.

Raman spectroscopy [32–36]. Laser Raman spectroscopy can be used widely for the investigation of surface species on oxides or supported metals. In some cases resonance Raman spectroscopy may also prove to be helpful. Raman shifts of surface species can be recorded in the low wavenumber region down to ca. 50 cm^{-1}. However, in some cases, sample sensitivity to the heating effects of the laser beam and background fluorescence from the inorganic oxide may be observed. The typical detection limits are of the order of 5–10% of a monolayer coverage for samples having specific surface areas of 100–200 m^2g^{-1}. Information to a depth of ca. 2 nm can typically be obtained by Raman spectroscopy for surface samples. The transformation of active structures or phases during the tailoring of catalysts and/or during their use in catalysis may be explored by means of the *in situ* Raman technique using a heated cell.

Electron Energy Loss Spectroscopy (EELS) [37–39]. This method provides very high sensitivity (depending on the surface dipole moment, as few as 10^{10} molecules may be detected within an area of only 1 mm^2). Spectral resolution, however, is restricted to the order of ca. 50 cm^{-1}, which is poor in comparison to IR and Raman spectroscopy which are readily capable of achieving resolutions of 2 cm^{-1} or better. Although EELS has the potential advantage of being able to open up the entire vibration region from below 100 cm up to 4000 cm^{-1}, its application is restricted to the study of geometrically flat surfaces since the method employs the specular reflection of electrons. Attempts have recently been reported of the application of this method to the study of evaporated Rh on an Al_2O_3 model surface.

Inelastic Electron Tunneling Spectroscopy (IETS) [40–42]. It is possible to obtain vibrational information through the use of IETS. The most attractive feature of this technique is that the 200–4000 cm^{-1} region of the spectrum is accessible, the sensitivity of the method being comparable to that of EELS while the resolution is much better (by several cm^{-1}) than EELS. However, to obtain high resolution measurements must be made at temperatures using equal to or less than 4.2 K, while special sample preparation is necessary for the metal–insulator–metal junction. Both IR and Raman active modes are observed in IET spectra.

Ultraviolet–visible (UV–vis) Diffuse Reflectance Spectroscopy (DRS) [43–46]. UV–vis spectra of surface metal complexes are usually obtained by means of the diffuse reflectance method. Electronic spectra of surface species, particularly ligand field spectra of transition metal complexes attached to inorganic oxides, provide information on the local environment (symmetry) and on the oxidation state of the particular transition metal ion. Information regarding the character of metal–metal bonding may also be obtained by this technique.

Photoluminescence spectroscopy [47–51]. This technique is seldom very useful despite its high sensitivity. Luminescence spectra may be obtained by illuminating a characteristic band of the surface species, and could well be applicable to the study of the electronic properties of surface OH groups and their coordination. Another application is in the study of the environments of supported metal centers. In addition to data on electronic states, vibrational information might be obtainable from low-temperature measurements.

Electron Spin Resonance (ESR) spectroscopy [52–55]. Considerable success in studies of the nature of paramagnetic transition metal ions, their surface bonding characteristics and symmetry has been achieved through the use of ESR spectroscopy. ESR spectra are highly sensitive to environmental changes in the vicinity of the paramagnetic center, this high sensitivity permitting the detection of very low concentrations of surface species.

Nuclear Magnetic Resonance (NMR) spectroscopy [56–59]. High-resolution NMR studies of solids have become possible through the use of multiple pulse techniques. In this way the line width of the absorption bands in solid samples can be reduced by several orders of magnitude, so that the chemical shifts of nonequivalent nuclei can be resolved. Another technique allowing line narrowing in solid samples is known as magic angle spinning. Such spinning of a sample effectively averages out the dipolar interactions and/or the anisotropy in the chemical shift tensor. Proton decoupled ^{13}C NMR spectroscopy has been applied to the study of functionalized inorganic oxide systems, while Fourier transform ^{31}P NMR spectroscopy may be used to establish the nature of phosphorus ligands attached to solid surfaces if particles of the solid are dispersed in a suitable organic solvent.

Mössbauer spectroscopy [60–64]. This technique, which involves the resonant emission–absorption of energetic γ-photons, has found considerable application both in catalytic systems and transition metal chemistry involving ^{57}Fe, ^{119}Sn, etc. Parameters such as isomer shift, quadrupole splitting and magnetic hyperfine splitting allow a detailed analysis of the chemical state of certain atoms within a material. Conversion electron Mössbauer spectroscopy recently developed has led to an enhancement of the surface sensitivity as well as enabling spectra to be obtained from various depths within the solid. An *in situ* cell for Mössbauer measurements of supported catalysts has been designed by several workers.

X-Ray Photoelectron Spectroscopy (*XPS*) [65–67]. Typically ca. 1% of a monolayer is detectable by this method if the element of interest is distributed on the surface. X-Ray photoelectron spectroscopy enables quantitative elemental analysis of surface groups, determination of the valency states of metal ions and characterization of their chemical environment. However, such quantitative analysis may be difficult for rough surfaces, porous materials or cavities (e.g. zeolites) where surface groups may not be 'seen' by the XPS technique. The charging of insulating materials can pose severe problems for the correct determination of binding energies; nevertheless, this surface-sensitive (< 2 nm) technique is a convenient and powerful method for catalyst characterization.

Extended X-Ray Absorption Fine Structure (*EXAFS*) *spectroscopy* [68–72]. EXAFS is a relatively simple and direct method for obtaining information on the local structures around a selected atom in a complex or in all phases of matter. By studying oscillatory variations in the X-ray absorption coefficient over a wide energy range beyond ca. 50 eV from the threshold to over 1000 eV, this technique provides three types of information concerning the elements present in the catalyst; bond distance, coordination number and Debye–Waller factor. Interatomic distances can be obtained accurately (to (±) 2 pm) from even quite sparse EXAFS information. Coordination numbers are, at present, somewhat more difficult to obtain with any certainty (here the accuracy is currently only (±) 30%) because of a lack of understanding of the phenomenon of 'shake up–shake off'. Debye–Waller or disorder factors obtainable by this method show bond length variations through the bulk of a material, and can give an indication of the nature of the surface. Two modifications extend the applicability of the EXAFS technique. These are surface (sensitive) EXAFS, which detects X-ray fluorescence arising from core-hole filling or measures the Auger electrons generated by core-hole filling, and the study of X-ray absorption near-edge structure (XANES) which provides structural information about the central metal atom and its oxidation state(s).

Secondary Ion Mass Spectroscopy (*SIMS*) *and Ion Scattering Spectroscopy* (*ISS*) [73–75b]. These techniques give valuable information on the composition, distribution and mutual location of heteroatoms in surface metal complexes and multimetal clusters. However, the application of SIMS and ISS to the characterization of such modified surfaces is not without complication since some uncertainty can arise regarding ion beam damage in the region of the bonded functional group.

Thermal Desorption Spectroscopy (*TDS*). Originally called temperature programmed desorption (TPD) [76], this technique possibly provides the most informative thermoanalytical method for the characterization of surface functional groups and chemisorbed species [77]. Modifications of TPD are temperature programmed reduction (TPR) [78] and temperature programmed decomposition (TPDE) [79]. The latter yields information regarding the composition, homogeneity and bonding character as well as on the reactivity of surface metal complexes.

Electron microscopy [80–85]. Electron microscopes come in two categories: scanning electron microscopes (SEM) which employ a ca. 10 nm electron beam at the specimen surface and transmission electron microscopes (TEM). Present TEMs can yield a resolution of 0.2–0.3 nm, providing extremely thin ($\leqslant$ 20 nm) specimens are available. With the normal kinds of inorganic oxide-supported metal catalysts, particles of 1.0–1.5 nm in diameter can be detected. The scanning transmission miscroscope (STEM) uses a high brightness dark-field emission gun to produce a probe ca. 0.3 nm in diameter and combines the techniques of SEM and TEM. The local composition and the distribution of chemical elements can be determined by electron probe microanalysis (EPMA), where the sweep of the electron beam is allied with an X-ray detector. When the overall loadings are very small, an averaged elemental analysis for supported metal catalysts may be performed by X-ray fluorescence spectroscopy instead of conventional elemental analysis.

5. Polymers and Inorganic Oxides as Supports [3, 86–88]

When used as supports for the attachment of transition metal complexes, organic polymers and inorganic oxides exhibit contrasting advantages and disadvantages. Typical examples of such inorganic and organic support materials are listed in Table II.

TABLE II
Support materials for attached metal catalysts

Inorganic	Organic
silica	polystyrene
glass	polyamines
alumina	polymethacrylate
zeolite	polybutadiene
clay	polyvinyls
other metal oxides	silicon resins
e.g. titania	cellulose
carbon	crosslinked dextrans
	polyamino acids

One of the most important differences between the chemistry of a catalyst attached to an inorganic support compared with that of the analogous catalyst affixed to a polymer is in its *temperature stability*. Polymeric supports are often subject to considerable degradation both in the presence and absence of oxygen at ca. 420 K, while the application of an inorganic-supported catalyst is generally limited by the thermal stability of the metal complex rather than that of the inorganic oxide.

Another striking difference arises from the *rigid* substrate structure of inorganic oxides in comparison to polymers. Control of the degree of coordination unsaturation of a metal center should be easier on the surface of an inorganic oxide than in a flexible polymer matrix. Polymer-supported catalysts are deactivated either through intermolecular aggregation or condensation, or by chelation of the metal center. Even with a highly crosslinked polymer a similar situation can arise because of the short-range flexibility of the polymer. Rigid inorganic matrices circumvent these two deactivation processes. Thus oxygen can adsorb reversibly below room temperature on an iron(II) porphyrin **11** attached to a silica gel [89].

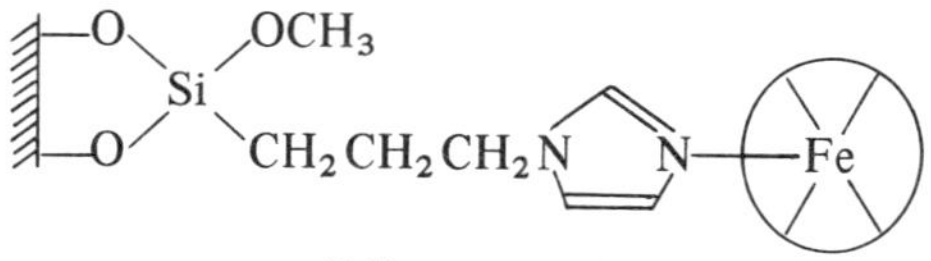

(11)

The rigid inorganic support prevents the irreversible dimerization forming $Fe^{III}-O-Fe^{III}$ which usually takes place on polymers.

Most inorganic supports are more *robust* to high pressures when used in columns or to the reduced pressure which arises during the rapid evacuation of solvents, and are somewhat more *resilient* with respect to sudden changes in solvent polarity. Although polymers can be stiffened by crosslinking, they become increasingly brittle as the extent of crosslinking increases with the result that attrition of the polymer particles becomes a serious problem.

Most macromolecular species, however, have the advantage of *inertness* over large pH ranges, although highly acidic environments can cause some damage. In contrast, many inorganic oxides actually dissolve in highly alkaline or acidic solutions.

Lipophilic polymers are less readily poisoned by metal impurities and by polar substances than *hydrophilic* inorganic oxide surfaces. Conversion of a hydrophilic to a hydrophobic surface can be achieved by increasing the chain length of the anchoring methylene groups or by treating the surface with hexamethyldisilazane in a toluene solution.

A potential application of inorganic-supported complexes could be in the area of bifunctional catalysis where one of the functions of the catalyst (for example, its *acid–base character*) is associated with the intrinsic properties of the support.

Conversely, this behavior could be a problem in an inorgnic catalyst if an undesirable side-reaction occurred on the support itself.

Another difference between polymer and inorganic oxide supports is in the ease and versatility of *chemical modification*, where macromolecular species undoubtedly have a tremendous advantage. Although well-characterized metal complexes containing anchoring ligands have been prepared in a homogeneous phase and then attached to an inorganic substrate under mild conditions, this preparative method has not been exploited for polymer-anchored catalysts because of synthesis limitations.

Since catalyst sites are generally distributed inside a polymer matrix, the choice of reaction medium with such supports is limited to solvents having good swelling or polymer-penetrating properties. However, such solvent compatibility is not necessarily the same as that required for chemical reaction. Thus, in the presence of a catalyst formed by equilibrating $RhH(CO)(PPh_3)_3$ with a phosphinated silica *surface*, the rate of hydrogenation of 1-pentene was six times that observed when the same complex was supported *inside* a phosphinated Amberlite ZAD-2 resin [90]. However, the steric effects observed with polymer-supported catalysts during the hydrogenation of substituted and bulky olefins are unlikely to be observed when the same catalysts are supported on silica.

The *swelling* of polymers under variable temperature and solution conditions can make the practical control of diffusional variables difficult. However, inorgnic substrates may be selected where diffusional changes are negligible for most reaction conditions of interest, and with many different types of inorganic oxides available the range of options is wide.

6. Surface Hydroxyl Groups on Inorganic Oxides

6.1. SILICA

The Si—O bond distance of SiO_2 is 0.161 nm and the Si—O—Si angle in, for example, β-cristobalite is 144°, but usually this value varies over 125–160° depending upon the degree of lattice disorder in the silica gel [91]. The density of silica gel, i.e. ca. 2.2 g cm^{-3}, which differs from that in cristobalite (2.27 g cm^{-3}) and tridymite (2.30 g cm^{-3}), supports a disordered structure.

The types of surface functional groups on silica have been reviewed many times [24b, 92–100], the most importance functional groups for the attachment of transition metal complexes being hydroxyl groups. The pK_a values of surface silanol groups lie in the range between 4 and 7, depending on the method of measurement [101]. Since siloxane bonds are generally much less reactive they have been used comparatively rarely for the binding of metal centers, although IR evidence suggests the involvement to some extent in reaction of strained siloxane bridges in unannealed silica [102].

Thermal treatment of silica at 473 K may remove all the physically bound water from the material, but hardly affects the surface silanol groups [103]. The resulting silica surface possesses ca. 5 OH groups nm^{-2} (values of 4.6 to 5.2 have been quoted in the literature [104–108]). At elevated temperatures the progressive elimination of water takes place between neighboring silanol groups; thus the silanol concentration was decreased to 1 OH group nm^{-2} by dehydroxylation at 1073 K, when only isolated free silanol groups were observed [109]. However, even after heat treatment at 1273 K some surface hydroxyl grups were retained. The variation in the average population of hydroxyl groups on the surface with the dehydroxylation temperature is illustrated in Figure 2 for a number of inorganic oxides. An estimation of the hydroxyl group population was obtained from studies of the thermal desorption of water and from surface reactions involving D_2, D_2O, Grignard reagents, halogen-containing molecules such as BCl_3, $AlCl_3$, $SiCl_4$ and chloromethylsilanes, alkylaluminum compounds, B_2H_6, etc. [92, 110–127]. It should be noted, however, that the initial number of surface hydroxyl groups on a surface may vary with the origin or nature of the material being studied.

By studying the stoichiometric reactions of $AlCl_3$ and $SiCl_4$ with the surface hydroxyl groups of silica gels, Peri and Hensley [128] found that about 90% of

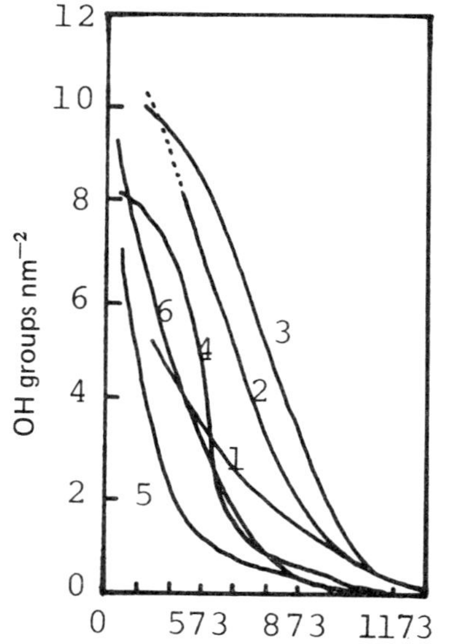

Fig. 2. The effect of temperature on the number of hydroxyl groups on the surfaces of various inorganic oxides. 1, SiO_2; 2, γ-Al_2O_3; 3, MgO; 4, ZnO; 5, TiO_2 (rutile); 6, α-Fe_2O_3.

such groups were paired after drying at temperatures below 870 K. With Aerosil or Cab-O-Sil pretreated at 623 K, only 61% pairing of surface hydroxyl groups was indicated by this method. Methyltrimethoxysilane, $MeSi(OMe)_3$, reacted both bifunctionally and monofunctionally with Cab-O-Sil in a 60:40 ratio [129]. Reactions of chloromethylsilanes with Aerosil dried at 573–773 K indicated a 55% pairing of surface hydroxyl groups [119]. Thus, on silica gel surfaces dehydroxylated below ca. 773 K, the surface hydroxyl groups appear to be sufficiently close together to interact bimolecularly with the reagents studied, even in those cases where adjacent hydroxyl groups are known not to hydrogen bond because their separation is greater than 0.31 nm [130]. This tendency diminishes with Aerosil or Cab-O-Sil.

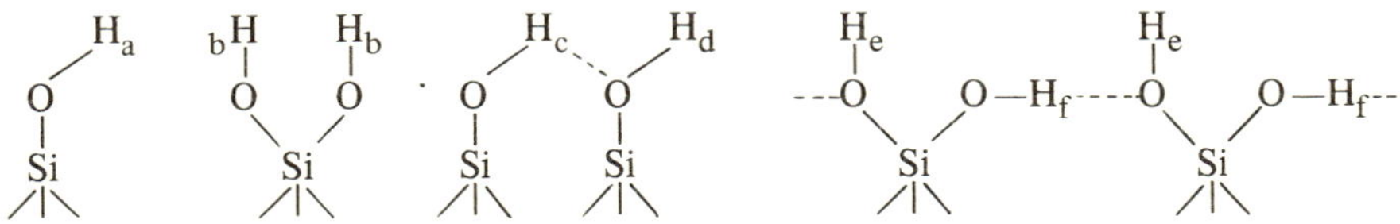

Van Rosmalen and Mol [131] have reported four absorption bands in the OH stretching region of the IR spectra of dry silica gels heated at 873 K. The bands were assigned as follows; ν_a (isolated single silanols): 3749 cm^{-1}, ν_b (very weakly coupled symmetric/asymmetric geminal pair): 3742 cm^{-1}, ν_c (hydrogen-bonded silanol of a vicinal pair): ca. 3600 cm^{-1} and ν_d (another silanol of a vicinal pair): 3720 cm^{-1}. Under similar pretreatment conditions, silica powders such as Cab-O-Sil and Aerosil exhibited only one structureless band at 3748 cm^{-1}. Hair and Hertl [132] suggested that a band at 3747 cm^{-1} is due to single OH groups (ν_a) while bands at 3751 and 3743 cm^{-1} are due to geminal OH groups (ν_b). Ghiotti *et al.* [133] also observed a sharp band at 3748 cm^{-1} in the IR spectrum of highly dehydroxylated Aerosil silica (1.5 OH groups nm^{-2}) corresponding to ν_a and at 3540 and 3720 cm^{-1} in the IR spectrum of a fully hydroxylated Aerosil corresponding to ν_c and ν_d, respectively. In addition to these bands, those at 3680 and 3450 cm^{-1} were assigned to ν_e and ν_f; such hydrogen bonding can occur on the (100) face of β-cristobalite [134].

Band splitting at ca. 3750 cm^{-1} has been contested by Van Cauwelaert *et al.* [135], Hockey [136], Morrow and Cody [137] and Klier *et al.* [138]; these authors observed no distinct shoulder formation nor splitting in this spectral region. There is still no general agreement as to whether isolated and geminal silanol groups can be distinguished via their stretching frequencies. However, three essentially different configurations of surface silanol groups are obviously present on silica surfaces pretreated above 473 K; i.e., isolated, geminal and vicinal silanols. The former two species (ν_a and ν_b) are observed at 3742–3751 cm^{-1} in the IR spectrum, while the band arising from the hydrogen-bonded vicinal pair (ν_c) appears at ca. 3540–3600 cm^{-1} [119, 131, 133, 139, 140]. The 3640–3660 cm^{-1} band has been ascribed to inaccessible internal OH groups, since this broad band exhibits no isotope shift on hydrogen exchange with D_2O. Annealing removes bulk hydroxyl groups without any significant alteration in the nature of the surface.

Line-narrowing studies of ^1H NMR spectra have demonstrated that the protons associated with silica surfaces migrate from oxygen to oxygen [141]. This proton hopping increases with temperature. The average hopping frequency was estimated to be of the order of 10^4 s^{-1} at 773 K which is two to three orders of magnitude less than that observed for the hopping frequency on a decationated Y-zeolite [142]. Through the use of multiple pulse experiments [143], the average proton–proton distance on a silica gel after calcination at 773 K was estimated to be greater than 0.4 nm; such a separation does not exclude the possible presence of geminal pairs.

Armistead and Hockey [144] have claimed that $(CH_3)_3SiCl$ and $(CH_3)_2SiCl_2$ react selectively with isolated free silanol groups, whereas CH_3SiCl_3 and $SiCl_4$ also react with hydrogen-bonded OH groups. Hertl *et al.* [129, 145] have also reported on the diminished reactivity of hydrogen-bonded silanol groups in their reaction with $Me_2Si(OMe)_2$ or with $(Me_3Si)_2NH$. The decreased reactivity of hydrogen-bonded OH groups has also been described by Yates *et al.* [126] in their reaction with trimethylaluminum. In contrast, an increased reactivity has been reported by Snyder and Ward [146] and by Kunawicz *et al.* [147]. Hair and Hertl [132] have suggested that geminal OH groups react more rapidly with CH_3SiCl_3 than do single OH groups on the basis of the geminal and single OH bands observed in the IR spectrum. The hydrogen-exchange reaction with D_2O also indicated a higher reactivity for geminal pairs than for single OH groups, but in this case the most reactive groups were hydrogen-bonded vicinal pairs [131]. For these various reasons, no general conclusions may be reached at present regarding the relative reactivities of surface hydroxyl groups.

6.2. ALUMINA

Since the bond distance of the Al—O bond in alumina is longer than that of the Si—O bond in silica (when linked tetrahedrally, for example, it is 0.178 nm), it follows that it is more ionic. This is supported by observations of proton migration on an alumina surface where the observed hopping frequency of 10^5 s^{-1} at 773 K is one order of magnitude larger than that for silanols [141]. It has been suggested that on average the protons are widely separated on Al_2O_3 dried at 773 K [148]. The ionization of surface hydroxyl groups on alumina, i.e. $Al{-}OH \rightleftharpoons Al{-}O^- + H^+$ (aq.), is characterized by the isoelectric point

(IEP) of the solid, which is the pH at which the net surface charge is zero in an aqueous suspension. The IEP of alumina lies in the pH range 7–9, with the result that the surface hydroxyl groups of alumina exhibit either acidic or basic behavior. Normally the acidic behavior of such groups is used for the attachment of transition metal complexes or surface anchoring groups.

Peri and Hannan [149] have assumed a density of 12.5 OH groups nm^{-2} for a fully hydroxylated alumina surface, whereas ca. 7.7 OMe groups nm^{-2} were formed on reaction with methanol in an autoclave [92]. As for silica, there is some deviation in the reported values for hydroxyl group populations on Al_2O_3 surfaces. The manner in which the average OH concentration depends on the pretreatment temperature of γ-Al_2O_3 is shown in Figure 2, which indicates that dehydroxylation of Al_2O_3 commences at ca. 473 K.

Peri [150] has considered the (100) crystal plane of γ-Al_2O_3 as being the most favorably exposed in the solid. He has suggested that the local environment of an OH group in the oxygen-defect plane generated by the high-temperature dehydroxylation of γ-Al_2O_3 is responsible for the occurrence of five distinct OH bands in the IR spectrum at 3800, 3780, 3744, 3733 and 3700 cm^{-1}, respectively [151]. Tsyganenko and Filimonov [152] have presented an alternative explanation for the three OH stretching frequencies at 3800, 3740 and 3700 cm^{-1} usually observed with alumina on the basis of the crystallography of Al_2O_3 spinel. A more detailed discussion employing essentially the same theoretical basis has been advanced by Knözinger and Ratnasamy [99]. They considered the following five configurations and assigned stretching vibrations to isolated OH groups which were correlated with the estimated net charges (and hence bond orders and force constants) of the five possible OH configurations. These assignments are compatible with those discussed previously [153–157].

3760–3780 cm^{-1}

Type Ia; (111) and (110) faces

3785–3800 cm^{-1}

Type Ib; (100), (111) and (110) faces

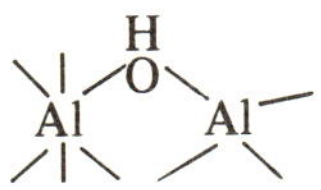

3730–3745 cm^{-1}

Type IIa; (111) face

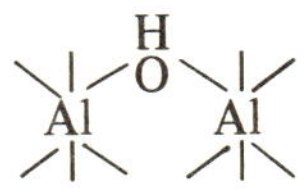

3740–3745 cm^{-1}

Type IIb; (111) and (110) faces

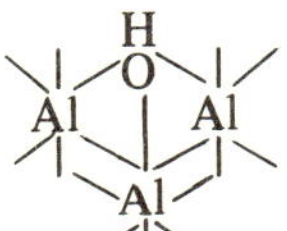

3700–3710 cm^{-1}

Type III; (111) face

Thus oxygen-defect surfaces pretreated at high temperatures may exhibit a heterogeneous character during the attachment of transition metal complexes, i.e. under circumstances where adequately hydroxylated surfaces would be most desirable. In general, surfaces dehydroxylated at 523–800 K possessing 2–6 OH groups nm^{-2} are used for supporting attached transition metal complexes. However, the use of an oxygen-defect surface exhibiting a Lewis acidity could provide a bifunctional metal catalyst under special circumstances.

6.3. SILICA–ALUMINA

The OH regions in the IR spectra of an amorphous silica–alumina (Al_2O_3 content: 30%) and a silica gel have been found to be virtually identical [134, 158, 159]. After treatment at 773 K, only a single strong band at 3750 cm^{-1} characteristic of isolated Si—OH groups could be observed. ^1H NMR spectroscopy indicated that the vast majority of OH groups on SiO_2–Al_2O_3 are not shared by an adjacent aluminum atom [110]. Schreiber and Vaughan [160] using multiple pulse NMR techniques determined the density of hydroxyl groups (both Si—OH and Al—OH) from the initial magnitude of the free induction decay observed after a 90° pulse; information on the local environment and on the arrangement of hydroxyl groups was obtained from an analysis of the free induction decay line shape.

TABLE III

Surface hydroxyl group densities on silica–aluminas of varying compositions pretreated at 773 K as determined by NMR spectroscopic techniques

SiO_2 content (wt. %)	Surface areas (m^2g^{-1})	OH group densities (OH groups nm^{-2})	
		Si—OH	Al—OH
100	576	1.6	0
90	693	1.5	0
75	615	1.2	0
50	727	1.0	2.7
25	839	0.2	2.4
10	768	0	3.4
0	571	0	3.1

The results listed in Table III indicate that Al—OH groups are apparently not detected in silica–alumina samples with a silica content above 75%, while nearly 75% of the protons on the 50% sample appear to be associated with Al—OH groups. Hence, a clear qualitative change appears to occur in the nature of the hydroxyl groups on the surface between samples containing 75% and 50% silica. A semiquantitative model has been proposed for the surface of a silica–

alumina by Peri [161], based on a previous model for the surface of SiO_2 where the resemblance to the (100) face of cristobalite was employed. Peri proposed eight possible types of site with different local composition (Si/Al ratio) and bonding features. After pretreatment at 873 K, the OH groups were considered to be located on the Si atoms with a predicted concentration of ca. 1.0, a value comparable with the observed value of ca. 0.8. Weiss *et al.* [162], however, concluded that the hydroxyl groups were bound to both silicon and aluminum, particularly on surfaces treated at lower temperatures.

Thus, although both Si—OH and Al—OH groups are present on fully hydroxylated or low temperature-treated silica–alumina surfaces, the hydroxyl groups combined with Al atoms appear to be more readily removed than the silanol groups on Si atoms, with the result that no Al—OH groups exist on silica–alumina with an Al_2O_3 content $< 50\%$ after drying at 773 K.

6.4. ZEOLITES

Zeolites A, X and Y, ZSM-5, synthetic mordenites, etc., consist of a network of SiO_4 and AlO_4 tetrahedra with different SiO_2/Al_2O_3 ratios in which the -1 charge on the AlO_4 tetrahedron is compensated by inclusion in the structure of cations such as H^+, NH_4^+, Na^+ and Ca^{2+}, thus maintaining electroneutrality. The more detailed properties of zeolites are described in the article by Howe in this book. Recently, the structures of zeolites have been studied using magic angle NMR techniques in association with ^{29}Si and ^{27}Al nuclei [163–166] and also by neutron diffraction methods [167]. Here, the nature of the hydroxyl groups is summarized in relation to their attachment to metal complexes.

Hydroxyl groups in H-zeolites may be schematically represented from a structural viewpoint as:

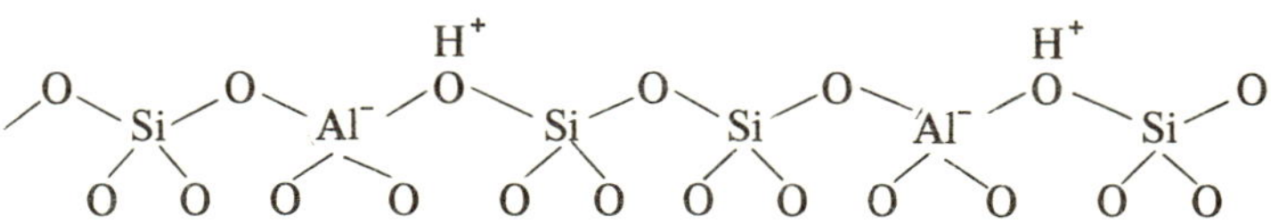

Uytterhoeven *et al.* [168] have reported that H-zeolites exhibit a sharp band at 3650 cm^{-1} and a broader one at 3550 cm^{-1} in their IR spectra. The higher frequency band was thought to arise from OH groups vibrating freely in the supercages in the structure, while the lower frequency band was associated with the withdrawal of electrons from the OH bond by adjacent Al^{3+} ions. Sodium (X and Y) faujasites after dehydration at temperatures below 523 K give rise to OH stretching bands in the IR spectrum at 3750 and 3690 cm^{-1} as described by Ward [169, 170]. The band at ca. 3690 cm^{-1} has been attributed to molecular water coordinated to the Na^+ ion. Divalent cation-exchanged zeolites give rise to complex bands in 3500–3800 cm^{-1} range of the spectrum which are derived from various hydroxyl groups. The mobility of the proton in zeolite cavities is much greater than that on silica or alumina surfaces. From NMR line shape and second moment calculations for various surface models of decationated Y-zeolites, Freude *et al.* [171] have suggested the presence of mainly pairs of hydroxyl groups in decationated Y-zeolite with a proton–proton distance of 0.37 nm rather than the random proton distribution which exists in cavities.

6.5. TITANIA

In the anatase, rutile and brookite forms of titania, each Ti^{4+} ions is surrounded by six O^{2-} ions and each O^{2-} ion is associated with three Ti^{4+} atoms placed at the apices of an isosceles triangle [91]. A packing density of 6.6 Ti^{4+} ions nm^{-2} has been estimated on a theoretical basis for this arrangement by Hollabaugh and Chessick [172]. The Ti—O single bond has a 63% ionic character in contrast to 50% for the Si—O bond in SiO_2, with the Ti—O bond in anatase being relatively more covalent than that in rutile because of the relatively distorted structure of the former solid. The TiO_2 surface is amphoteric having an IEP value of ca. 6. On fully hydroxylated titania surfaces, half the OH groups are acidic in character while the other half are mainly basic [108], thus prompting a study of the surface acidity of the oxide in connection with its catalytic activity [173, 174]. Surface dehydroxylation of TiO_2 begins at ca. 423 K, the variation in the population density of hydroxyl groups with pretreatment temperature being as depicted in Figure 2. From IR spectral studies it has been shown that the greatest proportion of OH groups are removed from TiO_2 surfaces at 573 K [175] with only a few isolated OH groups remaining at higher temperatures [176]. This feature of the dehydroxylation process may be readily appreciated from the OH configurations on the surface as deduced from crystallography and shown in Scheme 1.

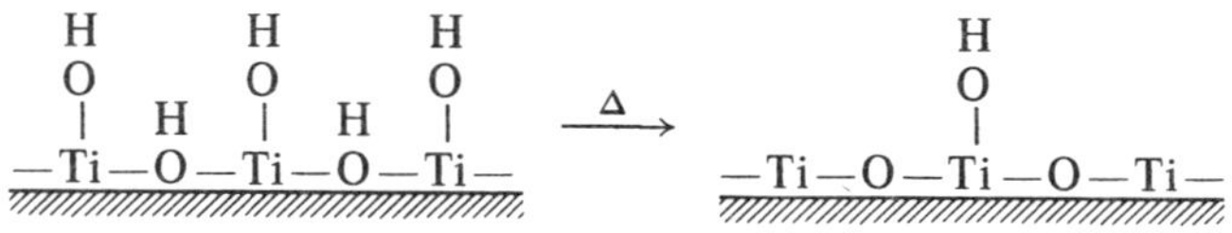

Scheme 1. Dehydroxylation of a TiO_2 surface.

Primet *et al.* [177] have suggested that two OH group configurations exist in the (001) face of anatase thus explaining stretching bands observed at 3715 and 3665 cm^{-1} in the IR spectrum. Jackson and Parfitt [178] and Munuera and Stone [179] have assigned the OH absorption bands arising from a rutile surface to the hydroxyl groups on the (110) face which is favored crystallographically. Tsyganenko and Filimonov [152] have suggested that Type I and II hydroxyl groups should be expected on the surface of anatase as well as rutile. Both the published data and Scheme 1 appear to be consistent with the following general assignments for the OH frequencies: on anatase, 3715–3728 cm^{-1} for single OH and 3645–3675 cm^{-1} for bridged OH groups; on rutile, 3700–3740 cm^{-1} for single OH and 3660–3685 cm^{-1} for bridged OH groups [152, 177–181].

6.6. OTHER INORGANIC OXIDES

Most work on SnO_2 has been carried out in relation to the modification of SnO_2 electrodes. Although the IEP of the SnO_2 surface (~5.5) is close to that for TiO_2, the Sn—O bond is probably more covalent than the Ti—O bond of TiO_2. Residual OH groups can still be detected on SnO_2 after heating to 773 K. These surface groups can serve as reactive sites for the attachment of transition metal complexes or for anchoring other functions.

The use of magnesia as a support for attaching transition metal complexes has been very limited although this oxide could be useful for supporting bifunctional metal catalysts. The surface of MgO generally has a relatively high concentration of hydroxyl groups with the exact amount being strongly dependent upon the origin of the sample and its preparation conditions. Some samples retain carbonate species on the surface even after high temperature calcination. However, only surface hydroxyl groups exist on MgO samples after treatment at temperatures above 573 K according to IR [182–184], luminescence [185, 186] and studies of the thermal desorption of water [187, 188]. The variation in the average population of hydroxyl groups on MgO surfaces with pretreatment temperature is recorded in Figure 2.

Cleavage of MgO crystals usually leads to the predominance of the (100) face in the cleavage planes. The two sharp bands observed for MgO at 3730–3750 cm^{-1} and 3610–3630 cm^{-1} in the IR spectrum have been assigned to single and bridged OH groups situated on the (100) plane [152]. However, the (111) face formed by annealing $Mg(OH)_2$ could provide a more reactive surface.

Transition metal catalysts attached to ZnO have not been widely studied although a tetrakis η^3-allyldichromium, $Cr_2(\eta^3\text{-}C_3H_5)_4$, has been supported through interaction with the surface hydroxyl groups of zinc oxide (Kadox-25) to form the average composition $[\text{≹}Zn—O]_4Cr_2(C_3H_5)$ (**12**); the catalyst obtained by reaction of the surface complex exhibited an activity for carbon monoxide hydrogenation [189]. The IEP of ZnO is in the pH range 8.7–9.7 with the result that the surface is basic in character. The attachment of transition metal complexes or functional groups to such a surface usually requires quite vigorous reaction conditions.

On heat treatment, surface hydroxyl groups start to be removed from about 473 K. Although the removal of physisorbed water is not complete at 423 K, with a large flat crystal the physisorbed water could be removed by evacuation at room temperature [190]. A large fraction of the hydroxyl groups may be removed by evacuation at 623 K as shown in Figure 2.

Dent and Kokes [191], Atherton *et al.* [192], Mattmann *et al.* [193] and Tsyganenko and Filimonov [152] have considered the (0001) and (000$\bar{1}$) faces of the wurtzite structure to be preferentially exposed at the surface. The narrow bands observed in the IR spectrum at 3660–3680 cm^{-1} and 3610–3630 cm^{-1} have been attributed to a singly coordinated hydroxyl group on a Zn atom or a hydroxyl group bridging between two adjacent Zn atoms on a (0001) plane, and to a triply bridging OH group on the (000$\bar{1}$) plane, respectively.

Although carbon is not an inorganic oxide, mention should be made of the fact that considerable quantities of foreign elements are usually found within its structure on chemical analysis [92]. These heteroatoms may be bound at the edges of the crystallites and form surface compounds, or they can be incorporated within the carbon layers. The most important and best known of the surface compounds of carbon are those with oxygen and with sulfur. The surface oxygen content varies over a wide range with the specific surface area and involves carboxyl groups, phenolic hydroxyl groups, carbonyl groups, etc. For this reason, it is not easy to distinguish the chemical attachment of a transition metal complex on a carbon surface from its physical adsorption since such surfaces have strong adsorption capabilities linked with the presence of many surface functional groups.

7. The Functionalization of Inorganic Oxide Surfaces

7.1. SILICA

In order to functionalize a silica surface, a molecule containing the ligand group, or a group that may readily be converted to a ligand group, is reacted with the surface silanols as shown in Scheme 2.

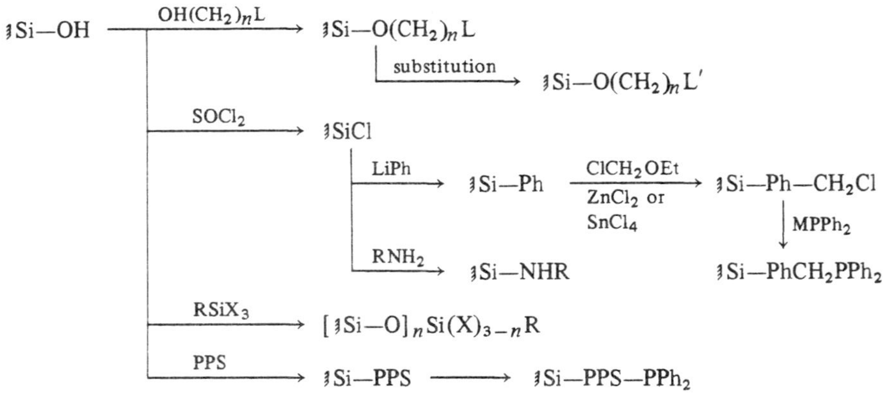

L = NH_2, PPh_2, SH, etc.; R = organofunctional group;
X = Cl, OR or O_2CCH_3; M = Li or K; PPS = polyphenylsiloxane

Scheme 2. Functionalization of a silica surface.

The Si—O—Si linkage so formed is very strong, with the Si—C bond being much more stable hydrolytically and thermally than either an Si—O—C or an Si—N—C linkage. Generally the Si—C (alkyl) bonding is more stable than an Si—Ph bonding. A wide range of compounds of the type $RSiX_3$ is available for a one-step reaction under mild conditions.

In the first method the silica surface is esterified using an appropriately substituted alcohol [86, 194–196]. Azrak and Angell [195] have investigated this process using $OH(CH_2)_2SH$ and $OH(CH_2)_3NH_2$ at 520 K. The resulting surface groups, $-(CH_2)_2SH$ and $-(CH_2)_3NH_2$, may be characterized by IR spectral analysis through SH and NH stretching frequencies observed at 2580 cm^{-1} and 3330 and 3390 cm^{-1}, respectively. These surface groups were stable at 520 K and resisted hydrolysis at 370 K. The second method involves initial chlorination followed by further reaction with LiPh or RNH_2 [197–199]. The aromatic rings thus obtained are functionalized like polystyrene. A third approach has been employed successfully by Allum *et al.* [200–202] to anchor PPh_2 [200], NH_2 [200, 203], C_5H_5 [18] and other ligands [200, 204, 205]. The convenient preparation of bifunctional phosphines with a choice in the length of the alkyl bridge, $X_3Si(CH_2)_nPPh_2$, can be performed by methods described by Eaborn [206] and Murrell [87]. These widely used reactions involve condensation with chlorosilanes or alkoxysilanes [207–209]. Chlorine analysis following the reaction of $Cl_2Si(CH_3)(CH_2)_2PPh_2$ or $Cl_3Si(CH_2)_2PPh_2$ in benzene solution with dehydrated silica at about one-half monolayer coverage indicated that bidentate bridges are formed predominantly [87]. However, in many cases it is not clear whether one or two such units are present. The Si—O—Si bond gives at least one very intense band in the 1100–1000 cm^{-1} spectral region due to the antisymmetric stretching mode. Unfortunately, this band masks the observation of the Si—O—C (alkoxy) group whose corresponding mode also appears in this region [210–212]. However, whether the Si—OEt group remained on the surface or not could be checked by observing the EtO mode, although estimation of the Si—Cl linkage which usually generates one band between 600–550 cm^{-1} proved impossible. Reaction of $Cl_3Si(CH_2)_2PPh_2$ with the hydroxyl groups of silica occurs under quite mild conditions, e.g. in refluxing benzene or xylene. In some cases the conversion may be more complete at elevated temperatures, e.g. 420–460 K [213]. Chemical analysis suggests that only a monolayer of phosphine ligands is covalently bonded to silica surfaces by the reaction of excess silylated phosphine with silica. At saturation coverage, two phosphine groups nm^{-2} exist on the silica surface, yielding about 1 mmol phosphine per gram of silica. The amount of phosphine covalently anchored was found to be 42% for OCH_3, 40% for O_2CCH_3 and 28% for OC_2H_5, compared to 100% for Cl, this sequence being in the approximate order of the steric requirement for silyl groups, i.e. $Cl < OCH_3 < O_2CCH_3 \lesssim OC_2H_5$. Expressed in another way, the following order of reactivities is conferred on silyl groups through the attachment of these functions, $Cl > O_2CCH_3 \gtrsim OCH_3 > OC_2H_5$ [87, 97, 214, 215]. The use of alkoxy compounds may be preferred over chlorosilanes when the generation of aggressive hydrogen chloride is to be avoided during the preparation of attached metal catalysts. A fourth preparative method involves coating the surface with a polyphenylsiloxane prior to the introduction of functional groups. This process is stated not to be completely reproducible [198, 216].

7.2. ALUMINA AND ZEOLITES

The use of these materials has been studies by Capka and Hetflejs [217]. Thus, PPh_2, CN, NMe_2 and C_5H_4N groups have been attached onto such surfaces through the types of reactions summarized in Scheme 3.

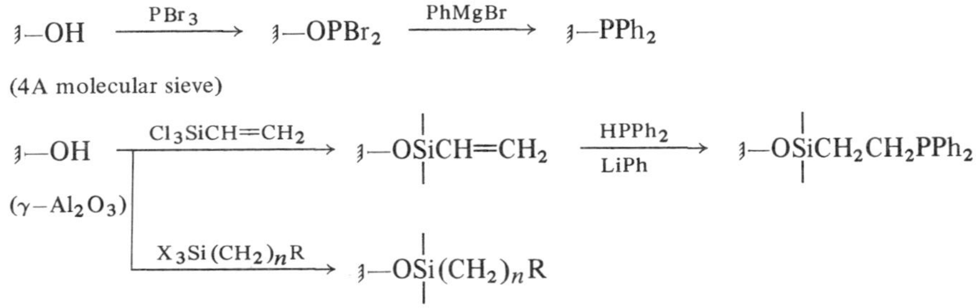

X = Cl or OEt; R = NMe_2, CN or C_5H_4N; n = 2 or 3

Scheme 3. Surface functionalization of alumina and zeolites.

Functionalization certainly occurs, but generally the reaction stoichiometry of surface OH groups and silyl compounds is not completely clear. The reaction pattern evidently depends upon the concentration and configuration of the surface hydroxyl groups.

7.3. STANNIC OXIDE AND CARBON

Miller and his associates [218, 219] have bound optically active (*S*)(−)-phenylalanine methyl ester via its amino groups to acyl chlorides attached to a carbon black surface, the latter being obtained by treating the carbon surface with thionyl chloride. This modified carbon has been used as a chiral electrode in electrochemical reduction or oxidation reactions. Tetra(aminophenyl)porphyrin has also been attached to a carbon surface via —COCl groups in studies undertaken by Murray and associates [220, 221].

$$\text{⊰C}-\text{COCl} \xrightarrow{(NH_2)_4TPP} [\text{⊰C}-\text{CO}]_2(NH)_2TPP(NH_2)_2$$

The surface hydroxyl groups of carbon have been interacted with $(MeO)_3Si(CH_2)_3NHCH_2CH_2NH_2$ followed by complexation with Cu^{II} ions to yield surface-stable Cu^{II} complexes [222]. Silane compounds have also been used for surface modification to enable the covalent bonding of ferrocene.

The hydrogen chloride liberated during the condensation of trichlorosilanes with the surface hydroxyl groups of SnO_2 is strongly adsorbed onto the surface unlike the situation with SiO_2 [224]. Ghosh and Spiro [225] have reported the attachment of a bipyridyl molecule on an SnO_2 surface via an alkylene bridge, followed by bonding with a tris(bipyridyl)ruthenium(II) complex.

$$\wr Sn{-}OH \xrightarrow{Cl_3Si(CH_2)_2bpy} [\wr Sn{-}O]_3Si(CH_2)_2bpy$$

8. Attached Metal Complex Catalysts

8.1. METAL COMPLEXES ATTACHED VIA FUNCTIONAL LIGANDS

Interest in this area has stimulated research into methods for attaching transition metal complexes to an inert support, the principal one studied being silica (Table IV). There are two main methods which differ only in the order of the steps involved [87, 200–202, 228, 236].

In method A the functional group is first attached to the support surface and then the transition metal complex is attached [87, 200–202, 237–240].

$$\wr{-}OH + X_3Si(CH_2)_nL \longrightarrow \wr{-}O\underset{|}{\overset{|}{Si}}(CH_2)_nL$$

$$\wr{-}O\underset{|}{\overset{|}{Si}}(CH_2)_nL + ML'_m \longrightarrow [\wr{-}O\underset{|}{\overset{|}{Si}}(CH_2)_nL]_yML'_{m-y}$$

where X = Cl, OR or O_2CR; $n = 0–18$; $y = 1–3$; and L = PPh_2, NR_2, $NH(CH_2)_2NH_2$, Cp, etc.

An alternative method of preparation is provided by method B in which a metal–silane complex is first formed and then condensed directly onto the surface [87, 200–202].

$$y\,X_3Si(CH_2)_nL + ML'_m \longrightarrow [X_3Si(CH_2)_nL]_yML'_{m-y}$$

$$\wr{-}OH + [X_3Si(CH_2)_nL]_yML'_{m-y} \longrightarrow [\wr{-}O\underset{|}{\overset{|}{Si}}(CH_2)_nL]_yML'_{m-y}$$

Method A has the disadvantage that the precise nature of the catalytic site can only be determined with difficulty. The possibility exists in this case that more than one surface product may be obtained due to inhomogeneity in the

TABLE IV

Transition metal complexes attached to inorganic oxides via anchoring functional groups

Reactions	Surface structures[a]	References
hydrogenation	[}Si–OSi$(CH_2)_2PPh_2$]$_3$RhCl	[200, 201, 226, 227]
	[}Si–OSi$(CH_2)_8PPh_2$]$_3$RhCl	[87]
	[}Si–OSi$(CH_2)_2PPh_2$]RhCl	[200, 201, 227, 228]
	[}Si–OSi$(CH_2)_2PPh_2$]$_3$RhH(CO)	[200, 201, 229]
	[}Si–OSi$(CH_2)_2PPh_2$]Rh(acac)(CO)	[200, 201]
	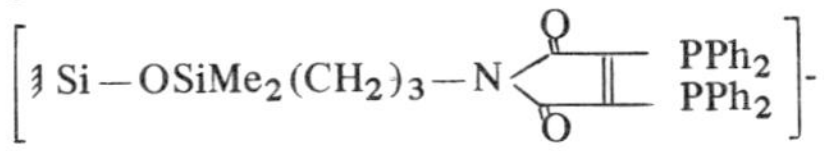 Rh^+(nbd)BF_4^-	[196]
	[}Si–OSi$(CH_2)_nPPh_2$]$_x$RhCl$(C_2H_4)_y$	[19]
	[}Si–OSi$(CH_2)_4C_6H_4PPh_2$]$_2$IrCl	[202, 226]
	[}Si–OSi$(CH_2)_2PPh_2$]$_3RuCl_2$	[202]
	[}Si–OSiCp]$TiCl_2$Cp	[230]
	[}Si–OSi$(CH_2)_3$Cp]$TiCl_2$Cp	[231]
hydroformylation	[}Si–OSi$(CH_2)_2PPh_2$]$_3$RhCl	[200, 202, 228]
	[}Si–OSi$(CH_2)_8PPh_2$]$_3$RhCl	[87]
	[}Si–OSi$(CH_2)_2PPh_2$]$_2$RhCl(CO)	[87, 200, 201, 227, 228]
	[}Si–OSi$(CH_2)_2PPh_2$]$_3$RhH(CO)	[200, 201, 229]
	[}Si–OSi$(CH_2)_2PPh_2$]Rh(acac)(CO)	[200, 201, 227]
	[}Si–OSi$(CH_2)_3NR_2$]RhCl$(CO)_2$	[200, 227, 229]
	[}Si–OSi$(CH_2)_2$CN]RhCl$(CO)_2$	[200, 227]
	[}Si–OSi$(CH_2)_{14}PPh_2$]RhCl(COD)	[87, 201, 228]
	[}Si–OSiMe_2Cp]Co$(CO)_2$	[18]
hydrosilylation	[}Si–OSi$(CH_2)_2PPh_2$]$PdCl_2$	[217]
	[}Al–OSi$(CH_2)_2PPh_2$]$PdCl_2$	[217]
	[}Si–OSi$(CH_2)_nPPh_2$]$_x$RhCl$(C_2H_4)_y$	[19]
oligomerization	[}Si–OSi$(CH_2)_2PPh_2$]$_2NiCl_2$	[201]
	[}Si–OSi$(CH_2)_2PPh_2$]$_2$Ni(COD)	[201]
other reactions	[}Si–O]$_2$Si[CpFeCp]	[232]
	[}C–CO]$_2(NH)_2$TPP$(NH_2)_2$	[233]
	[}Si–PPh_2]Pd(C_3H_5)Cl	[4]
	[}Si–OSi$(CH_2)_3NH_3^+$] [N$(PPh_3)_2$]$_{5-x}$- [$(CN)_5$Co–OO–Mo(O)$(H_2O)(CN)_5$]	[234]
	[}Si–OSi($CH_2CH_2CH_2PPh_2$)($CH_2CH(CH_3)CH_2PPh_2$)]- $Pd_n(OAc)_2$	[235]

[a] Cp = cyclopentadiene; acac = acetylacetonate; nbd = norbornadiene; COD = cycloocta-diene; TPP = tetraphenylporphyrin.

distribution of ligands. However, the range of ligands can be extended readily by standard organic reactions even if the analogous homogeneous reactions are precluded by the presence of reactive SiX_3 groups, thus:

$$\}Si{-}OSi(CH_2)_3Cl + KPPh_2 \longrightarrow \}Si{-}OSi(CH_2)_3PPh_2$$

Method B provides a valuable means of preparing relatively well-characterized surface complexes with a known ligand to metal ratio. The analogous method is not available when organic polymers are used as supports. The attachment of $Cl_3Si(CH_2)_8PPh_2PdCl_2(PhCN)$ to silica has been shown to yield only the monophosphine palladium complex over a wide range of Pd concentration, while a mixture of mono- and di-phosphine complexes was formed by the reaction of $(PhCN)_2PdCl_2$ with $\}Si{-}OSi(CH_2)_8PPh_2$ [87].

Complexes that cannot be prepared by method A may however be obtained at surfaces. Thus, the reaction of $[RhCl(CO)_2]_2$ with phosphinated silica gives the monophosphine surface complex, $\}Si{-}OSi(CH_2)_2PPh_2RhCl(CO)_2$ (**13**), whereas $[(EtO)_3Si(CH_2)_2PPh_2]_2RhCl(CO)$ (ν_{CO} 1770 cm^{-1}) reacts with silica to give the attached diphosphine complex [200]. Smith *et al.* [241] have reported that $Ni(CO)_4$ gives a monophosphine complex on reaction with phosphine–silica, but that a dicomplex is obtained by direct reaction of $[(EtO)_3Si(CH_2)_2PPh_2]_2Ni(CO)_2$ with silica. In practice, however, many complexes of this type cannot be isolated as crystalline solids but only as oils that must be purified chromatographically using nonhydroxylic phases; e.g. $(EtO)_3Si(CH_2)_3(acac)Rh(CO)_2$ (oil) [200], $[(EtO)_3SiC_5H_4Fe(CO)_2]_2$ (wax) [18], $[(EtO)_3Si(CH_2)_2PPh_2]_2RhCl(CO)$ (solid) [200]. Such complexes are often used without purification.

If the ligand attaching the metal to the inorganic support is involved in dissociative equilibrium with the metal under catalytic conditions, metal leaching can be serious [242, 243]. The covalent Si—O—Si bonding obtained by such attachment is extremely stable, the problem of metal leaching thus arising from the phosphine ligand attached to silica. Loss of metal can also occur with some bidentate phosphine ligands. On the other hand, Allum *et al.* [244] claim that the terdentate complex $[(EtO)_3Si(CH_2)_2PPh_2]_3RhH(CO)$ when linked to silica shows no loss of activity and no detectable loss of rhodium when tested for 75 h under continuous flow conditions during the liquid-phase hydroformylation of 1-hexene at 80–150 °C. Bidentate phosphine complexes are less likely to be reduced to elemental metal particles than unidentate complexes [196].

If oxidation of the phosphine ligand of an attached metal–phosphine complex occurs in the presence of low levels of oxygen contamination, the resulting phosphine oxide could generate a more labile phosphine oxide–metal complex; metal leaching would then occur from the labile phosphinated oxide complex.

Poisoning of a relatively flexible attached catalyst by physical blocking of the active sites may be less severe than with a directly attached catalyst.

$$[\not\!\mathrm{Si{-}O\overset{|}{\underset{|}{Si}}(CH_2)_2PPh_2}]_3\mathrm{RhCl}$$

(14)

Thus, the attached catalyst **14** which catalyzes the hydrogenation of 1-hexene and isoprene at 393–413 K and 15 atm H_2 pressure, retains substantial activity in the presence of mercaptans like *n*-butylmercaptan [202]. The same species attached to an organic polymer was deactivated by a sulfur species. The conversion of a catalyst surface to a lipophilic surface by reaction of the residual OH groups with hexamethyldisilazane could prevent poisons from adsorbing onto the surface [227]. Hancock *et al.* [227] have also found that such hexamethyldisilazane treatment has positive effects on aldehyde yields as well as on the selectivity in the hydroformylation of 1-hexene over an attached Rh catalyst.

8.2. TRANSITION METAL COMPLEXES ATTACHED DIRECTLY TO INORGANIC OXIDE SURFACES

Metal complexes have been attached heterogeneously onto inorganic oxide surfaces through simple adsorption, e.g. $[PdCl_2(PhCN)_2]/SiO_2$ [245], etc. [246–249], by means of nonvolatile liquid-phase catalyst adsorption (for example, using benzylbutyl phthalate), e.g. $[RhCl(CO)(PPh_3)_2]/SiO_2$ [250], etc. [251–258], or by ion exchange, e.g. $[Rh(COD)PNNP(+ \text{ or } -)]^+[ClO_4]^-$ [259], etc. [260–262]. Simple adsorption on inorganic oxides is a traditional method which has few controllable and uniform chemical features. This type of catalyst is not likely to contribute any new chemical aspects toward catalyst preparation and fundamental research. With supported liquid-phase metal complexes the way in which chemical attachment to the support occurs, and the situation and environment of the supported metal complex, remain ambiguous. Ion exchange cannot always be used to prepare surface complexes of uniform composition. Usually, part of the supported compounds precipitate on the surface with the formation of a dispersed phase. Ion exchange involves long-range forces incapable of controlling the local structure with definite and specific bond direction. In addition, the exchange capacity of the usual oxide supports is limited, thus preventing control of the concentration of supported ions over a wide range. Catalysts of these types, which were popular in the early stages of the development of hybrid catalysts, are outside the scope of this book.

Organometallic complexes and certain metal complexes are capable of reacting directly with the surface hydroxyl groups on inorganic oxide surfaces to form definite covalent bonds [12a, 263–265]. Direct attachment of this kind provides another valuable approach to the heterogenization of homogeneous metal complexes. The metal center in this type of catalyst should have little mobility and would be very close to the support surface. Whilst on the one hand, such limited mobility could place these catalysts at a large disadvantage in comparison with homogeneous catalysts possessing a flexible metal center, on the other hand attachment to an inorganic oxide may produce a new active metal center with no counterpart as far as homogeneous catalytic behavior is concerned. However, as with traditional heterogeneous catalysts, characterization of the catalyst when the latter is directly attached to the surface remains a major problem.

Ballard [12a] has found that benzyl, η^3-allyl and hydroborate compounds of transition metals such as Ti, Zr, Hf and Cr can react in more than one way with the OH groups of silica or alumina, the product obtained depending upon several factors such as the metal content, the nature of the support, the reactivity of the metal complex, the support pretreatment temperature, the attachment temperature, etc. Candlin and Thomas [263] have also reported the direct reaction of η^3-allyl and benzyl complexes of many transition metals such as Ti, Zr, Nb, Cr, Mo and W with surface OH groups. Yermakov [4a, 13, 235] has developed this approach widely, and in particular investigated surface metal complexes attached to silica in detail. Iwasawa [266–268, 325, 331–339], through the use of spectroscopic methods, has also demonstrated the possibility of tailoring at a surface molecular level using metal η^3-allyl complexes. Generally, the attaching reaction may be expressed as:

$$\wr\text{—OH} + \text{MR}_n \longrightarrow [\wr\text{—O}]_x\text{MR}_{n-x} + x\,\text{RH}\ (x = 1\text{–}3) \tag{15}$$

During the interaction of η^3-allyl complexes of Ni, Pd and Pt with silica dehydrated at 573 K or titania dehydrated at 473 K, 1 mol of propene is evolved per metal atom M ($x = 1$, $n = 2$) and a residual allyl ligand on M retains its π-allyl character; e.g. ν_{C-C} 1490 cm^{-1} for Pt/SiO_2 [269, 270] and 1485 cm^{-1} for Pt/TiO_2 [271a]. The surface Pd– and Pt–allyl complexes are decomposed by hydrogen at 293 K with the formation of metal particles.

$$n\ \wr\text{—OM}(\eta^3\text{-C}_3\text{H}_5) + \text{H}_2 \longrightarrow n\,[\wr\text{—OH}] + [\text{M}]_n + n\,\text{C}_3\text{H}_8 \tag{16}$$

Carturan *et al.* [272] have measured the Pd and Pt metal particle sizes on a glass ($SiO_2/Al_2O_3/Na_2O$ = 71 : 18 : 11) by small-angle X-ray scattering (SAXS). They found that catalytic activity for the hydrogenation of 1-hexene and for the selective hydrogenation of PhC≡CH to PhCH=CH_2 at 298 K was determined essentially by clusters with a diameter smaller than 0.3 nm [272, 273]. The reason was attributed to the preferential formation of α-Pd hydride on the smaller size particles. Physical adsorption of $M(\eta^3\text{-}C_3H_5)_2$ without chemical bonding to the support generated larger metal particles. Very small Pt particles are also formed through the interaction of $Pt(\eta^3\text{-}C_4H_7)_2$ and a silica surface with Re^{II} ions, followed by hydrogen treatment at 873 K. The binding energy of the $4f_{7/2}$ electrons of Pt in highly dispersed [Pt(2.9 wt.%) + Re(6.5 wt.%)] / SiO_2 is 0.5 eV higher than in Pt/SiO_2, indicating some interaction between Re^{II} ions and Pt particles [274]. An increase in specific catalytic activity for the hydrogenation of benzene, the dehydrogenation of cyclohexane and the hydrogenolysis of ethane and neopentane has been reported [275]. Reduction of $\equiv Si-ONi(\eta^3\text{-}C_3H_5)$ (**17**) also produces small metal particles of diameter $\leqslant 0.2$ nm according to magnetic measurements [275].

Matsuo and Klabunde [276] have prepared supported dispersed Ni catalysts **18** from toluene-solvated nickel atoms as follows:

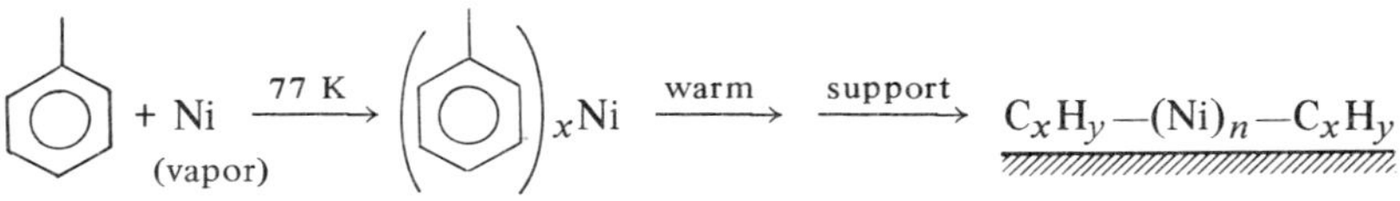

(18)

where the support was SiO_2, Al_2O_3, MgO or carbon. The data of Scott *et al.* [277] and Klabunde *et al.* [278] indicate that the crystallites are on average less than 0.2 nm and that they agglomerate to form chains of larger particles.

Highly dispersed catalysts containing nickel, molybdenum or tungsten sulfides on SiO_2, i.e. M_xS_y (**19**), are obtained by exposing the Ni^0, Mo^{II} or W^{II} species derived from the attached metal–allyl complexes to H_2S vapor [13]. Electron microscopy studies have suggested a high dispersion of sulfide particles; e.g. < 0.1 nm for $[NiS]_x/SiO_2$. In thiophene hydrogenolysis the catalytic activity was found to be higher than that of a conventional catalyst.

The bidentate bonding geometry in **15** is clearly possible if two surface hydroxyl groups are located adjacent to each other, when x is 2 and n is 3 or 4, as has been indicated by ESR spectroscopy [263], by the amount of RH evolved during the attachment reaction [4, 12a], by the ligand exchange of the residual ligands (R_{n-2}) with butanol [12a] or by means of the quantitative TPDE chromatogram [326, 335, 336]. The surface benzyl zirconium complex is one of the most active catalyst for ethene polymerization of attached organometallic complexes. Because the polymeric product is itself insoluble, one of the main advantages of the attached catalyst, viz. its ease of separation after reaction, would appear to be lost during polymerization unless a gain occurs through a dramatic increase in catalytic activity on attachment. Their definite composition and structure and the versatility of their direct control may provide another advantage for the use of attached catalysts in fundamental research.

Ballard [12a] has found that the polymer produced when styrene is polymerized using $Zr(^{14}CH_2Ph)/Al_2O_3$ as a catalyst contains a radioactive benzyl ligand:

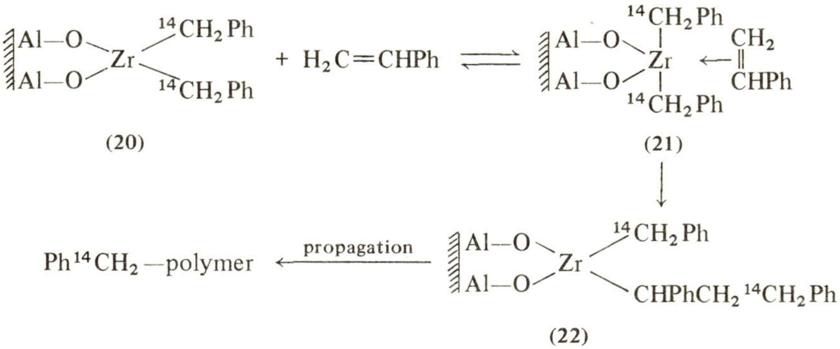

However, not all surface benzyl complexes follow this polymerization scheme, only those containing special coordinatively unsaturated species. Additional activation of the catalyst by heating *in vacuo*, hydrogen treatment or UV irradiation in an atmosphere of hydrogen is observed, probably as a result of the partial formation of Zr^{III} hydrides. The formation of surface titanium(III) hydrides in a $Ti(benzyl)_4/SiO_2$ system has been demonstrated by Yermakov *et al.* [279] through the use of ESR spectroscopy. The observation of ν_{Ti-H} at 1560 and 1630 cm^{-1} in the IR spectrum [279] indicates that at least two different hydride species are present. Active surface hydrides are also produced by the decomposition of a surface zirconium–hydroborate complex by heating to 373–523 K (ν_{Zr-H} 1925 cm^{-1}) [13a].

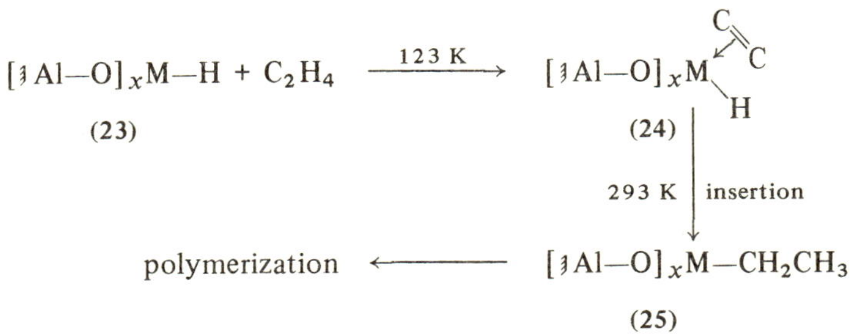

In contrast to ethene polymerization depicted above, the activity and stereospecificity (up to 60% crystallinity) of catalysts in the polymerization of propene increase in the series [12b]: $Ti(benzyl)_4/Al_2O_3 < Zr(benzyl)_4/Al_2O_3 < Hf(benzyl)_4/Al_2O_3$.

An important difference between homogeneous and heterogeneous transition metal complexes is that the hydride species functions as an active center, whereas species such as $[PhCH_2]_3ZrH$ are inactive in solution as a result of dimerization. However, +2, +3 and +4 oxidation states are present on the surfaces of these active polymerization catalysts, their relative abundance depending upon the nature of the organometallic complexes, the activation conditions and the reaction conditions. The surface heterogeneity of Al_2O_3-attached $Zr(BH_4)_4$ catalyst has been revealed through IETS studies by Evans and Weinberg [280], who observed non-uniform features for the vibrational spectra over the $B{-}H_t$, $B{-}H_b$ and $Zr{-}O$ stretching regions.

An active catalyst is obtained through the use of chromocene, $CrCp_2$, which is inactive in solution. Thus the SiO_2-attached chromocene catalyst **(26)** has been developed for a novel technological process allowing the preparation of high-density polyethylene via polymerization in the gas phase at 363–373 K and 35 atm pressure (Union Carbide Co.) [14, 15]. In contrast to other commercial polyethylenes, the chromocene catalyst produces polyethylenes with a relatively narrow molecular weight distribution.

$$\}Si{-}OH + CrCp_2 \longrightarrow \}Si{-}OCrCp$$

(26)

A full range of polymer molecular weights can be readily attained through control of only the hydrogen/ethene ratio. It is believed that the active catalytic site is provided by surface complexes with the composition $Si{-}OCrCp$. Polarographic measurements have shown that the attached chromium remains largely bivalent ($\geqslant 85\%$) after contact with ethene. The precise mechanism of the initiation reaction is not clear since direct insertion of ethene into a $Cr{-}Cp$ bond is hardly likely. The catalyst activity may be increased by increasing the dehydroxylation temperature of silica from 473 K to 1073 K. Cab-O-Sil has been used in view of the important role of isolated single silanol groups in the formation of active surface species.

Bis(triphenylsilyl)chromate is also active for ethene polymerization. In this case the catalytic activity is markedly increased if the complex is deposited on silica–alumina or silica with a pore size > 6.7 nm thus allowing the chromium complex to penetrate [281]. The surface active structure in this catalyst has not been clearly defined. Such catalysts may also be used for the oxidative double-bond scission of olefins [282].

$Mo(\eta^3\text{-}C_3H_5)_4$ reacts with surface hydroxyl groups in inorganic oxides with the following approximate order of readiness: $SiO_2{-}Al_2O_3 \approx Al_2O_3 \approx TiO_2 > SiO_2 >> ZnO \gtrsim Bi_2O_3$. The Mo complex can react with all types of surface hydroxyl groups but $Si{-}O{-}Si$ and $Al{-}O{-}Al$ bonds were found to be much less reactive. The bidentate form of the attached complex **27** may be obtained by reaction with silicas pretreated at 473–773 K or aluminas pretreated at 573–823 K. Yermakov [264] has observed an absorption band in the IR spectrum at 1650 cm^{-1} with the Mo complex attached onto SiO_2. This band has been attributed to a C=C stretching mode, suggesting that the allyl ligands may be attached to the molybdenum in the surface complex via σ-allyl rather than π-allyl bonds (which exhibit an absorption band at ca. 1500 cm^{-1}). Figure 3 depicts a typical TPDE chromatogram for the surface Mo complex **27** attached to γ-Al_2O_3 in the presence of H_2 [283].

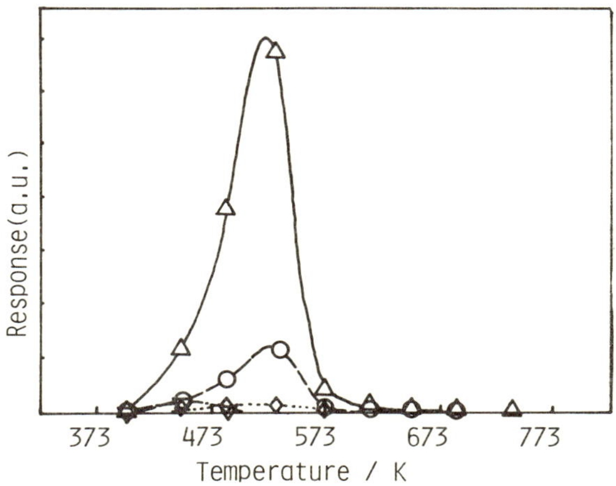

Fig. 3. Temperature-programmed decomposition chromatogram observed in the presence of H_2 for the Al_2O_3-attached Mo–allyl complex **27**. Experimental conditions: Mo/Al_2O_3, 0.73 wt.%; heating rate, 4 K min^{-1}; H_2 flow, 70 cm^3 min^{-1}. △: ethene; ○: ethane; ▽: propane; ◇: butene.

A major peak attributed to ethene evolution and a minor peak attributed to ethane evolution were observed at 520 K. The peak pattern and position remained unchanged despite a variation in the Mo content over the range 0.2–2.0 wt.%, indicating that the individual surface Mo species are distributed almost uniformly.

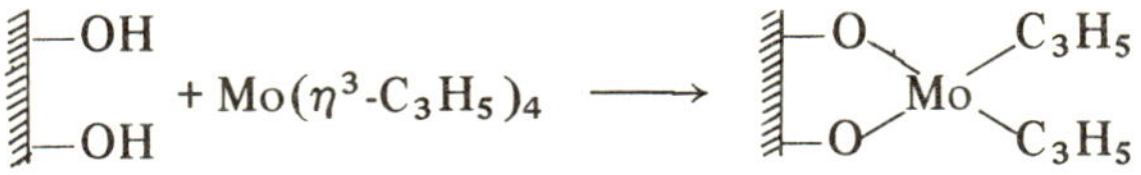

(27)

W(2-methallyl)$_4$ has been attached chemically to a silica surface without affecting its structure and compisition [284]. These Mo and W complexes on

SiO_2 [263, 264, 284, 285] and Al_2O_3 [263, 285] have been shown to be very active for olefin metathesis under reaction conditions where the corresponding soluble complexes are inactive. $[Mo(OEt)_5]_2$ [4] and $Mo_2(RCO_2)_4$ (R = H, CH_3, CF_3, C_6H_5) [286, 287] can also be interacted with the surface OH groups of SiO_2 or γ-Al_2O_3. However, the reactivity of these ligand groups is much less than that of η^3-allyl groups and their interaction with the surface involves uncontrollable and non-uniform adsorption processes in addition to reaction with OH groups. The use of $[Cr_2(CH_3CO_2)_4] \cdot 2H_2O$ [288] or $[Ph_2V_2(RCO_2)_4]$ [289] is subject to the same limitations.

Schwartz and his associates [290–293] have reacted triallyl rhodium, $Rh(\eta^3$-$C_3H_5)_3$, with silica and zeolite at room temperature and observed the evolution of propene during the attachment process:

$$\}\text{—OH} + Rh(\eta^3\text{-}C_3H_5)_3 \longrightarrow \}\text{—ORh}(\eta^3\text{-}C_3H_5)_2$$

(28)

The surface Rh complexes may be transformed by H_2 treatment to Rh hydride species which are active for olefin hydrogenation. Hjortkjaer *et al.* [294] have examined the interaction of $RhH(CO)(PPh_3)_3$ with alumina by IR spectroscopy. They propose a direct reaction with the surface hydroxyl groups as a result of observing carbonyl bands in the spectrum although the evolution of H_2 was not measured.

$$\}\text{Al—OH} + RhH(CO)(PPh_3)_3 \longrightarrow \}\text{Al—ORh}(CO)(PPh_3)_3$$

(29)

(ν_{CO} 1975 cm^{-1})

$$\downarrow 375\text{ K},\ CO\ (30\text{ atm})$$

$$\}\text{Al—ORh}(CO)_2$$

(30)

(ν_{CO} 2080 and 2000 cm^{-1})

The supported complex **29** has been used as a catalyst for propene hydroformylation although it exhibited a smaller activity than the species obtained via the adsorption of $Rh(\eta^3$-$C_3H_5)(CO)(PPh_3)_2$ on Al_2O_3 [295].

$Al(C_2H_5)_3$, which acts as a catalyst for ethene oligomerization but at a slow rate [296], reacts with the OH groups of alumina to give mainly $[\}\text{Al—O}]_2$ $Al(C_2H_5)$ wihich strongly promotes the production of high molecular weight polyethylene as the sole product [12a]. The corresponding SiO_2-attached compound, ʒSi—OAl$(C_2H_5)_2$, (2.8 Al atoms nm^{-2}, SiO_2 pretreated at 673 K), has been further reacted with $TiCl_2Cp_2$. The resulting surface possesses various species according to ESR spectroscopy but its specific activity is comparable with the activity of the corresponding homogeneous system [297].

Metal chloride attachment to a surface occurs simulataneously with partial chlorination of the surface [121]. Thus Soga *et al.* [298] have attached $TiCl_4$ to SiO_2 or γ-Al_2O_3, in a similar manner to that described by Grubbs [299], for the study of ethene and propene polymerization.

ʒ—OH + $TiCl_4$ ⟶ [ʒ—O]$TiCl_3$ + [ʒ—O]$_2TiCl_2$

(31) **(32)**

Species **31** was the main product (ca. 80%) when the support was pretreated at 573 K, and this was converted to ʒ—$OTiCl_2$ (**33**) by evacuation at 1073 K, although species **32** was not reduced to Ti^{III} [300]. Ti^{4+} and Ti^{3+} ions were suggested as being active for both ethene and propene polymerization while the Ti^{II} species was only active for ethene polymerization [301].

Research into polymerization catalysts following the original work of Ziegler [302] has been devoted toward the development of new highly efficient catalysts so that the need to remove residues from the final polymer product is eliminated. A second aim has been to develop catalysts which yield either a narrow or broad molecular weight distribution via a single-step process. This is not possible with conventional Ziegler catalysts. Thus $TiCl_4$ has been attached to MgCl(OH) and $MgCl_2$ or to SiO_2 and $MgCl_2$ [303]. Such studies have been conducted in some instances in the presence of additives, e.g. $PhCO_2C_2H_5/Al(C_2H_5)_3$ [304]. These catalysts are among the most active known for ethene polymerization. The formation of highly active catalysts through the interaction of $TiCl_4$ with $MgCl_2$ may be associated with structural defects derived from a disordering of the $MgCl_2$ crystal structure. The real structures of the active surface complexes are as yet unclear and further investigation is required.

Silica has been exposed to excess $TiCl_4$ vapor at temperatures within the range 423–573 K, which leads to the saturation of the surface with Ti; this was then followed by hydrolysis at 473 K. McDaniel *et al.* [305] have proposed the attachment structure **34** for the resulting surface species.

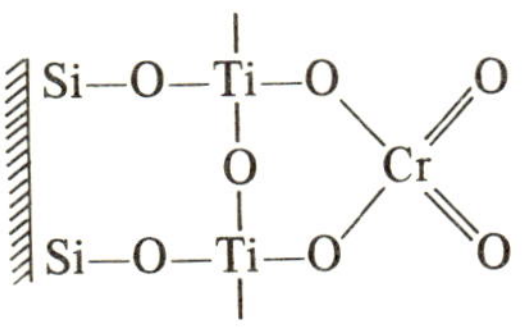

(34)

[ʒAl—O]$_2VCl_2$

(35)

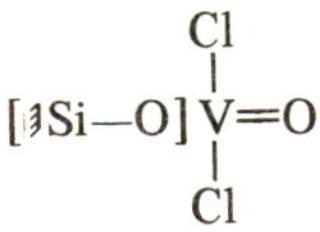

(36)

These workers found that the catalyst polymerized ethene without the need for high-temperature activation. The activity was clearly enhanced by the presence of a titania phase [305, 306]. The active species formed under the polymerization conditions have not been identified, but could be associated with the small fraction of Cr ions present. The surface oxychloride complexes of vanadium **35** and **36** have been prepared through the interaction between VCl_4 [307, 308] and $VOCl_3$ [309–312] and the OH groups of silica or alumina. Hydrolysis followed by subsequent calcination of the surface vanadium species produced dispersed particles of V_2O_5 containing vanadium atoms in different environments. This vanadium pentoxide phase was active and selective in the oxidation of 1-butene to maleic anhydride at 593–693 K [310], while the catalyst derived from $VOCl_3$ and Al_2O_3 showed a comparable activity to an NH_4VO_3-impregnation catalyst in the oxidative dehydrogenation of 1-butene.

The direct attachment of molybdenum may be achieved through the use of $MoCl_5$ [313–316]. When SiO_2 is preheated to 353 K and then subsequently treated with a CCl_4 solution of $MoCl_5$, surface compounds with the average composition $[\!\!\rightarrow\! Si{-}O]_3MoCl_2$ (**37**) are formed. Oxidation of the sample at temperatures > 573 K followed by hydrolysis at 353 K led to the formation of a proposed dimeric Mo^{VI} species [315]. When the concentration of the supported molybdenum was small, it was found that the surface contained Mo^{5+} and Mo^{6+} ions mainly in a tetrahedral coordination [314]. Che *et al.* [317] have reported the formation of a tetrahedral Mo^{VI} structure through the reaction of $MoCl_5$ and SiO_2 followed by hydrolysis and calcination. This tetrahedral structure was found to be more selective for the partial oxidation of propene to acrolein than the conventional supported molybdenum catalyst. It is probable that the formation of chemically attached species and dispersed MoO_3 particles occurs during the course of hydrolysis and calcination of the $MoCl_5/SiO_2$ system. This $MoCl_5/SiO_2$ system has been applied to olefin metathesis in the presence of a cocatalyst, e.g. an organomagnesium compound [316]. Similarly, the attachment of WCl_6 to SiO_2 followed by activation with organotin compounds has provided an efficient catalyst for olefin metathesis, which exhibits a greater stability than its homogeneous analogs [318].

One problem in the attachment of Group VIII metal halides to inorganic oxides is the difficulty of retaining the transition metal–surface oxygen bonds under the reaction conditions or during activation. This results in the generation of ill-defined heterogeneous surface phases. Surface structures with definite compositions obtained via the direct attachment of transition metal complexes to inorganic oxides are listed in Table V. The structures formed from the initial surface complexes which are described in the next section are also given in the second column.

TABLE V

Surface structure obtained through the direct attachment of transition metal complexes to inorganic oxides

Initial surface complex[a]	Transformed structure[a]	Catalytic reaction	References
$[\nmid \mathrm{Si-O}]_2\mathrm{Ti(benzyl)_2}$		polymerization	[12]
	$[\nmid \mathrm{Si-O}]_x\mathrm{Ti-H}$	polymerization	[279, 319, 320]
$[\nmid \mathrm{Si-O}]_2\mathrm{Zr}(\eta^3\text{-}\mathrm{C_3H_5})_2$		polymerization	[12, 263]
	$[\nmid \mathrm{Si-O}]_2\mathrm{Zr}\begin{smallmatrix}\diagup H\\ \diagdown H\end{smallmatrix}$	polymerization	[4, 319]
		isomerization	[321]
$[\nmid \mathrm{Al-O}]_2\mathrm{Zr}(\eta^3\text{-}\mathrm{C_3H_5})_2$		polymerization	[12, 263]
$[\nmid \mathrm{Al-O}]_x\mathrm{Zr(BH_4)}_{4-x}$			[12, 41, 319]
	$[\nmid \mathrm{Al-O}]_x\mathrm{Zr-H}$	polymerization	[4, 319]
$[\nmid \mathrm{Al-O}]_2\mathrm{Hf}(\eta^3\text{-}\mathrm{C_3H_5})_2$		polymerization	[4, 12, 319]
$[\nmid \mathrm{Si-O}]\mathrm{CrCp}$		polymerization	[14, 15]
$[\nmid \mathrm{Si-O}]_2\mathrm{Cr}(\eta^3\text{-}\mathrm{C_3H_5})$		polymerization	[12a, 13b, 263]
$[\nmid \mathrm{Al-O}]_2\mathrm{Cr}(\eta\text{-}\mathrm{C_3H_5})$		polymerization	[263]
	$[\nmid \mathrm{Si-O}]_2\mathrm{Cr}$	polymerization	[22]
		hydrogenation	[22]
	$[\nmid \mathrm{Al-O}]_2\mathrm{Cr}$	hydrogenation	[322]
		isomerization	[323]
	$[\nmid \mathrm{Si-O}]_2\mathrm{Cr}\begin{smallmatrix}\diagup\!\!\diagup O\\ \diagdown\!\!\diagdown O\end{smallmatrix}$	oxidation	[324]
	$[\nmid \mathrm{Al-O}]_2\mathrm{Cr}\begin{smallmatrix}\diagup\!\!\diagup O\\ \diagdown\!\!\diagdown O\end{smallmatrix}$		[322, 324]

Table V (continued)

Initial surface complex[a]	Transformed structure[a]	Catalytic reaction	References
$[(∤Si-O)Cr(\eta^3\text{-}C_3H_5)]_2$		polymerization	[22]
$[(∤Al-O)_2Cr(\eta^3\text{-}C_3H_5)]_2$		polymerization	[22]
	$[(∤Al-O)_2Cr]_2$	hydrogenation	[325]
	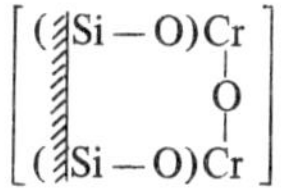	hydrogenation polymerization	[325] [22]
	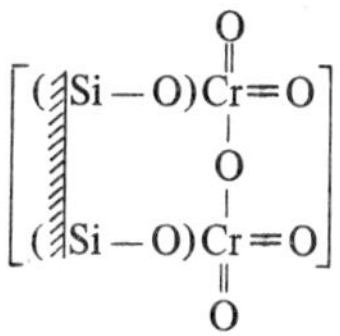	oxidation	[325, 326]
	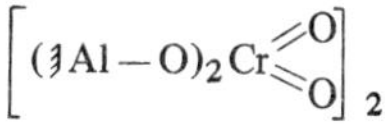	oxidation	[324, 325]
$[(∤Si-O)_2Cr(C_3H_5)_{0.5}]_{FM}$			[22]
	$[(∤Si-O)_2Cr]_{FM}$	hydrogenation	[22]
		polymerization	[22]
	$[∤Si-O]_x[Cr^{III}]_{FM}$	polymerization	[22, 326]
		oxidation	[22]
		hydrogenation	[22]
$[(∤Al-O)_2Cr]_{FM}$		hydrogenation	[22]

Table V (continued)

Initial surface complex[a]	Transformed structure[a]	Catalytic reaction	References
	$[(⦚Al-O)_2Cr-O-Cr(O-Al⦚)_2]_{FM}$	hydrogenation	[22]
	$[(⦚Al-O)_2Cr(=O)_2]_{FM}$	oxidation	[22]
$[(⦚Zn-O)_2Cr(\eta^3\text{-}C_3H_5)_{0.5}]_2$			[189]
	$[(⦚Zn-O)_2Cr]_2$	hydrogenation	[189]
$[⦚Si-O]_2Mo(\eta^1\text{-}C_3H_5)_2$		metathesis	[263, 264, 285]
$[⦚Al-O]_2Mo(C_3H_5)_2$		metathesis	[285, 327]
$[⦚Ti-O]Mo(C_3H_5)_3$			[189]
$[⦚Bi-O]_2Mo(C_3H_5)_2$			[189]
	$[⦚Si-O]_2Mo$	hydrogenation	[328, 329]
		isomerization	[323]
		H_2-D_2 exchange	[323]
	$[⦚Al-O]_2Mo$	hydrogenation	[329]
		H_2-D_2 exchange	[323, 330]
	$[⦚Si-O]_2Mo=O$	metathesis	[264, 285, 330]
	$[⦚Al-O]_2Mo=O$	metathesis	[285, 327, 330]
		$C_3H_6-C_3D_6$, H-exchange	[330, 338]
	$[⦚Si-O]_2Mo(=O)_2$	oxidation	[266, 331, 332]
	$[⦚Al-O]_2Mo(=O)_2$	oxidation	[333]

Table V (continued)

Initial surface complex[a]	Transformed structure[a]	Catalytic reaction	References
	$[⫽\mathrm{Ti-O}]_2\mathrm{Mo}(=\mathrm{O})_2$	oxidation	[189]
	$[⫽\mathrm{Bi-O}]_2\mathrm{Mo}(=\mathrm{O})_2$	oxidation	[189]
$[(⫽\mathrm{Si-O})_2\mathrm{Mo}(\mathrm{C_3H_5})]_2$		metathesis	[330, 334]
$[(⫽\mathrm{Si-O})_2\mathrm{Mo}(\mathrm{C_3H_5})_2]_2$		metathesis	[330, 334]
$[(⫽\mathrm{Al-O})_2\mathrm{Mo}(\mathrm{C_3H_5})_2]_2$		metathesis	[330, 334]
	$[(⫽\mathrm{Si-O})_2\mathrm{Mo}]_2$	hydrogenation	[329, 335]
	$[(⫽\mathrm{Al-O})_2\mathrm{Mo}]_2$	hydrogenation	[329, 336]
	$[(⫽\mathrm{Si-O})_2\mathrm{Mo}]_2(\mathrm{O_2})$	oxygen carrier	[335, 337]
		oxidation	[337]
	$[(⫽\mathrm{Si-O})_2\mathrm{Mo=O}]_2$	metathesis	[330, 334, 335, 338]
	$[(⫽\mathrm{Al-O})_2\mathrm{Mo=O}]_2$	metathesis	[330, 334, 336, 338]
		C_3H_6–C_3D_6, H-exchange	[330, 338]
	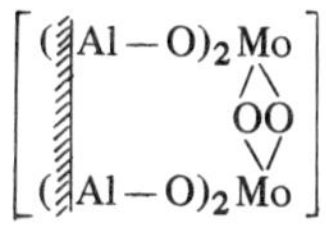	C_3H_6–C_3D_6, H-exchange	[330, 338]
		metathesis	[330, 334, 336, 338]
	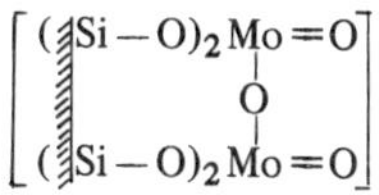	metathesis	[330, 334, 335]
	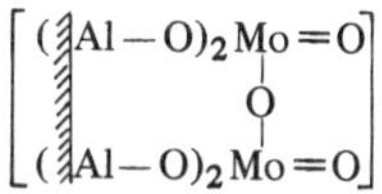	metathesis	[330, 334, 336]

Table V (continued)

Initial surface complex[a]	Transformed structure[a]	Catalytic reaction	References
	$[(‖Si-O)_2Mo(=O)_2]_2$	oxidation	[335, 339]
	$[(‖Al-O)_2Mo(=O)_2]_2$	oxidation	[189, 336]
	$[(‖Al-O)_2Mo(=O)(O)_2Mo(=O)(O-Al‖)_2]$	oxidation	[189, 336]
$[‖Si-O]_2W(\eta^3\text{-}C_4H_7)_2$		metathesis	[340–342]
$[‖Al-O]_2W(\eta^3\text{-}C_3H_5)_2$		metathesis	[263, 343]
$[‖Mg-O]_3W(\eta^3\text{-}C_3H_5)$		metathesis	[343]
	$[‖Si-O]_2W$		[340–342]
	$[‖Al-O]_2W$	hydrogenation	[343]
	$[‖Si-O]_2W{=}O$		[340–342]
	$[‖Al-O]_2W{=}O$	metathesis	[343]
	$[‖Mg-O]_2W{=}O$	metathesis	[343]
	$[‖Si-O]_2W(=O)_2$		[340–342]
	$[‖Al-O]_2W(=O)_2$		[343]
$[‖Si-O]Rh(\eta^3\text{-}C_3H_5)_2$			[291, 292]
$[‖Ti-O]_2Rh(\eta^3\text{-}C_3H_5)$			[271a]

Table V (continued)

Initial surface complex[a]	Transformed structure[a]	Catalytic reaction	References
[FM–O] Rh (η^3-C_3H_5)$_2$ (zeolite)			[293]
	[FM Si–O] Rh (η^3-C_3H_5) (H)	hydrogenation	[291, 292]
	[FM–O] Rh (η^3-C_3H_5) (H) (zeolite)	hydrogenation	[293]
	[FM–O] Rh(H)(H) (zeolite)	hydrogenation	[293]
	[FM Ti–O]$_2$ Rh–H	hydrogenation	[271a]
		hydroformylation	[271a]
	[FM TiO_2] [Rh]$_n$	hydrogenation	[271a]
[FM Al–O] Rh (CO) (PPh_3)		hydroformylation	[241]
[FM Si–O] Ni (η^3-C_3H_5)			[13a, 275]
[FM Si–O] Pd (η^3-C_3H_5)			[269, 270]
[FM Si–O] Pt (η^3-C_3H_5)			[269, 270]
[FM Ti–O] Pt (η^3-C_3H_5)			[271a]
	[FM SiO_2] [Ni]$_n$	hydrogenation	[275]
	[FM SiO_2] Ni_xS_y	dehydrosulfurization	[13a]
	[FM SiO_2] [Pd]$_n$	hydrogenation	[269, 270, 272, 273]
	[FM SiO_2] [Pt]$_n$	hydrogenation	[269, 270, 272, 273, 275]
	[FM TiO_2] [Pt]$_n$	hydrogenation	[271a]

[a] FM = full monolayer.

9. Tailored Metal Catalysts

9.1. MULTISTEP ATTACHED METAL CATALYSTS

Heteroatomic surface complexes can be produced via a two-stage attachment procedure [345]. Initially $SnCl_4$ is contacted with the surface OH groups of SiO_2 to form **38** and then an organometallic complex is attached by reaction of its sodium salt with **38**. The following steps have been proposed:

$$\wr Si{-}OH + SnCl_4 \longrightarrow [\wr Si{-}O]_2SnCl_2 \quad \mathbf{(38)}$$

$$[\wr Si{-}O]_2SnCl_2 + Na[MCp(CO)_x] \longrightarrow [\wr Si{-}O]_2Sn[MCp(CO)_x]_2 \quad \mathbf{(39)}$$

(M = Ni or Fe)

This method may be developed to provide a route for the surface preparation of heteronuclear metal species with definite compositions.

Polynuclear Pd complexes (1–5 Pd atoms) have been prepared by the stepwise attachment of a mononuclear Pd complex, $Pd(OAc)_2$, onto phosphinated silica using the following procedure [235]:

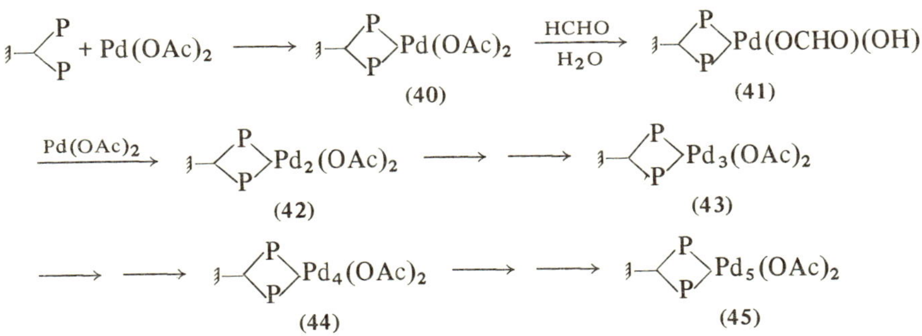

where 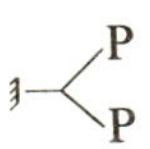 depicts $[\wr Si{-}O]_2Si\genfrac{}{}{0pt}{}{CH_2CH_2CH_2PPh_2}{CH_2CH(CH_3)CH_2PPh_2}$.

Yermakov [235] has studied the relationship between catalytic properties and the number of palladium atoms in the active site. As the number n of Pd atoms in the polynuclear complex $[Pd]_n$ increases, the binding energy of the Pd 3d level decreased, and when $n = 5$ it approached the energy of metallic palladium as depicted in Figure 4. During the hydrogenation of cyclopentadiene the activity increased monotonously as the value of n in the attached complexes increased, being virtually identical at $n = 5$ to the specific catalytic activity for metallic palladium. The ratio of the rates of hydrogenation of cyclopentadiene

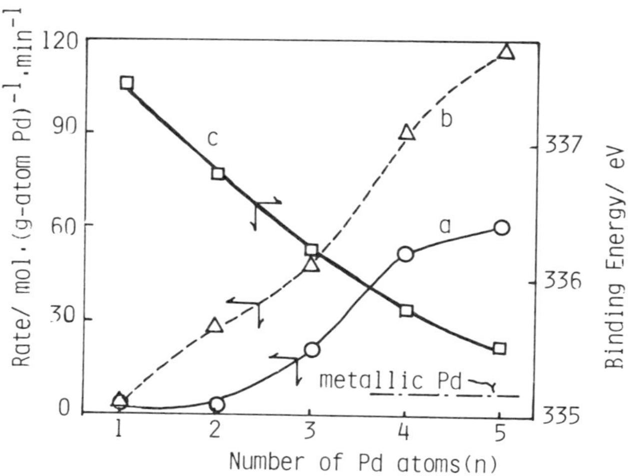

Fig. 4. Variation in the binding energy of the Pd 3d5/2 level and the catalytic activities of attached polynuclear $[Pd]_n/SiO_2$ catalysts with varying nuclearity (n) for different reations. Data (a) for the rate of disproportionation of cyclohexene at 373 K; (b) for the rate of hydrogenation of cyclopentadiene at 303 K; and (c) for the binding energies as measured for the Pd 3d emission line by X-ray photoelectron spectroscopy (XPS).

and cyclopentene was also close to the typical value for metallic palladium. Disproportionation of cyclohexene only occurred when $n > 3$ in the attached catalysts, with the rate attaining a maximum value at $n = 5$ (Figure 4). Thus, if the character of the metallic crystals is to be reproduced, a minimum of five Pd atoms is necessary in the attached polynuclear complex. The main problem with this system, however, is the paucity of evidence regarding the structures and compositions of the attached $[Pd]_n$ species.

9.2. TRANSITION METAL HYDRIDE CATALYSTS

Surface organometallic compounds may undergo various transformations as a result of gas-phase chemical treatments. Most of the surface complex obtained by the reaction of $Zr(\eta^3\text{-}C_3H_5)_4$ with SiO_2 is $[\ni Si-O]_2Zr^{IV}(\eta^3\text{-}C_3H_5)_2$ (**46**) with ca. 10% Zr^{III} ions in the material [320]. The surface complex **46** can be transformed into the zirconium dihydrides **47** on treatment in an H_2 atmosphere at 333 K. Schwartz and Ward [321] have shown that this SiO_2-attached ZrH_2 species exhibits high activity toward 1-butene, 1-hexene, allylbenzene and methylenecyclohexane isomerizations under mild conditions. Hydrolysis of the catalyst after completion of the isomerization reaction gave the respective alkane. The alkane/Zr ratio approached unity and H_2 was also evolved.

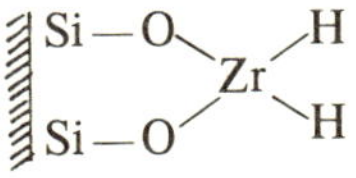

(47)

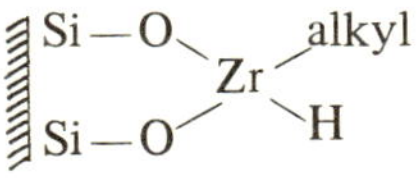

(48)

This may indicate that the active zirconium species present during isomerization is a Zr^{IV} alkyl hydride (**48**). Analysis of the IR spectrum led to the assignment of two peaks observed at 1620 and 2190 cm^{-1} to the Zr dihydride species but the first peak could also possibly be attributed to the bridged hydride Zr—H—Zr [320]. The species associated with the first peak readily reacted with D_2 and H_2O and also catalyzed the polymerization of ethene [235].

The silica-attached diallylrhodium complex, ≹Si—ORh(η^3-C_3H_5)$_2$ (**49**), may be formed by the reaction of Rh(η^3-C_3H_5)$_3$ and the hydroxyl groups of silica [292]. Subsequent treatment of the complex with H_2 leads to the formation of the silica-attached rhodium hydrides (ν_{Rh-H} 2010 cm^{-1}, $\nu_{Rh-H-Rh}$ 1800 cm^{-1}) [291]. The dihydride species (ν_{Rh-H} 2080 and 2020 cm^{-1}) is obtained by further treatment with H_2. Data for the catalytic hydrogenation of alkenes in the presence of these attached complexes reveal that both provide highly active surfaces. The intermediacy of a rhodium(alkyl)hydride species has been suggested. However, the surface exhibits some heterogeneity; further investigation is required for distinct story [292a].

Triallylrhodium reacts with surface hydroxyl groups located in the zeolite cavities of partially (15%) proton-exchanged X-type zeolite to form the attached diallylrhodium complex **50** at room temperature [293]. Subsequent treatment of the surface complex with H_2 at room temperature for 48 h led to the formation of a dark gray zeolite-attached rhodium hydride (**51**) (ν_{Rh-H} 2010 cm^{-1}) in conjunction with the evolution of 1 equiv of propane.

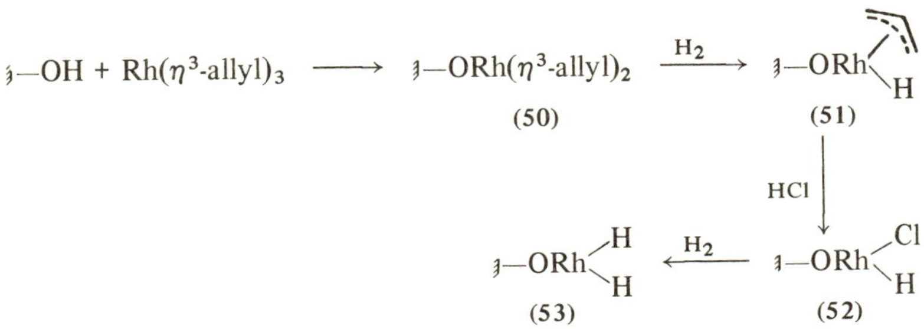

Scheme 4. Formation of zeolite-attached rhodium hydrides.

No bridging hydride ligands could be detected by IR analysis. The rhodium dihydride species **53** was obtained from the species **51** via **52**, and exhibited twin peaks at 2098 and 2029 cm^{-1}, respectively. Species **51** was shown to have a high catalytic activity for olefin hydrogenation, when size/shape selectivity for the substrate was also noted with the rates decreasing in the order 1-butene > 1-hexene > cyclohexene >> cycloheptene $\gtrsim$ cyclooctene $\gtrsim$ 1-methylcyclohexene. Poisoning experiments were employed to substantiate the suggestion that the catalytically active rhodium complex is located inside the zeolite cavity.

When TiO_2 (Degussa P-25) was pretreated at 473 K for 2 h and used as a support for the attachment of $Rh(\eta^3\text{-}C_3H_5)_3$, the bidentate complex **54** (ν_{C-H} 3050 cm^{-1}, ν_{C-C} 1462 cm^{-1}) was formed [271a]. Species **54** was converted to

$$\underset{(54)}{[\wr Ti{-}O]_2Rh(\eta^3\text{-}C_3H_5)} \xrightarrow{H_2} \underset{(55)}{[\wr Ti{-}O]_2Rh{-}H} \xrightarrow{H_2} \underset{(56)}{\wr [Rh]_n}$$

the hydride species **55** (ν_{Rh-H} 2048 cm^{-1}) by H_2 at 290–323 K. No bridging hydrides were observed in a sample with Rh/TiO_2 = 2. Further H_2 treatment of **55** at 773 K led to the formation of a small $[Rh]_n$ cluster (**56**) (1.4 nm in diameter) [271a]. In contrast to a conventional TiO_2-supported catalyst, for which a typical SMSI(strong metal–support interaction) [271b] is observed, the attached $[Rh]_n$ cluster catalyst did not show any distinct SMSI effect.

9.3. ATTACHED METAL OXIDE CATALYSTS

9.3.1. *Surface molybdenum monomers*

Ballard [12a], Candlin and Thomas [263] and Yermakov [264] have demonstrated that transition metal η^3-allyl complexes readily react with the surface OH groups of SiO_2 or Al_2O_3 under mild conditions. As shown in Figure 5, the decrease in the intensity of the OH stretching vibration on reaction of tetrakis η^3-allylmolybdenum with SiO_2 indicates that the Mo ions combine with the silica surface through the oxygen atoms of the OH groups.

Yermakov *et al.* [4, 328] on the basis of an IR spectrum have suggested that the ligands in the obtained surface complex **57** are σ-allyl type (ν_{C-C} 1650 cm^{-1}). Well-defined surface transformations of the initial metal–allyl complexes attached to SiO_2 or Al_2O_3 have been demonstrated by Yermakov and also by Iwasawa for Al_2O_3 surfaces. A reversible reduction–oxidation cycle involving the SiO_2-attached catalysts **58**, **59** and **60** as shown in Scheme 5 has been observed [4, 264, 331, 333]. The diffuse reflectance spectrum of **60** which exhibits absorption peaks at 290 ($t_1 \longrightarrow 3t_2$) and 225 ($2t_2 \longrightarrow 3t_2$) nm demonstrates that the Mo ions are situated in a tetrahedral position. An absorp-

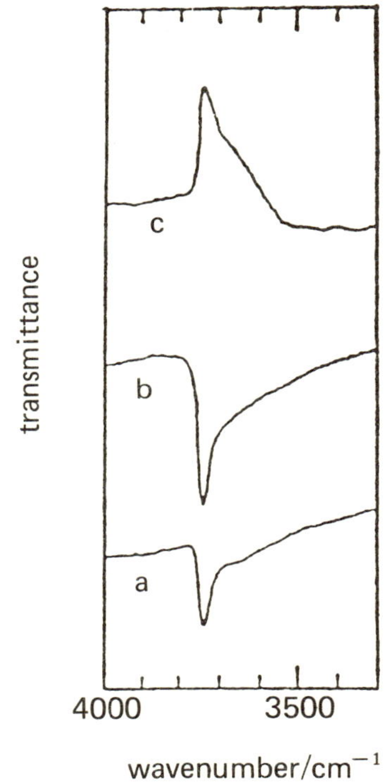

Fig. 5. Changes in the intensity of the OH vibration band in the IR spectrum during the attachment of a molybdenum complex to a silica surface. Curve (a): the attached catalyst 57; curve (b): SiO_2; and curve (c): the difference spectrum (a)–(b).

tion band in the IR spectrum at 916 cm^{-1} was assigned to a stretching frequency of the Mo—O bond in the oxomolybdenum(VI) species with the two terminal oxygen atoms in tetrahedral symmetry [331]. The emission energy from the lowest T_1 triplet state of the Mo^{VI} entity in species **60**, induced by illumination of the Mo=O charge-transfer band, was found to be independent of a change in the excitation energy, reflecting the homogeneity of the molybdenum species. Conventional catalysts obtained by an impregnation method using inorganic acids or their salts, such as ammonium paramolybdate in aqueous solution, give rise to a non-uniform distribution of active sites and with a resulting catalyst surface exhibiting heterogeneous properties being generally produced during immersion and calcination.

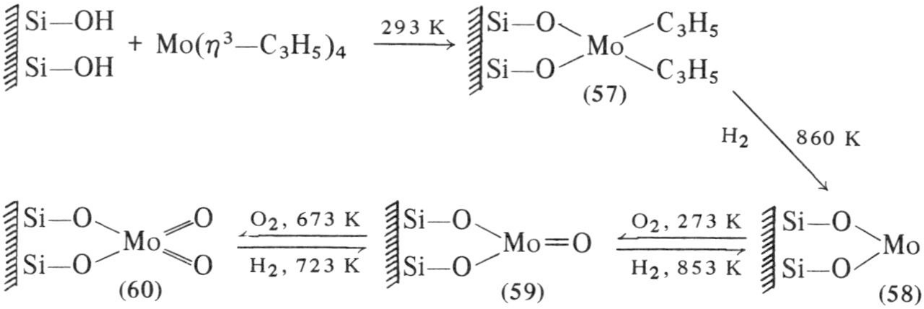

Scheme 5. Surface transformations of SiO_2-attached Mo monomers.

Catalyst **60** was found to be 11 times more active than the corresponding impregnation catalyst for the allyl-type oxidation of propene. Enhancement of allyl hydrogen abstraction by the attached Mo catalyst **60** leads to the disappearance of the kinetic isotope effects usually observed in acrolein formation [332]. The attached catalyst **60** is also active for the oxidation of ethanol to acetaldehyde. This reaction may be explained quite well in terms of a two-stage redox mechanism:

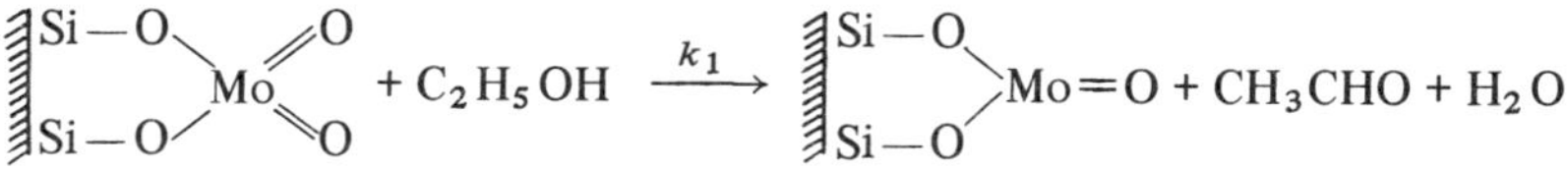

(a)

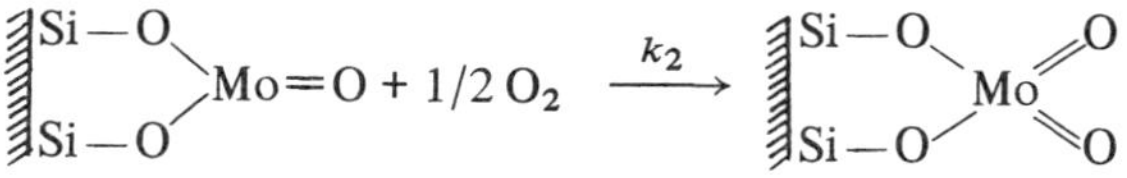

(b)

The rate of C—H breaking at the α-carbon of ethanol on the attached Mo catalyst **60** was approximately one order of magnitude larger than the corresponding rate over the impregnation catalyst [331]. A distorted molybdenum (VI) structure with a large difference between the Mo=O and Mo—O bond distances was more active than the species with nearly ideal T_d symmetry.

From this it may be concluded that in the selective oxidation of ethanol and propene a coordinatively unsaturated, distorted dioxomolybdenum(VI) structure with covalent Mo=O bonds is an essential factor for the abstraction of an $H^{\delta -}$ atom from a C—H bond.

The advantage of the use of a tailored metal catalyst in elucidating a reaction mechanism is the possibility that each surface structure which may be involved in the reaction sequence can be formed. Thus, the rate constants, k_1 and k_2, for steps (a) and (b) above in the oxidation of ethanol may be determined independently, with the rate for the steady-state reaction being expressed by the equation $P_E/r = 1/k_1 + P_E/2k_2\sqrt{P_{O_2}}$, where r is the rate of oxidation, P_E is the partial pressure of ethanol and P_{O_2} the partial pressure of oxygen. Good agreement between the values obtained from this steady-state equation and the individual rate constants evaluated for the two steps confirms the redox nature of the mechanism involving the Mo^{VI} and Mo^{IV} species [331].

The Al_2O_3-attached surface complex **61** may also be transformed to different oxidation states of molybdenum, i.e. **62–65** [327, 333]. These transformations are performed via two sequences, $Mo^{VI} \rightleftharpoons Mo^{IV} \rightleftharpoons Mo^{II}$ or $Mo^{VI} \rightleftharpoons Mo^{II} \rightleftharpoons Mo^{IV}$, depending upon the nature of the alumina.

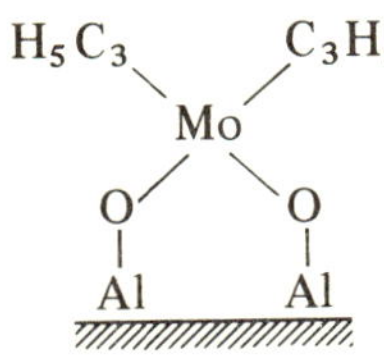
(61)

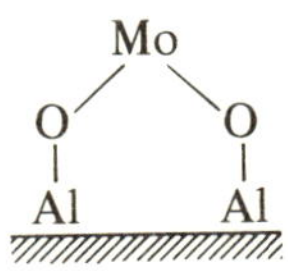
(62)

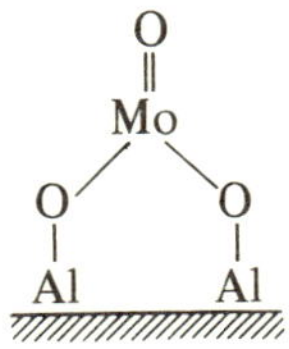
(63)

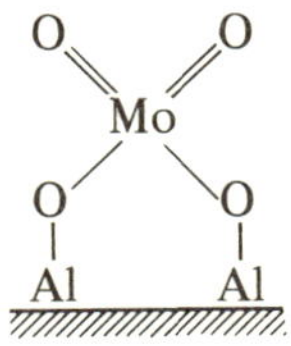
(64)

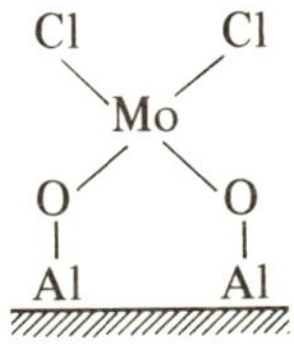
(65)

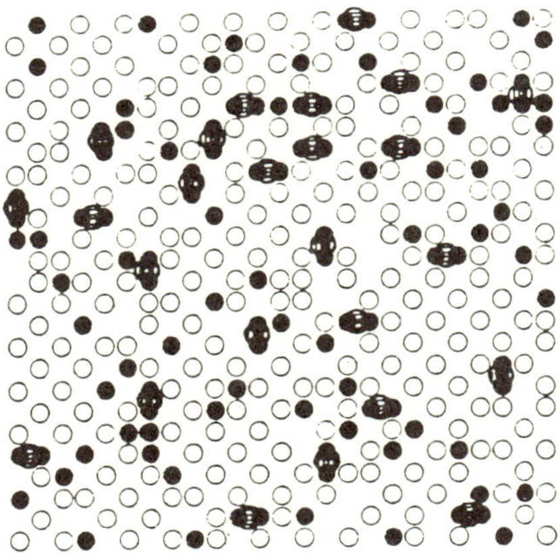

Fig. 6. The molybdenum distribution at an Al_2O_3 surface after removal of 82.1% of the OH groups; ○, oxygen atoms on the top layer; •, surface OH groups; ⊜, Mo atoms.

The Mo distribution on the surface is as illustrated in Figure 6 [327], which depicts the situation on an alumina surface when 82.1% of the full monolayer of OH groups has been removed. This figure was constructed on the basis of Peri's result [150] using an OH population of 2.2_4 OH groups nm^{-2} and a surface area of 190 $m^2 g^{-1}$. Structure **64** containing tetrahedral Mo^{VI} ions is characterized by a DRS band at 281 nm ($t_1 \longrightarrow 3t_2$) [346, 347]. The surface allyl complex **61** and the oxomolybdenum(IV) structure **63** have been shown to be very active for propene metathesis, i.e. $2\,CH_3CH{=}CH_2 \rightleftharpoons H_2C{=}CH_2 + CH_3CH{=}CHCH_3$ [285, 327]. The constant turnover frequency observed in the composition range Mo/Al_2O_3 = 0.73–1.98 demonstrates the uniform distribution of the molybdenum centers. The Mo^{VI} species (XPS binding energy for Mo $3d_{5/2}$ = 233.5 eV) and the Mo^{II} species (230 eV) showed no catalytic activity. The binding energies for the Mo $3d_{3/2}$ and $3d_{5/2}$ levels in the active catalysts **63** and **65** were 235.3 and 232.5 eV and 235.4 and 232.6 eV, respectively [285, 327].

A variety of values within the range 0.7–3.3 eV has been reported for the difference between the Mo 3d binding energies of Mo^{VI} and Mo^{IV} for molybdenum oxide and within the range 0.3–1.7 eV for some Mo complexes depending upon the type of ligand involved [348–351]. By comparison with the spectrum of the Mo^{VI} catalyst and on the basis of the uniformity of the species, these values suggest that the molybdenum ion is in a quadrivalent state in the active catalysts. More recently Yermakov [235] has also confirmed the presence of Mo^{IV} species in metathesis catalysts by means of XPS studies. The uptake of one oxygen atom per Mo atom during oxidation of the Mo^{II} with O_2 and of two chlorine atoms per Mo atom also demonstrates the formation of quadrivalent oxo- and dichloro-molybdenum(IV) structures. The metathesis activity is markedly affected both by the structure and oxidation state of the molybdenum, and by the nature of the supports to which the molybdenum ions are coordinated [285]. A general order of activity molybdenum structures in propene metathesis is as follows (the preparation of the Mo dimers listed is described below) [330, 334]:

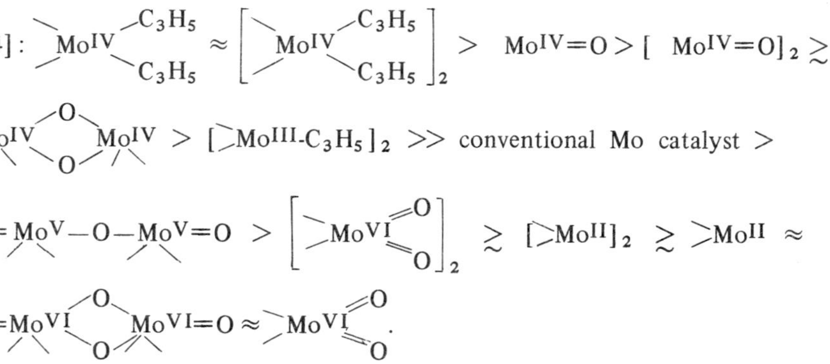

The metathesis reaction proceeds essentially on one molybdenum site, the most active catalysts being allyl-type catalysts in which a catalytically active carbene species (i) may be produced *in situ* from C_3H_5 ligands by thermal evacuation [330]. A carbene chain mechanism including a metallacyclobutane intermediate (ii) is currently accepted for olefin metathesis with both homogeneous and heterogeneous catalysts [352–361].

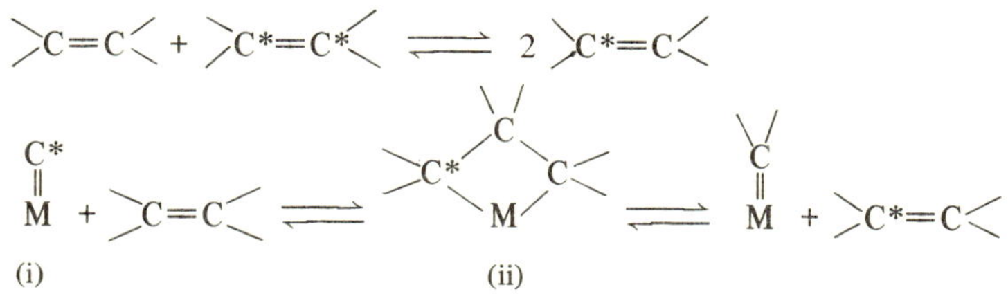

On the other hand, with oxomolybdenum(IV) catalysts which have no carbene sources, the initial rates of propene metathesis, d[ethene]/dt, were found to depend upon the position of deuteriation of the deuteriopropene [338]. Large isotope effects were observed in the metathesis of $D_2C{=}CHCH_3$ or $D_2C{=}CDCD_3$. The absence of an isotope effect during the metathesis of C_3D_6 or C_3H_6 on a catalyst where the carbene species had been produced in advance implies that the hydrogen isotope effects may be associated with the first step of the metathesis process, i.e. the formation of the initial carbene. It therefore follows that the breaking of a carbon–hydrogen bond of the methylene group, and not of the methyl group (allylic hydrogen), in a propene molecule must be involved in the initial stage of the reaction sequence [287]. Elimination of the methylene hydrogen atom is also indicated by the selective methylene–hydrogen exchange between C_3H_6 and C_3D_6 under metathesis conditions. Furthermore, the metathesis of ethene, $H_2C{=}CH_2 + D_2C{=}CD_2 \rightleftharpoons 2\,H_2C{=}CD_2$, proceeds readily on both Al_2O_3- and SiO_2-attached Mo catalysts, which excludes the involvement of a π-allylic intermediate in the mechanism [362–364]. It should be noted that Al_2O_3-attached molybdenum catalysts provide an almost selective methylene–hydrogen exchange, while SiO_2-attached molybdenum catalysts catalyze olefin metathesis without virtually any intermolecular hydrogen exchange. During metathesis of the C_3D_6–H_2 system, no hydrogen exchange occurred between C_3D_6 and H_2 and no incorporation of hydrogen atoms was observed into the metathesis products (C_2D_4 and C_4D_8). It is clear that the surface hydrogen atoms originally retained in the catalyst or derived from the addition of H_2 play no important role in metathesis, and for this reason a mechanism involving a half-hydrogenated propyl species [365, 366] is most unlikely. Consequently, on the basis of current tracer studies with active attached Mo monomer and dimer catalysts having well-defined reaction sites, the formation of the carbene complex (vi) from ethene or propene may be represented by a mechanism involving a molybdenum n-alkenyl complex (v) (Scheme 6). The reverse reaction of the sequence given in the scheme, i.e. (vi) $\longrightarrow$ (iii), has been demonstrated by ^{1}H and ^{13}C NMR spectroscopic studies [367].

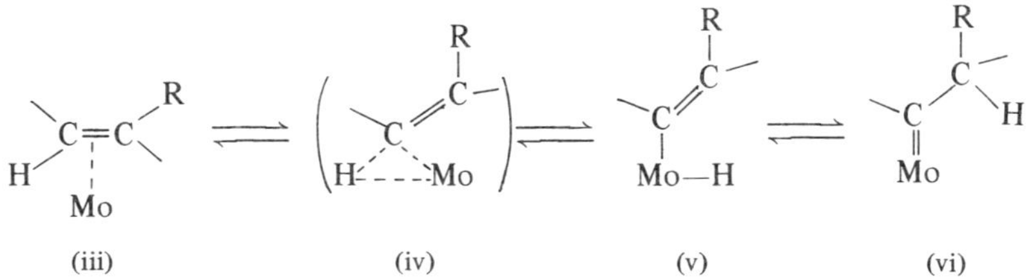

Scheme 6. Mechanism for initial carbene formation in olefin metathesis over attached oxomolybdenum catalysts.

9.3.2. *Surface molybdenum dimers*

Attached Mo dimer catalysts with well-defined surface structures may be synthesized molecularly employing the reaction between $Mo_2(\eta^3\text{-}C_3H_5)_4$ and the surface OH groups of SiO_2 or Al_2O_3, followed by suitable chemical treatment [329, 334–337]. Such surface Mo dimers must obviously rank among well-tailored metal catalysts.

$Mo_2(\eta^3\text{-}C_3H_5)_4$, which has an Mo—Mo bond distance of 0.2183 nm [368, 369], was reacted exclusively with the surface OH groups of silica as a 2.5–5 mmol dm^{-3} solution in pentane at 273 K. The intensity of the OH stretching band for SiO_2 in the IR spectrum diminished as a result of such Mo_2 attachment. When a silica sample of surface area 120 $m^2 g^{-1}$ and containing 3.5 OH groups nm^{-2}, but with Na^+ as an impurity, was used the surface composition $[(\text{⫶}Si{-}O)_2Mo(C_3H_5)]_2$ **(66)** was obtained, as shown in Scheme 7.

In the presence of H_2, regardless of the Mo quantity over the range 0.2–1.0 wt.%, almost the same TPDE pattern was observed indicating the nearly uniform nature of the surface Mo–allyl complexes **66** [335]. The elimination of the allyl ligands leads to the formation of a species containing Mo^{II} in a bidentate form **(67)**. The Mo^{II} ion may be oxidized quantitatively by molecular O_2 to yield the hexavalent molybdenum form. The Mo^{VI} species **70** thus obtained exhibited an XPS doublet peak which arises from emissions from the 3d level similar to those of the Mo^{VI} ions in pure MoO_3. The Mo^{VI} species **70** gave a $t_1 \longrightarrow 3t_2$ absorption at 290 nm assigned to the charge-transfer band ($Mo^{6+}{=}O^{2-}$) of a tetrahedral Mo^{VI} ion, similarly to an absorption at 285 nm for the tetrahedral dioxomolybdenum monomer. Those Mo^{VI} ions are located adjacent to each other as demonstrated by the complete quenching by one Mo^{VI} ion of the photoluminescence of an adjacent Mo^{VI} ion which is normally observed at 22.9×10^3 cm^{-1} and arises from the lowest triplet T_1 state through illumination of the charge-transfer band.

The tetrahedral dioxomolybdenum(VI) dimer **70** is reduced in a stepwise manner to the Mo^{IV} dimer **69** at 723 K and then to the Mo^{II} dimer **67** at

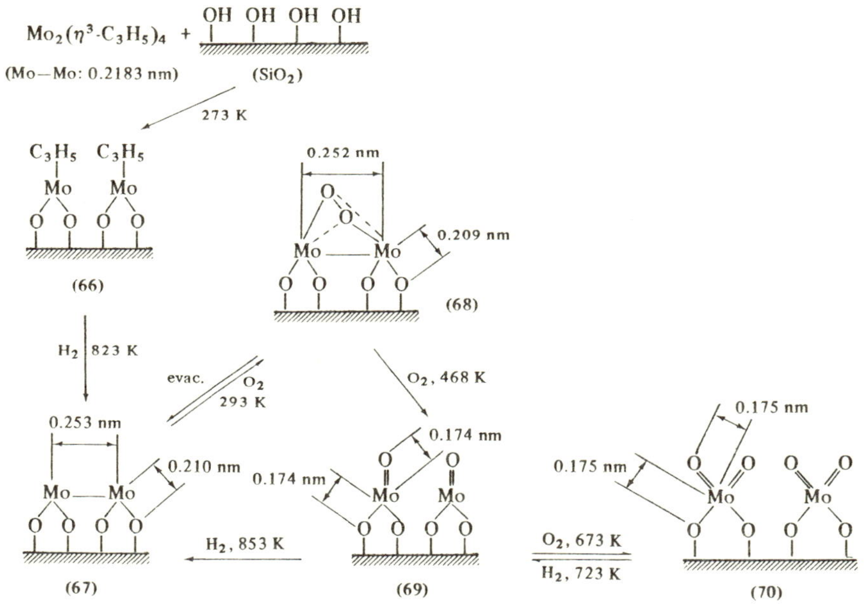

Scheme 7. Attachment of the Mo dimer onto silica and its structure transformations.

853 K, such reduction being accompanied by the uptake of a stoichiometric amount of H_2 and the quantitative evolution of water [335]. Conversely, the Mo^{II} dimer may be converted back to the Mo^{VI} dimer by the stoichiometric uptake of O_2 at 673 K. An unusual property was found with the Mo^{II} dimer **67** in that it binds dioxygen reversibly over the temperature range 273–428 K, as demonstrated by the results depicted in Figure 7 [335, 337]. Although it has been known previously that single Mo^{II} ions are instantaneously and irreversibly oxidized to Mo^{IV} ions in the presence of O_2 at 273 K, this attached Mo dimer–oxygen complex is the first reversible oxygen carrier reported among the Group 6A metals for which there is no analogy in solution.

These various surface transformations may be monitored directly by means of the EXAFS technique using an *in situ* cell. Typical Fourier transform of the EXAFS spectra of the SiO_2-attached Mo dimer catalysts during such surface molybdenum structure changes are shown in Figure 8(A) [370]. Even when the Mo loadings are as small as 0.5 wt.% good EXAFS spectra obtained, demonstrating an immediate advantage of well-tailored metal catalysts. The bond distances as determined by theoretical and empirical curve-fitting methods are given in Scheme 7. The Mo^{VI} dimer **70** exhibited an average Mo—O bond length of 0.175 nm which is typical for a tetrahedral MoO_4 structure, the coordination

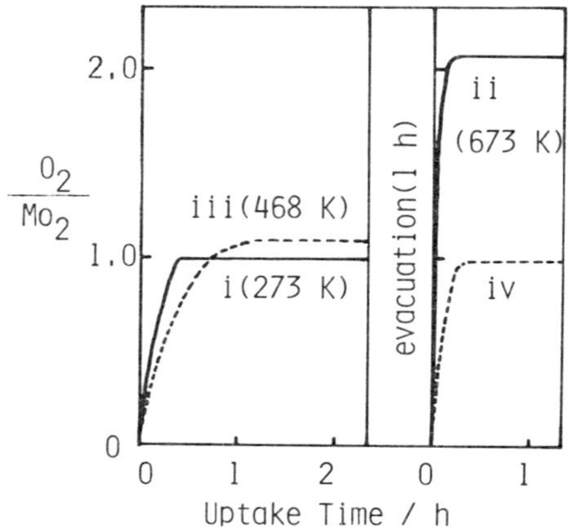

Fig. 7. Oxygen uptake on the Mo^{II} dimer structure **67** (Mo/SiO_2 = 0.68 wt.%). Step a (—): (i) exposure to O_2 (13.3 kPa) at 273 K followed by evacuation at 273 K for 1 h, then (ii) exposure to O_2 (9.3 kPa) at 673 K. Step b (- - -): (iii) exposure to O_2 (9.3 kPa) at 468 K followed by evacuation at 468 K for 1 h, then (iv) exposure to O_2 (9.3 kPa) at 673 K.

number of 4.2 obtained also supporting the T_d symmetry of the structure. A drastic change in structure is observed, however, when the Mo species are reduced to the Mo^{II} level (**67**). A new peak which may be attributed to an Mo—Mo bond appears at a bond distance of 0.253 nm whilst at the same time the Mo—O single bond distance is lengthened by ca. 0.03 nm. It was concluded that in **67** the number of molybdenums adjacent to a given Mo atom was unity. When this structure was exposed to oxygen, no essential change in the Mo—Mo distance occurred, suggesting that the reversible oxygen binding may be attributed to the unusual resistance of the Mo—Mo bond to breakage. However, irreversible oxidation of Mo^{II} ions to the oxo-Mo^{IV} structure **69** was accompanied by such Mo—Mo bond breaking. The essential features of this molybdenum–oxygen bonding may be affected by the Mo—Mo separation, its local stereochemistry and electronic structure, and also basic impurities in the support surface. A theoretical description of reversible oxygen bonding onto the Mo^{II}—Mo^{II} species **67** is not as yet available. The reversibly bound oxygen is active in the oxidation of carbon monoxide, cyclohexene and NO [337].

A surface molybdenum–allyl complex **71** of composition $[(\equiv Si-O)_2Mo(C_3H_5)_2]_2$ can be prepared via reaction between $Mo_2(C_3H_5)_4$ and a silica gel with 4.6 OH groups nm^{-2} and a surface area of 285 m^2g^{-1} (Scheme 8) [334, 335].

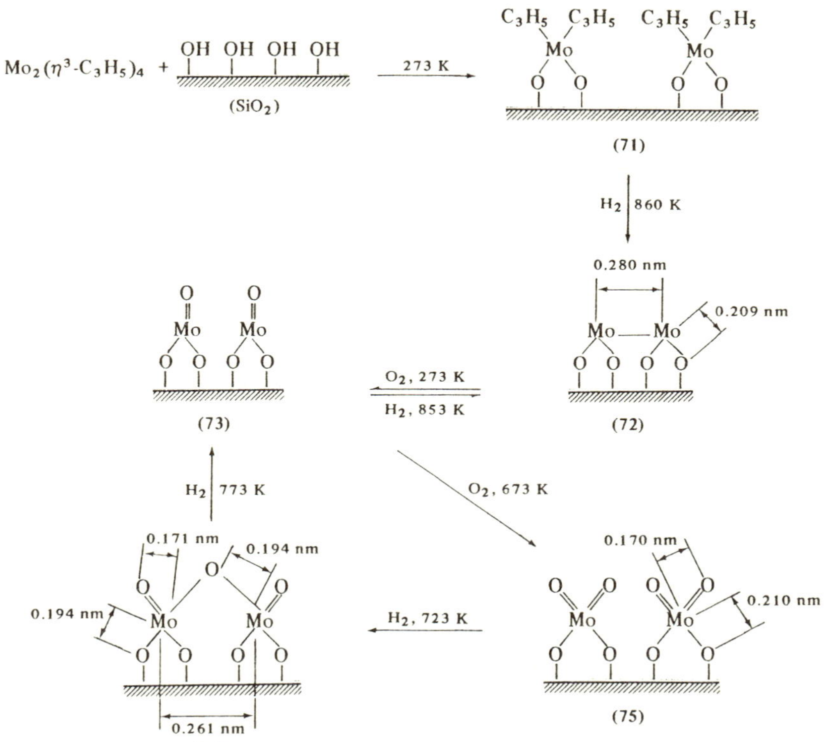

Scheme 8. Attachment of the Mo dimer onto silica and its surface transformations.

The Mo^{II} dimer structure on this support (**72**) exhibits an Mo—Mo bond distance of 0.280 nm which is much larger than the value of 0.253 nm reported for structure **67**. Species **72** was sensitive to oxygen and was oxidized in the nomal fashion to the Mo^{IV} dimers **73** at 273 K and to the Mo^{VI} species **75** at 673 K with a stoichiometric uptake of oxygen. This Mo^{VI} species exhibits considerable distortion as demonstrated by EXAFS studies [370]; the Fourier transforms of the spectra obtained are shown in Figure 8(B). In this case Mo=O double bonds are observed with a bond distance of 0.170 nm while the Mo—O single bond distance is as large as 0.210 nm. The DRS spectrum (charge-transfer band at 306 nm) suggests that Mo^{VI} ions are associated in 5- or 6-coordination, while the XANES spectrum excludes the possibility of octahedral symmetry. A direct Mo—Mo bond was not observed at room temperature.

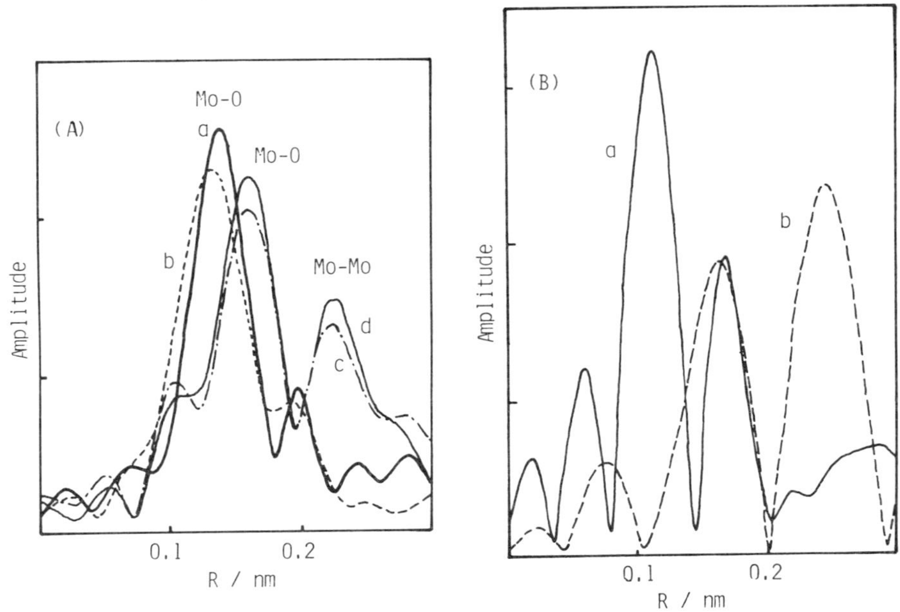

Fig. 8. Fourier transforms of EXAFS spectra arising from SiO_2-attached Mo dimer catalysts. (A): curve a, $[(\equiv Si-O)_2Mo(=O)_2]_2$ (**70**); curve b, $[(\equiv Si-O)_2Mo{=}O]_2$ (**69**); curve c, $[(\equiv Si-O)_2Mo]_2$ (**67**); curve d, $[(\equiv Si-O)_2Mo]_2(O_2)$ (**68**).
(B): curve a, $[(\equiv Si-O)_2Mo(=O)_2]_2$ (**75**); curve b, $[(\equiv Si-O)_2Mo]_2$ (**72**).

In contrast to species **70**, the Mo^{VI} dimer **75** was first reduced to the Mo^{V} level on treatment with H_2 at 723 K. The DRS of the resulting diamagnetic Mo^{V}-level species **74** (with peaks at 385 nm (main) and 442 nm (sh)) is not a superimposition of the Mo^{VI} spectrum (peak at 306 nm) and the Mo^{IV} spectrum (peak at 369 nm). EXAFS and XANES data also supported the oxygen-bridged structure **74**, where the Mo—Mo separation was shown to be 0.261 nm. The Mo—O bond distance decreased, the Mo=O bond distance being unchanged.

The attached Mo^{VI} dimers **75** function much more efficiently as a catalyst for the oxidation of ethanol over the temperature range 400–450 K than a conventional impregnation catalyst [339]. The oxidation of ethanol on such Mo^{VI} dimers takes place simultaneously with the cooperative redox cycle $Mo^{VI}_2 \rightleftharpoons Mo^{V}_2$ as illustrated in Scheme 9 where the two adjacent molybdenum(VI) species provide an active site.

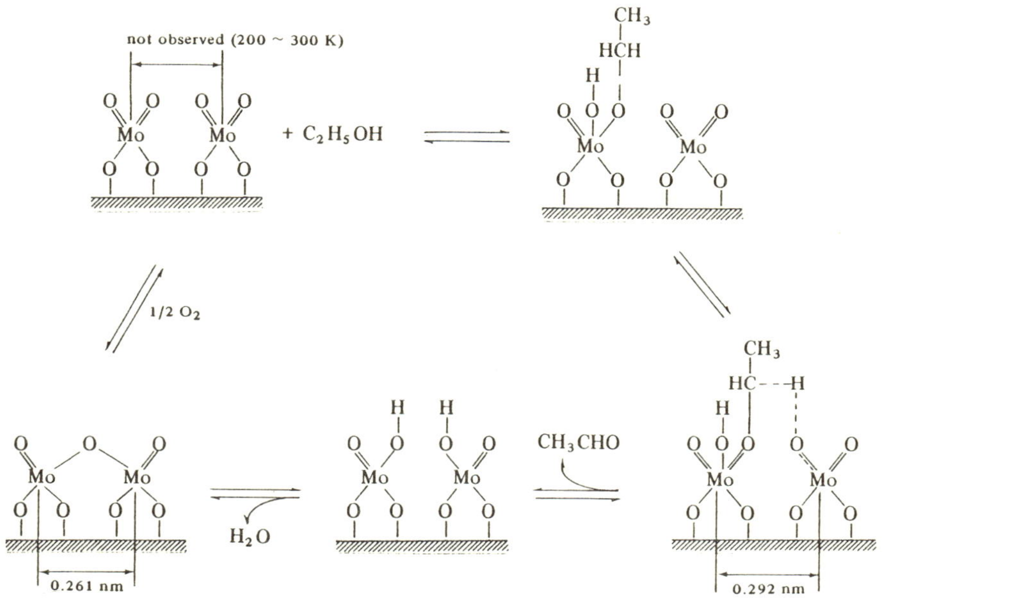

Scheme 9. Mechanism for the oxidation of ethanol on an attached Mo^{VI} dimer catalyst.

Molybdenum species with a lower valency such as $Mo^{V}{}_2$ (74), $Mo^{IV}{}_2$ (73) and $Mo^{II}{}_2$ (72) were much less active or virtually inactive. The dissociative adsorption of ethanol on the Mo^{VI} ion under the reaction conditions employed is rapid as demonstrated by DRS (shift of the 307 nm peak to 336 nm due to an increase in the coordination number) and by IR (1360–1450 cm^{-1} absorption band corresponding to CH_3 and CH_2 deformation and a C—H stretching band at 2980–2900 cm^{-1}). It was estimated from EXAFS analyses after ethanol adsorption in the absence of oxygen that the Mo—Mo separation corresponds to a distance of 0.292 nm brought about by hydrogen bonding as shown in Scheme 9. Most of the Mo^{VI} dimers were reduced to Mo^{V} dimers (**74**) with an Mo—Mo separation of 0.261 nm as a result of the oxidative dehydrogenation of ethanol for 60 min at 433 K in the absence of oxygen. Such changes were followed by EXAFS and DRS methods, the results suggesting that $H^{\delta-}$ abstraction of the methylene hydrogen occurs through cooperative catalysis involving two adjacent Mo^{VI} species in the transition state associated with the dehydrogenation step. The monomer is more efficient per Mo weight than the dimer as far as the rate of reaction is concerned, but the dimer has a greater influence on the turnover frequency (rate/site). It should be noted that each individual molybdenum ion associated with a dimer site acts in consort with the adjacent molybdenum ion of the dimer located at a suitable distance from it rather than acting independently. This leads to the loss of the original character of the ion.

The Al_2O_3-attached Mo dimer catalysts may be prepared by taking advantage of the reaction between $Mo_2(\eta^3\text{-}C_3H_5)_4$ and the surface OH groups of $\gamma\text{-}Al_2O_3$ (surface area, 190 $m^2 g^{-1}$). The pattern of the synthesis is similar to that in Scheme 8 although the temperatures required for each transformation step are different [334, 336]. The resulting surface complex, i.e. $[(\equiv Al{-}O)_2Mo(C_3H_5)_2]_2$ (**76**), exhibited a fairly uniform TPDE chromatogram over the composition range $Mo/Al_2O_3 = 0.2{-}1.0$ wt.% [336]. The surface distribution of Mo dimers in such Al_2O_3-attached catalysts is as illustrated in Figure 9 where the distribution of hydroxyl groups (2.6 OH groups nm^{-2}) was obtained in a similar way to that depicted in Figure 6 and the Mo dimers (0.6 wt.%) were distributed as assemblies involving four OH units on the basis of the stoichiometric reaction between $Mo_2(\eta^3\text{-}C_3H_5)_4$ and four OH groups.

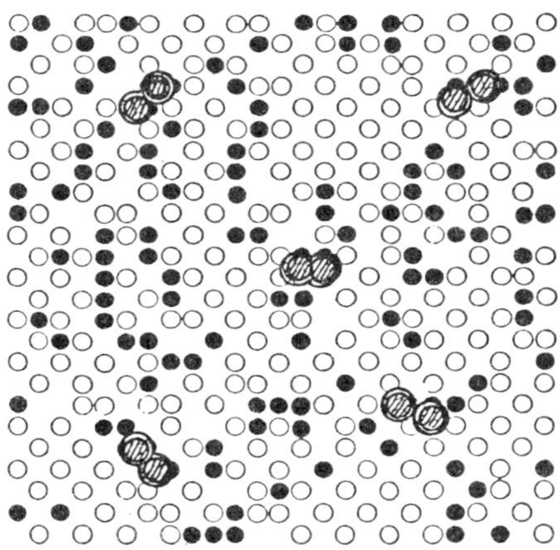

Fig. 9. The distribution of molybdenum dimers on an Al_2O_3 surface after removal of 79% of the OH groups; ○, surface lattice oxygen atoms; •, surface OH groups; (dimer symbol), molybdenum dimer units. Average $Mo_2{-}Mo_2$ separation, 3.0 nm.

The species $[(\equiv Al{-}O)_2Mo(=O)_2]_2$ (**77**) is characterized by $2t_2 \longrightarrow 3t_2$ (225 nm) and $t_1 \longrightarrow 3t_2$ (283 nm) absorptions, similar to the monomer **64**, which may be assigned to the charge-transfer bands of a tetrahedral Mo^{VI} ion, suggesting that each Mo^{VI} ion of the 'paired' structure **77** has a similar structure to that enjoyed in **64**. Two characteristic absorption peaks at 310 and 405 nm attributed to the bismolybdenyl(V) structure developed during the course of the stoichiometric reduction of Mo^{VI} to Mo^V. The Mo^V ion could be further reduced in a stepwise manner to Mo^{IV} and then to the Mo^{II} level, taking up during this process a stoichiometric amount of H_2 and evolving the stoichiometric (100%) amount of water [336]. It should be noted that this reduction step does not proceed quantitatively if a conventional Al_2O_3-supported catalyst is employed. The Mo^{II} dimer also showed a much higher activity during ethene hydrogenation at 200 K (turnover frequency > 5 s^{-1} at a C_2H_4 pressure of 20.3 kPa and an H_2 pressure of 81.1 kPa) than the usual Mo catalyst [329].

When the alumina surface is rougher (surface area 385 $m^2 g^{-1}$, 4.5 OH groups nm^{-2}), the 5-coordinated oxygen-bridged structure **83** can be produced in the manner depicted in Scheme 10. This Mo^{VI} species exhibits a charge-transfer (Mo=O) transition peak at 353 nm, quite different in position from the 333 nm absorption peak for an octahedral Mo^{VI} ion attached to an identical Al_2O_3 surface and also from the 283 nm band for a tetrahedral Mo^{VI} species. In contrast to these other two cases, reduction of the Mo^{VI} dimer **83** and oxidation of the Mo^{II} dimer **79** did not lead to the formation of the same Mo^{IV} species, the spectroscopically and catalytically distinct oxomolybdenum(IV) dimer **81** and the di-μ-oxodimolybdenum(IV) species **80** being formed in these respective steps.

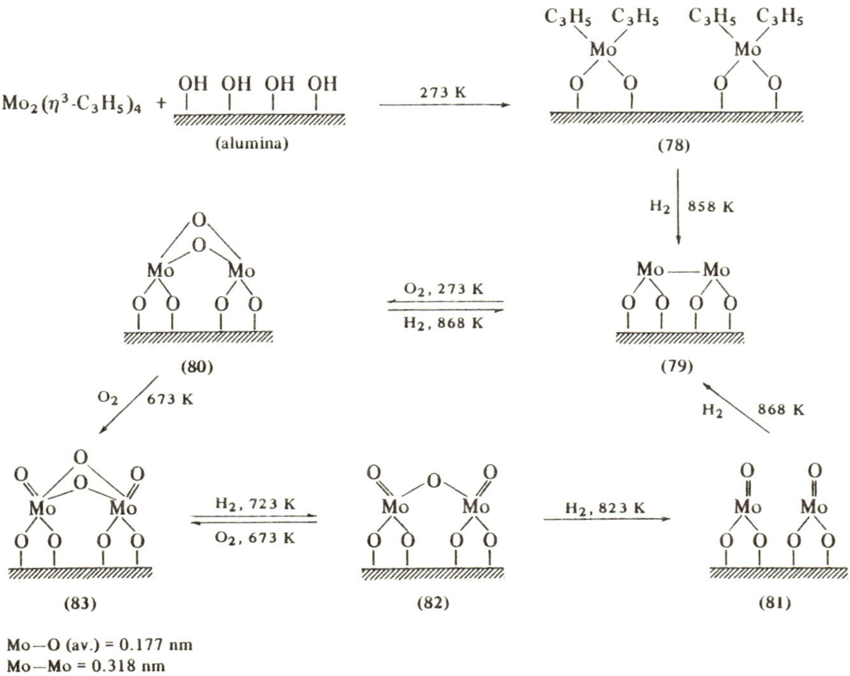

Scheme 10. Attachment of the Mo dimer onto alumina and its structure transformations.

The existence of structure **83** is supported by EXAFS data [371]. Both empirical and theoretical curve-fitting analyses have been performed with the oxygen atoms in the first shell and molybdenum atoms in the second shell using a k^3-weighting scheme over the k-range 40–120 nm^{-1}. From this it was concluded on the basis of the bond distance and the coordination number that one Mo atom is located 0.318 nm away from an absorbing Mo atom which is surrounded by five oxygen atoms at an average distance of 0.177 nm [371]. The surface structures **79**–**83** depicted in Scheme 10 may be mutually transformed via stoichiometric reduction and oxidation reactions as the results shown in Figure 10 demonstrate [336].

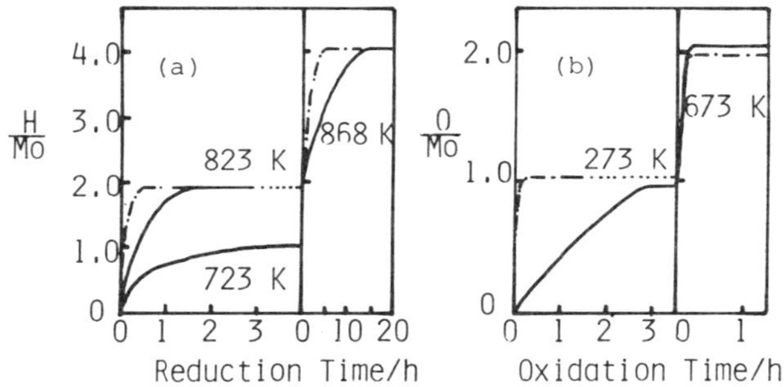

Fig. 10. (a) Reduction curves for the Al_2O_3-attached Mo^{VI} monomer (**64**) (-··-) and the Al_2O_3-attached Mo^{VI} dimer species (**83**) (Mo/Al_2O_3 = 0.56 wt.%) (—) in the presence of H_2 at 8.0 kPa pressure. (b) Oxidation curves for the Al_2O_3-attached Mo^{II} monomer (**62**) (-··-) and the Al_2O_3-attached dimer species (**79**) (—) in the presence of O_2 at 8.0 kPa pressure.

The homogeneous dinuclear Mo^{II} complexes with a metal–metal bond between the two adjacent Mo atoms exhibit new absorption bands associated with d_σ, d_π or δ orbitals [372–377]. The d–d absorption which occurs in the > 500 nm range for the Al_2O_3-attached Mo^{II} dimer **79**, but which is absent for the single Mo^{II} ion, indicates direct Mo—Mo bonding. It has been suggested that the Mo^{II} dimer catalyst **79** exhibits cooperative catalysis via the two adjacent Mo^{II} ions in the hydrogenation of 1-butene and 1,3-butadiene as shown in Table VI. The Mo^{II} dimer **79** has an activity which is three orders of magnitude greater than that of the reduced Mo catalyst produced by conventional methods and is also ca. one order of magnitude more active than the Mo^{II} monomer catalyst **62**. The activation energy for the hydrogenation decreases if direct bonding between the Mo atoms is effected at a suitable distance, e.g. ca. 0.28 nm, but a much shorter distance, e.g. ca. 0.25 nm, is unfavorable.

TABLE VI
Activities and selectivities of Al_2O_3-attached molybdenum catalysts for the hydrogenation of 1,3-butadiene and 1-butene at 273 K.

Reactants	Products	Selectivities (%)					
		Mo dimer		Mo monomer		Impreg. cat.[a]	
1,3-butadiene	*trans*-2-butene	20	40	30	60	78	96
	cis-2-butene	20		30		18	
	1-butene	60		40		4	
Relative activity		3280		660		1	
1-butene	butane	100		100		—	
Relative activity		21		1		—	

[a] A conventional molybdenum catalyst obtained by an impregnation technique.

9.3.3. *Oriented MoO_3 phases on graphite*

Various metal complexes can insert between the layers in graphite and can be used as active centers for catalysis. By the reduction of transition metal compounds, lamellar graphite compounds containing metals can be obtained. Such intercalation compounds have been shown to be catalytically active in various reactions such as *ortho–para* hydrogen conversion, hydrogenation, dehydrogenation, olefin isomerization, dimerization, ammonia synthesis and Fischer–Tropsch synthesis [378–383]. In general, transition metal–graphite intercalation compounds do not preserve their well-defined structures during catalysis and hence the active structures have not been characterized.

Oxychlorination of the $MoCl_5$–graphite intercalation compound (ν_{Mo-Cl} 351 cm^{-1}) leads to the formation of MoO_3 (ν_{Mo-O} 989, 873 and 605 cm^{-1}) probably via $MoOCl_3$ and MoO_2Cl_2 as intermediates [384]. When the reaction was performed at 673–773 K, MoO_3 oriented phases were produced [385–387] with the MoO_3 (010) crystalline faces parallel to the (001) graphite sheets. Electron diffraction microscopy and scanning electron microscopy confirmed that MoO_3 crystal growth is epitaxial on graphite, and that a silica support does not induce such preferential orientation. The specific influence of oriented MoO_3–graphite on the catalysis of propene oxidation has been observed [388, 389], selectivity for acrolein and CO_2 formation being correlated with the development of the (100) side face and (010) MoO_3 basal face, respectively. Active sites for acrolein formation may be formed at the edge of the (100) crystale face. Such epitaxial orientation may be applied to the well-controlled preparation of layer structures attached to inorganic supports.

9.3.4. *Surface chromium monomers and dimers*

Silica-attached Cr dimer catalysts may be synthesized using the dinuclear complex $Cr_2(\eta^3\text{-}C_3H_5)_4$ as illustrated in Scheme 11 [325, 326].

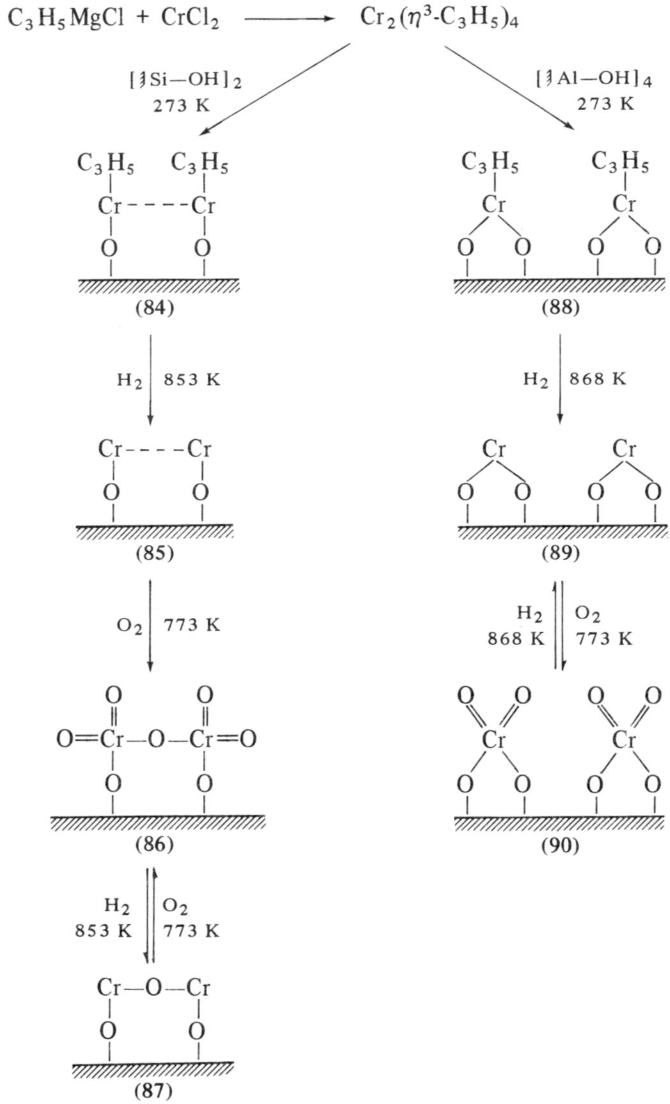

Scheme 11. Attachment of the dinuclear Cr complex onto silica or alumina and its subsequent surface transformations.

The attachment of $Cr_2(\eta^3\text{-}C_3H_5)_4$ to the surface OH groups of SiO_2 (1.3 OH groups nm^{-2}) yields the new surface complex **84** with a C_3H_5/Cr ratio of 1.0, which is characterized by an absorption peak at 553 nm which may be assigned to the $\delta \longrightarrow \delta^*$ ($^1A_{1g} \longrightarrow {}^1A_{2u}$) transition corresponding to the dinuclear Cr^{II}—Cr^{II} structure [325]. The bond distance between two adjacent bivalent chromium atoms has been estimated as 0.222 nm, assuming a linear relationship between the metal–metal bond distance and the energy (18.1×10^3 cm^{-1}) of the $^1A_{1g} \longrightarrow {}^1A_{2u}$ transition [373–376]. This value is much larger than that of 0.1975 nm for the same distance in the $Cr_2(\eta^3\text{-}C_3H_5)_4$ crystal [390–393].

The surface complex **84** exhibits a simple TPDE peak as shown in Figure 11, indicating the uniform nature of the species. This allyl-attached complex is remarkably thermostable in comparison to $Cr_2(\eta^3\text{-}C_3H_5)_4$. The Cr^+ (d^5) dimer **85** (gray color) obtained from **84** via allyl elimination gave no ESR signal and its DRS was very different from other bi to hexa-valent species. The surface species **85** could be oxidized by O_2 at 773 K to the dichromate structure **86** with a characteristic peak at 485 nm. Species **86** is readily reduced with H_2 at 853 K to the Cr^{II} dimer **87** with stoichiometric hydrogen uptake [326]. The rapid oxidation of **87** to **86** at 673 K also proceeds stoichiometrically.

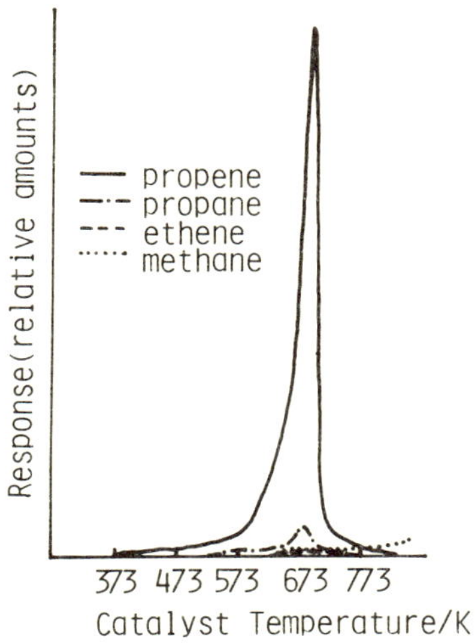

Fig. 11. Temperature programmed decomposition curve of the attached dinuclear complex **84** in the presence of hydrogen. Heating rate, 4 K min^{-1}; catalyst weight, 0.35 g.

Alumina-attached Cr dimer catalysts may also be obtained via the reaction between $Cr_2(\eta^3\text{-}C_3H_5)_4$ and the surface OH groups of $\gamma\text{-}Al_2O_3$ in a 1 : 4 stoichiometric ratio [325]. The C—C stretching frequency (1563 cm^{-1}) of the surface complex **88** thus obtained is intermediate between the values of ca. 1650 cm^{-1} for the σ-allyl and ca. 1500 cm^{-1} for the π-allyl type, suggesting a dynamic σ-allyl or static π-allyl (blue shift on attachment) type of ligand. The dependence of the ESR signal intensity of **88** ($g = 2.03_5$, $\Delta H = 1300$ G) upon temperature suggests a weak, direct exchange interaction between two adjacent Cr^{III} ions.

Complex **88** has been converted quantitatively into the Cr^{II} dimer **89** in which there is no direct bonding between the two adjacent Cr^{II} ions situated at a distance < 0.24 nm unlike the situation in complex **84**. Species **89** was oxidized to the paired monochromate(VI) structure **90** with a charge-transfer band at 386 nm. The Cr^{VI} ions in **90** could be reduced back to Cr^{II} (**89**) at 868 K with a stoichiometric uptake of H_2. In contrast, conventional Al_2O_3-supported catalysts show nonstoichiometric and heterogeneous behavior in the redox steps; the reduction of Cr^{6+} ions at 868 K led to $Cr^{2.15+}$ as an average composition.

The corresponding Cr monomer catalysts were similarly prepared using a mononuclear $Cr(\eta^3\text{-}C_3H_5)_3$ complex and the following steps:

$$\underset{(91)}{[\wr Al{-}O]_2Cr(C_3H_5)} \xrightarrow{H_2} \underset{(92)}{[\wr Al{-}O]_2Cr} \underset{H_2}{\overset{O_2}{\rightleftharpoons}} \underset{(93)}{[\wr Al{-}O]_2Cr(=O)_2}$$

The Cr^{II} monomer (**92**) is remarkably active for propene hydrogenation at temperatures as law as 195–260 K. However, a problem arises with the highly active, attached Cr^{II} monomer catalyst in that the catalytic activity shows a drastic decrease as a result of the unfavorable environmental change which occurs during the oxidation–reduction cycles necessary for catalyst retreatments, i.e. **92** $\rightleftharpoons$ **93**; during this procedure, the number of unsaturated coordination sites on a given Cr atom decreases from 2.6 to 0.9. In contrast, the activity of the Cr^{II} dimers **89** are not affected in this way. Although the average number of vacant sites decreases marginally from 1.9 to 1.5, this change is compensated by a decrease in the activation energy for the reaction from 17.5 to 14.6 kJ mol^{-1} [325]. The Cr–allyl species is not transformed to an active Cr—H species on treatment with H_2. Of the structures **84**–**93**, the coordinatively unsaturated, Al_2O_3-attached Cr^{II} dimer **89** is the most active and stable structure for alkene hydrogenation at low temperatures. Both the Cr^{II} monomer **92** and the Cr^{II} dimer **89** have also been shown to be about two orders of magnitude more active than conventional Cr catalysts in effecting the double-bond shift of 1-butene. Reaction on the attached catalysts proceeds via a π-allyl anion intermediate whereas conventional catalysts utilize a carbenium cation intermediate.

9.3.5. *Two-dimensional surface* Cr^{III} *phase attached onto* SiO_2 [22, 326]

A two-dimensional Cr^{III} layer attached onto an SiO_2 surface **(100)** may be produced via the reaction between excess $Cr(\eta^3\text{-}C_3H_5)_3$ and a full monolayer of OH groups on SiO_2, followed by stepwise transformations of the type shown in Scheme 12.

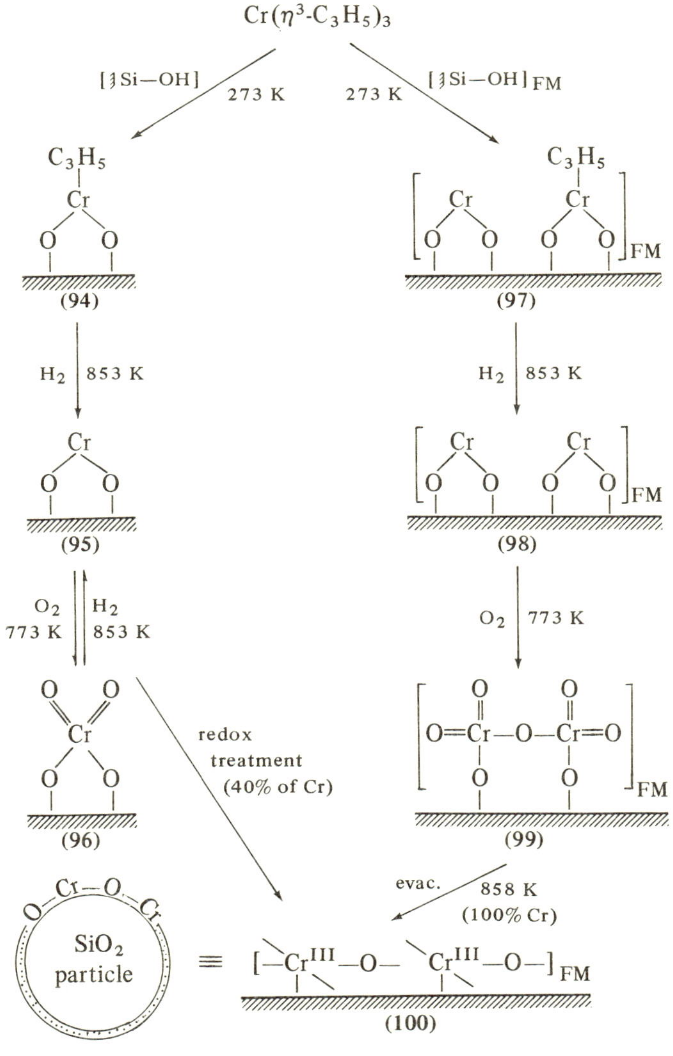

(FM = full monolayer)

Scheme 12. Stepwise transformation of attached Cr complexes to form a two-dimensional Cr^{III} layer.

$Cr(\eta^3\text{-}C_3H_5)_3$ can react with a full monolayer of OH groups on surfaces with pores $\geqslant 3.0$ nm in diameter. The two-dimensional Cr^{II} phase **98** may be oxidized to the dichromate(VI) structure **99** which is characterized by a charge-transfer band at 485 nm. This Cr^{VI} monolayer (**99**) can be quantitatively converted by evacuation at 858 K to the Cr^{III} layer **100** as has been demonstrated by DRS (463 nm ($^4A_{2g} \longrightarrow {}^4T_{2g}$) and 607 nm ($^4A_{2g} \longrightarrow {}^4T_{1g}$)) and by ESR spectroscopy (β phase Cr^{III}, antiferromagnetic at low temperatures). Adsorption measurements of CO at 293 K and of O_2 at 200 K indicate one vacant site on each Cr atom contained in the two-dimensional phase after evacuation of the latter at 858 K. It is almost impossible to prepare the two-dimensional Cr^{III} layer **100** by the usual impregnation methods; a traditional supported Cr catalyst involves a three-dimensional Cr_2O_3 phase, Cr^V and Cr^{VI} ions, in addition to a two-dimensional Cr^{III} assembly.

The coordinatively unsaturated Cr^{III} layer attached onto an SiO_2 surface (**100**) has been shown to be very active for the catalytic oxidation of CO at ca. 470 K, with the catalytic activity being two orders of magnitude greater than that of a conventional Cr catalyst. The existence of a $Cr^{3+}—O^{2-}$ bond in Cr_2O_3 has been well characterized via its Racah parameter (B_{35}) as employed in ligand field theory [394, 395];

$$B_{35} = (\delta E/15)\,[(\Delta - \delta E)/(0.8\Delta - \delta E)]$$

where Δ and δE represent the energy of the first d–d transition of the Cr^{III} ions and the energy difference between first and second absorption bands, respectively. Smaller values of B_{35} indicate a stronger exchange interaction between d^3 ions, i.e. a shorter $Cr^{III}—Cr^{III}$ distance. The value of 481 cm^{-1} observed for species **100** suggests that the $Cr^{3+}—O^{2-}—Cr^{3+}$ layer possesses suitable exchange interaction and is structurally more relaxed than that of crystalline α-Cr_2O_3 which has a Cr—Cr distance of 0.265 nm. In Figure 12 the activation energies for the oxidation of CO with O_2 in the presence of various catalysts have been plotted against B_{35} where a 'reversed volcano' relationship is shown. From this it follows that the structurally relaxed surface structure **100** containing square-pyramidal Cr^{III} ions and capable of dissociating O_2 molecules to generate active oxygen provides the most active structure for catalysis of CO oxidation. The oxidation of CO with O_2 over the two-dimensional layer **100** is well expressed by a two-stage redox mechanism

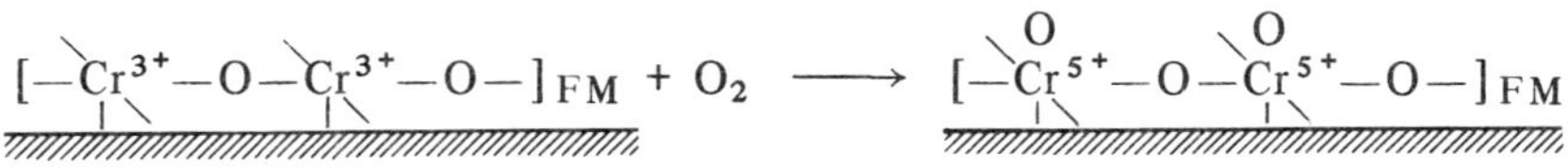

(**100**) (**101**)

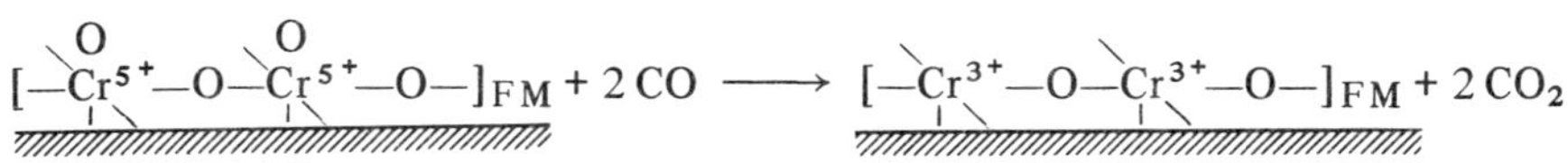

(**101**) (**100**)

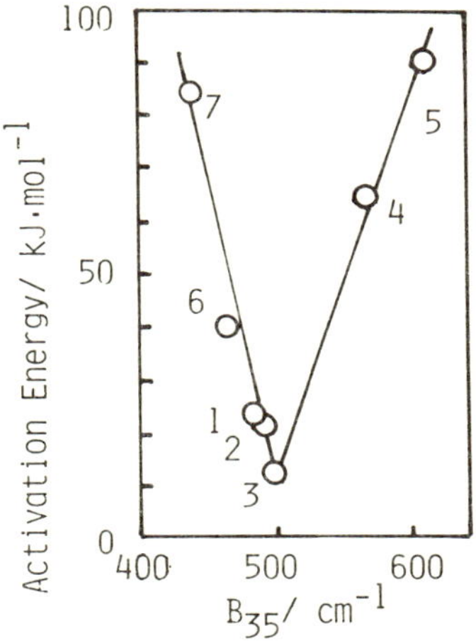

Fig. 12. Plot of the Racah parameter B_{35} and activation energy for CO oxidation with various catalysts. 1, Cr^{III} monolayer (100%) (**100**); 2, Cr^{III} in the dinuclear catalyst; 3, Cr^{III} obtained by regeneration of the surface of the mononuclear catalyst **96**; 4, impregnation catalyst (2.72 wt.% Cr); 5, impregnation catalyst (0.9 wt.% Cr); 6, Cr_2O_3; 7, Cr_2O_3.

where FM depicts full monolayer coverage. The existence of the intermediate Cr^V species **101** has been confirmed by DRS in which peaks at 705 and 540 nm were obtained and assigned to $^2B_{2g} \longrightarrow {}^2E_g$ and $^2B_{2g} \longrightarrow {}^2B_{1g}$ transitions, respectively, for the shortened octahedral Cr^V species [22]. The Cr^{III} monolayer catalyst **100** has also been shown to be remarkably active for the gas-phase polymerization of ethene at 273 K. In this case the sequence of activity for the various structures is Cr^{III} monolayer > [⫶Si—O]$_2$Cr^{II} > [⫶Si—O]$_2$Cr^{III}(C_3H_5) >

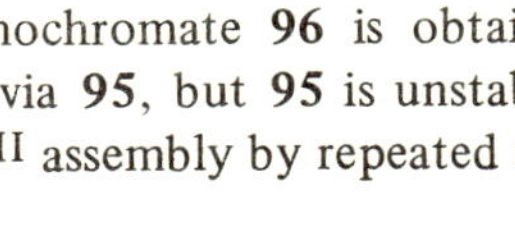

≈ impregnated Cr catalyst (reduced) > [⫶Si—O]$_2$Cr(=O)$_2$.

The SiO_2-attached monochromate **96** is obtained from the initial surface mononuclear complex **94** via **95**, but **95** is unstable being partially rearranged to the two-dimensional Cr^{III} assembly by repeated redox treatments.

9.3.6. *Attached tungsten monomers*

W(η^3-C_4H_7)$_4$ reacts with SiO_2 after dehydration of the latter at 673 K to form mainly [⫶Si—O]$_2$W(η^3-C_4H_7)$_2$ (**102**) but the C_4H_7/W ratio in the complexes changes from 1 to 3 as the initial dehydration temperature of SiO_2 is increased from 473 K to 873 K. Reduction of the surface complex **102** with H_2 at 673–

873 K leads to the formation of two kinds of W^{II} species in different environments. One type of species is oxidized stepwise to W^{IV} and then to W^{VI}, while th other is oxidized directly to the W^{VI} level [342]. The surface complex $[\equiv Al{-}O]_2W(\eta^3\text{-}C_3H_5)_2$ (**103**), formed by attachment to $\gamma\text{-}Al_2O_3$, was highly active for propene metathesis at 298 K, while the oxo species $[\equiv Al{-}O]_2W{=}O$ (**104**) also catalyzed the metathesis but with a much lower activity than **103** [343]. A supported tungsten catalyst prepared by traditional impregnation methods was inactive under similar conditions. Much work remains to be undertaken on tungsten catalysts, e.g. details of surface characterization, preparation of the dimer and tetramer, etc.

9.3.7. *Surface cobalt species*

$Co_2(CO)_8$ supported on SiO_2 or $\gamma\text{-}Al_2O_3$ can be transformed into various structures and cobalt oxidation states as shown in Scheme 13 [396, 397].

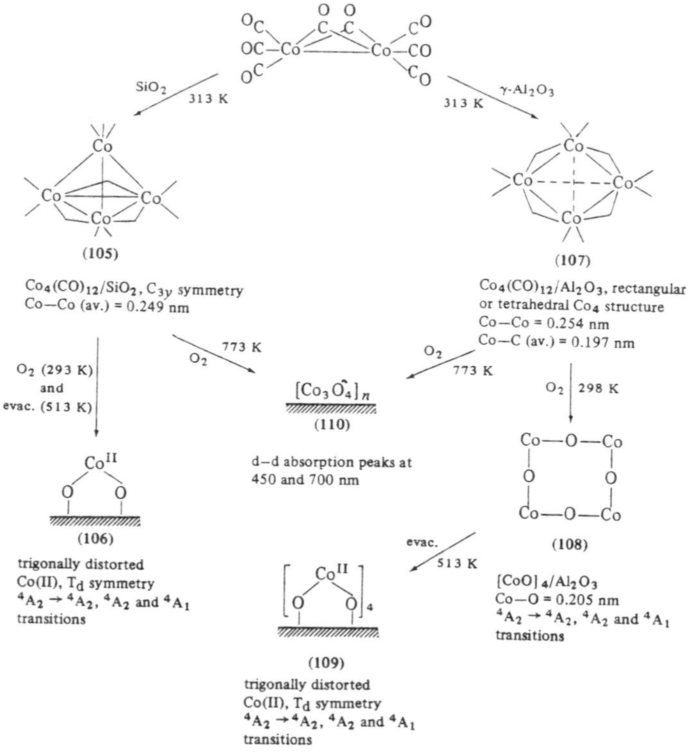

Scheme 13. Transformations of surface cobalt species.

When $Co_2(CO)_8$ was supported on an SiO_2 surface pretreated at 473 K, dimerization occurred to form a $Co_4(CO)_{12}$ cluster with C_{3v} symmetry, exhibiting a $\sigma \longrightarrow \sigma^*$ transition at 375 nm and liberating one CO molecule per Co atom. On exposing the cluster to O_2 at 293 K for 4 h and subsequent heating to 513 K under vacuum, the cobalt was oxidized to a Co^{II} species (**106**) by reaction with the surface OH groups of SiO_2. The Raman spectrum of this species exhibits a new peak at 975 cm^{-1} attributed to the Co—O—Si bond formed by the surface reaction. The DRS showed distinct peaks at 675, 596 and 520 nm which were assigned to the transitions $^4A_2(F) \longrightarrow {}^4A_2(P)$, $^4A_2(F) \longrightarrow {}^4A_2(P)$ and $^4A_2(F) \longrightarrow {}^4A_1(P)$, respectively, associated with a Co^{II} ion situated in T_d coordination with low symmetry perturbation. Transformation of a supported $Co_2(CO)_8$ complex provides a new way for preparing Co^{II} ions in T_d coordination dispersed on SiO_2 over a wide range of Co loadings up to nearly complete monolayer coverage; such species exhibit almost uniform characteristics in the loading range 1–5 wt.%.

Supporting $Co_2(CO)_8$ on γ-Al_2O_3 pretreated at 573 K also led to the formation of $Co_4(CO)_{12}$. The Co—Co bond distance in the resulting surface cluster **107** was shown to be 0.254 nm by EXAFS studies. This value is larger than that of 0.249 nm for $Co_4(CO)_{12}$ of C_{3v} symmetry on SiO_2, close to the value of 0.253 nm for pure $Co_2(CO)_8$. However, the large increase in the coordination number of the Co atoms surrounding a given Co atom (EXAFS) when $Co_2(CO)_8$ is supported on γ-Al_2O_3, the appearance of a new absorption peak at 413 nm similar to the peak for a γ-Al_2O_3-supported $Co_4(CO)_{12}$ cluster and the value of 0.197 nm for the average Co—C bond distance suggest the presence of a Co tetramer rather than a Co dimer. $Co_4(CO)_{10}(PPh_3)_2$, which has a rectangular Co_4 structure, has a Co—Co bond (bridging CO) distance of 0.252 nm; the bridging CO ligand leads to a stretching vibration in the IR spectrum at 1866 cm^{-1} [398]. Both these values are close to the respective values (0.253 nm and 1855 cm^{-1}) for the $Co_4(CO)_{12}$ cluster supported on γ-Al_2O_3. The possibility of a mutual transformation between structures with T_d (no bridging CO ligand), C_{3v} (three bridging CO ligands) and D_{2d} (tetrahedral Co_4 unit with four bridging CO ligands) symmetries has been suggested for $Co_4(CO)_{12}$ [399]. Thus the formation of a D_{4d} rectangular or a D_{2d} tetrahedral Co_4 carbonyl species with four bridged carbonyls (**107**) on pretreated γ-Al_2O_3 has been suggested [396]. In this case it is probable that $Co_4(CO)_{12}$–Al_2O_3 surface interaction occurs leading to expansion of the Co—Co bond distance, thus preferring structure **107** to the C_{3v} symmetry of **105**.

Species **107** is very sensitive to O_2. It is rapidly oxidized at 298 K to form a new species $[CoO]_n$ (probably with $n = 4$) (**108**) with simultaneous breaking of the Co—Co bond and formation of a new Co—O bond at a distance of 0.205 nm which is shorter than the 0.213 nm distance observed for the CoO crystal.

On heating to 513 K *in vacuo*, species **108** reacted with the surface OH groups of γ-Al_2O_3 to form a Co^{II} assembly (**109**) of trigonal distorted T_d symmetry with the accompanying elimination of H_2O. Treatment of species **105** weakly adsorbed on SiO_2 and of species **107** on γ-Al_2O_3 with O_2 at 773 K produced the Co_3O_4 clusters (< 1.5 nm) **110**, characterized by peaks at 450 and 700 nm. The Co_3O_4 species **110** has been suggested on the basis of TEM, Raman, IR, DRS and the stoichiometry in its reaction with CO or H_2 [397] to be a thin, small cluster $[Co_3O_4]_n$ (n = 3–5) with an active structure resembling the (110) plane of the Co_3O_4 spinel. These coordinatively unsaturated $[Co_3O_4]_n$ clusters exhibited an activity which was two to three orders of magnitude higher for the catalytic oxidation of CO than that of the usual supported Co_3O_4 catalysts prepared using aqueous $Co(NO_3)_2$ solution, as the results depicted in Figure 13 demonstrate.

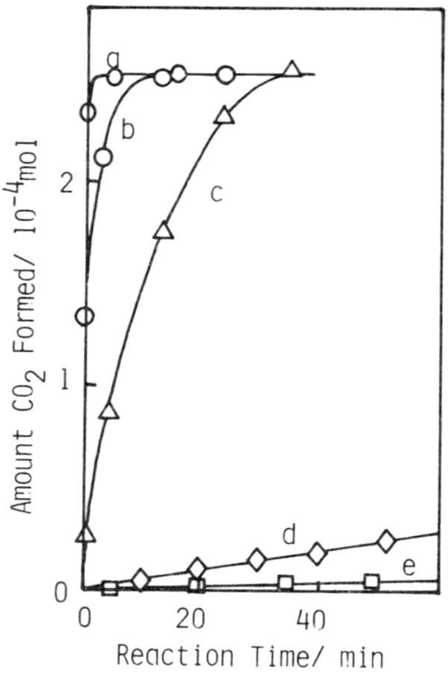

Fig. 13. The reaction of CO with O_2 catalyzed by various attached and impregnated cobalt catalysts. a, $[Co_3O_4]_n/SiO_2$; b, $[Co_3O_4]_n/\gamma$-Al_2O_3; c, $[Co^{II}]_4/\gamma$-Al_2O_3; d, $[CoO]_4/\gamma$-Al_2O_3; e, Co_3O_4/γ-Al_2O_3 catalyst obtained by conventional impregnation (0.287 g catalyst with Co/support = 5 wt.%). Reaction temperature throughout was 273 K.

The activity of the attached Co_3O_4 clusters **110** apparently exceeds that of the most active industrial Pt or Pd catalysts. Indeed, the observed activity for CO oxidation is apparently markedly affected by the nature of the surface cobalt structures; $[Co_3O_4]_n/SiO_2$ > $[Co_3O_4]_n/\gamma\text{-}Al_2O_3$ > $[Co^{II}]_4/\gamma\text{-}Al_2O_3$ >> traditional Co_3O_4/SiO_2 ~ $[CoO]_4/\gamma\text{-}Al_2O_3$ > Co^{II}/SiO_2 > traditional $Co_3O_4/\gamma\text{-}Al_2O_3$ > ion-exchanged Co^{II}/SiO_2. The reaction can be well explained by means of the following two-stage redox mechanism:

$$\underline{[Co_3O_4]_n} + CO \longrightarrow \underline{[Co_3O_{4-x}]_n} + CO_2$$

$$\underline{[Co_3O_{4-x}]_n} + x/2\ O_2 \longrightarrow \underline{[Co_3O_4]_n}$$

(SiO_2: $n = 5, x = 1/5$; $\gamma\text{-}Al_2O_3$: $n = 3, x = 1/3$)

Such $[Co_3O_4]_n$ clusters have also been shown to be highly active for ethene hydrogenation without any activation, whereas conventional cobalt catalysts show no activity unless the samples are reduced with H_2 at high temperature [397].

10. Concluding Remarks

A wide variety of organometallic molecules which react with inorganic oxides and functionalized surfaces to give well-characterized surface compounds have been illustrated in this chapter. This area of science is still in its infancy and obviously provides both a novel surface chemistry and an unique opportunity for the development of new and exciting molecular catalyst technologies. The fundamental principles are now obvious and the foundation has been laid for a systematic approach to the tailoring of new attached metal catalysts with specific structures, compositions, environments and properties through an appropriate choice of inorganic supports, transition metal complexes, types of attaching reaction and selective transformation steps. Exploratory studies are under way at present to build synthetic tailored metal catalysts to allow the development of a thorough working knowledge of the origin of these excellent catalysts. Once the essential ingredients of activity, selectivity and specificity have been identified, it should be possible to incorporate this knowledge into the rational design and synthesis of practical catalysts which are deliberately tailored for specific reaction chemistries. There is still much work to be done in this area of surface organometallic chemistry and surface molecular catalytic chemistry.

Department of Chemistry,
Faculty of Science,
The University of Tokyo,
Hongo, Bunkyo-ku,
Tokyo 113, Japan

References

1. J. P. Collman: *Acc. Chem. Res.* **1**, 136 (1968).
2. R. Ugo (ed.): *Aspects of Homogeneous Catalysis*, Vol. 2, Reidel, Dordrecht/Boston (1974).
3. F. R. Hartley and P. N. Vezey: *Adv. Organometal. Chem.* **15**, 189 (1977).
4. (a) Yu. I. Yermakov, B. N. Kuznetsov and V. A. Zakharov: *Catalysis by Supported Complexes*, Elsevier, Amsterdam (1981); (b) M. S. Scurrell: in *Catalysis* (eds. C. Kemball and D. A. Dowden), The Chemical Society, London (1978).
5. M. Mazzei, W. Marconi and M. Riocci: *J. Mol. Catal.* **9**, 381 (1980).
6. L. L. Murrell: in *Advanced Materials in Catalysis* (eds. J. J. Burton and R. L. Garten), p. 235, Academic Press, New York (1977).
7. C. H. Brubaker, Jr., E. S. Chandresekaran, C. Gibbons, R. H. Grubbs and L. C. Kroll: *J. Am. Chem. Soc.* **97**, 2128 (1975).
8. R. Jackson, J. Raddlesden, D. J. Thompson and R. Whelan: *J. Organometal. Chem.* **125**, 57 (1977).
9. D. E. Leyden and W. Collins (eds.): p. 333, Gordon and Breach, New York/London/Paris (1980).
10. Yu. I. Yermakov, B. N. Kuznetsov, Yu. P. Grabovski and A. N. Startzev: *J. Mol. Catal.* **1**, 93 (1975/76).
11. W. Skupinski and S. Malinowski: *J. Mol. Catal.* **4**, 95 (1978).
12. (a) D. G. H. Ballard: *Adv. Catal.* **23**, 263 (1973); (b) *J. Polym. Sci., Polym. Chem. Ed.* **13**, 2191 (1975).
13. (a) Yu. I. Yermakov: *J. Mol. Catal.* **21**, 35 (1983); (b) Yu. I. Yermakov and V. Zakharov: *Adv. Catal.* **24**, 173 (1975).
14. F. J. Karol, G. L. Karapinka, C. Wu, A. W. Dow, R. N. Johnson and W. L. Carrick: *J. Polym. Sci., Part A-1* **10**, 2621 (1972).
15. F. J. Karol, C. Wu, W. T. Reichle and N. J. Maraschin: *J. Catal.* **60**, 68 (1979).
16. T. S. Piper, F. A. Cotton and G. Wilkinson: *J. Inorg. Nucl. Chem.* **1**, 165 (1955).
17. A. Davison, J. A. McCleverty and G. Wilkinson: *J. Chem. Soc.* 1133 (1963).
18. F. R. W. P. Wild, G. Gubitosa and H. H. Britzinger: *J. Organometal. Chem.* **148**, 73 (1978).
19. M. Czaková and M. Čapka: *J. Mol. Catal.* **11**, 313 (1981).
20. L. k. Przhevalskaya, D. B. Furman, U. A. Shrets, S. S. Zhokovskii, T. M. Kharitonova, G. N. Bondarenko, A. M. Taber, V. B. Kazanskii, O. V. Bragin, I. V. Kalechits and V. E. Vasserberg: *Kinet. Katal.* **19**, 1036, 1283 (1978).
21. G. V. Bakulina, Zh. L. Dykh, L. I. Lafer, V. I. Yakerson, A. M. Taber, Yu. S. Mardashev, I. V. Kalechits and A. M. Rubinshtein: *Izv. Akad. Nauk SSSR, Ser. Khim.* 251 (1977).
22. Y. Iwasawa and N. Ito: to be published.
23. D. W. Studer and G. L. Schrader: *J. Mol. Catal.* **2**, 223 (1977).
24. (a) T. Castrills, H. Knözinger, J. Lieto and M. Wolf: *Inorg. Chim. Acta* **44**, L239 (1980); (b) H. P. Boehm and H. Knözinger: in *Catalysis*, Vol. 4, (eds. J. R. Anderson and M. Boudart) pp. 39–128, Springer Verlag, Berlin (1983); (c) W. N. Delgass, G. L. Haller, R. Kellerman and J. H. Lunsford: *Spectroscopy in Heterogeneous Catalysis*, Academic Press, New York (1979).
25. L. H. Little: *Infrared Spectroscopy of Adsorbed Species*, Academic Press, New York (1966).
26. M. L. Hair: *Infrared Spectroscopy in Surface Chemistry*, Marcel Dekker, New York (1967).

27. C. H. Amberg: in *Infrared Spectroscopy and the Solid–Gas Interface* (ed. E. A. Flood), p. 869, Marcel Dekker, New York (1967).
28. A. V. Kiselev and V. I. Lygin: *Infrared Spectra of Surface Compounds*, Wiley, New York (1975).
29. J. Pritchard and T. Catterick: 'Infrared Spectroscopy', in *Experimental Methods in Catalytic Research* (eds. R. B. Anderson and P. T. Dawson), p. 281, Academic Press, New York (1976).
30. W. N. Delgass, G. L. Haller, R. Kellerman and J. H. Lunsford: *Spectroscopy in Heterogeneous Catalysis*, Academic Press, New York (1979).
31. C. L. Angell: in *Fourier Transform Infrared Spectroscopy* (eds. J. R. Ferraro and L. J. Basile), pp. 1–36, Academic Press, New York (1982).
32. T. A. Egerton and A. H. Hardin: *Catal. Rev. Sci. Eng.* **11**, 1 (1975).
33. R. P. Cooney, G. Curthoys and T. T. Nguyen: *Adv. Catal.* **24**, 293 (1975).
34. P. J. Hendra and M. Fleischmann: 'Raman Spectroscopy at Surfaces', in *Topics in Surface Chemistry* (eds. E. Kay and P. S. Bagus), Plenum Press, New York (1978).
35. B. A. Morrow: in *Raman Spectroscopic Studies of Surface Species, ACS Symp. Ser. No. 137*, p. 119, Am. Chem. Soc., Washington, DC (1980).
36. E. Payen, M. C. Dhamelincourt, P. Dhamelincourt, J. Grimblot and J. P. Bonnelle: *Appl. Spectrosc.* **36**, 30 (1982).
37. H. Ibach, H. Hopster and B. Sexton: *Appl. Surf. Sci.* **1**, 1 (1977).
38. H. Ibach and D. L. Mills: *Electron Energy Loss Spectroscopy and Surface Vibrations*, Academic Press, New York (1982).
39. L. H. Bubois, P. K. Hansma and G. A. Somorjai: *Appl. Surf. Sci.* **6**, 173 (1980).
40. W. H. Weinberg: *Annu. Rev. Phys. Chem.* **29**, 115 (1978).
41. H. E. Evans and W. H. Weinberg: *J. Am. Chem. Soc.* **102**, 2548 (1980).
42. D. G. Walmsley: 'Inelastic Electron Tunneling Spectroscopy', in *Vibrational Spectroscopy of Adsorbates* (ed. R. F. Willis), p. 67, Springer Verlag, Berlin (1980).
43. W. N. Delgass, G. L. Haller, R. Kellerman and J. H. Lunsford: *Spectroscopy in Heterogeneous Catalysis*, Academic Press, New York (1979).
44. K. Klier: *Catal. Rev. Sci. Eng.* **1**, 207 (1967).
45. C. E. Jones and K. Klier: *Annu. Rev. Mater. Sci.* **2**, 1 (1972).
46. W. W. Wendlandt and H. G. Hecht: *Reflectance Spectroscopy*, Interscience, New York (1966).
47. E. J. Bowen (ed.): *Luminescence in Chemistry*, Van Nostrand, London (1968).
48. G. T. Pott and W. H. J. Stork: *Catal. Rev. Sci. Eng.* **12**, 163 (1975).
49. S. Coluccia, M. Deane and A. J. Tench: *Proc. 6th Int. Congr. Catal. London (1976)*, Vol. 1, p. 171, The Chemical Society, London (1977).
50. A. M. Gritscov, V. A. Shvets and V. B. Kazansky: *Chem. Phys. Lett.* **35**, 511 (1975).
51. Y. Iwasawa and S. Ogasawara: *Bull. Chem. Soc. Jpn.* **53**, 3709 (1980).
52. D. E. O'Reilly: *Adv. Catal.* **12**, 31 (1960).
53. R. J. Kokes: 'Electron Spin Resonance Spectroscopy', in *Experimental Methods in Catalytic Research* (ed. R. B. Anderson), pp. 436–475, Academic Press, New York (1968).
54. J. H. Lunsford: *Adv. Catal.* **22**, 265 (1972).
55. B. D. Flockhart: 'Electron Spin Resonance Studies of Adsorbed Species', in *Surface and Defect Properties of Solids*, Vol. 2, pp. 69–96, The Chemical Society, London (1973).
56. E. R. Andrew: in *Nuclear Magnetic Resonance–Rapidly Rotating Solids in Magnetic Resonance* (eds. C. K. Coogan, N. S. Ham, S. N. Struart, J. R. Pilbrow and G. V. H. Wilson, pp. 163–176, Plenum, New York (1970).

57. U. Haeberlen: *High Resolution NMR in Solids*, Academic Press, New York (1976).
58. M. Mehring: *High Resolution NMR Spectroscopy in Solids*, Springer Verlag, Berlin (1976).
59. D. Slotfeldt-Ellingsen and H. A. Resing: *J. Phys. Chem.* **84**, 2204 (1980).
60. R. L. Mössbauer: *Z. Phys.* **151**, 124 (1958).
61. H. Sinfelt: *Acc. Chem. Res.* **10**, 15 (1977).
62. H. M. Garger and M. C. Hobson: *Catal. Rev. Sci. Eng.* **11**, 117 (1975).
63. W. Jones: in *Characterization of Catalysts* (eds. J. M. Thomas and R. M. Lambert), p. 114, Wiley, Chichester/New York (1980).
64. J. A. Dumesic and H. Topsoe: *Adv. Catal.* **26**, 121 (1977).
65. M. W. Roberts: *Adv. Catal.* **29**, 55 (1980).
66. T. Edmonds: in *Characterization of Catalysts* (eds. J. M. Thomas and R. M. Lambert), p. 30, Wiley, Chichester/New York (1980).
67. T. A. Carlson: *Photoelectron and Auger Spectroscopy*, Plenum Press, New York/London (1975).
68. D. E. Sayers, E. A. Stern and F. W. Lytle: *Phys. Rev. Lett.* **27**, 1204 (1971).
69. G. H. Via, J. H. Sinfelt and F. W. Lytle: *J. Chem. Phys.* **71**, 690 (1979).
70. R. B. Greegor and F. W. Lytle: *J. Catal.* **63**, 476 (1980).
71. A. Bianconi, L. Incoccia and S. Stipcich (eds.): *EXAFS and Near Edge Structure*, Springer Verlag, Berlin (1983).
72. R. W. Joyner, A. D. Cox and R. F. Pettifer: in *Characterization of Catalysts* (eds. J. M. Thomas and R. M. Lambert), pp. 237–283, Wiley, Chichester/New York (1980).
73. A. Benninghoven: *Critical Reviews in Solid State Sciences*, CRC Press, Cleveland, Ohio (1978).
74. W. L. Baun: in *Industrial Applications of Surface Analysis* (eds. L. A. Casper and C. J. Powell), *ACS Symp. Ser. No. 199*, pp. 121–141, Am. Chem. Soc., Washington, DC (1982).
75. (a) C. R. Brundle: ref. 74, pp. 13–32; (b) J. M. Poate and T. M. Buck: 'Ion Scattering Spectroscopy', in *Experimental Methods in Catalytic Research* (eds. R. B. Anderson and P. T. Dawson), pp. 175–210, Academic Press, New York (1976).
76. R. J. Cvetanovic and Y. Amenomiya: *Adv. Catal.* **17**, 103 (1967).
77. E. E. Ibok and D. F. Ollis: *J. Catal.* **66**, 391 (1980).
78. S. D. Robertson, B. D. McNicol, J. H. de Baas, S. C. Kloet and J. W. Jenkins: *J. Catal.* **37**, 424 (1975).
79. A. Brenner, D. A. Huucul and S. J. Hardwick: *Inorg. Chem.* **18**, 1478 (1979).
80. P. B. Hirsch, A. Howie, R. B. Nicholson, D. W. Pashley and M. J. Whelan: *Electron Microscopy of Thin Crystals*, Krieger, New York (1977).
81. D. B. Holt, M. D. Muir, P. R. Grant and I. M. Boswarva: *Quantitative Scanning Electron Microscopy*, Academic Press, London (1975).
82. A. Howie: in *Characterization of Catalysts* (eds. J. M. Thomas and R. M. Lambert), p. 89, Wiley, Chichester/New York (1980).
83. S. J. B. Reed: *Electron Microprobe Analysis*, Cambridge University Press, Cambridge (1975).
84. R. Woldseth: *X-ray Energy Spectrometry*, Kevex Corporation, Burlingame (1973).
85. R. Jenkins and J. L. De Vries: *Worked Examples in X-ray Analysis*, Springer Verlag, New York (1970).
86. P. Hodge and D. C. Sherrington (eds.): *Polymer-Supported Reactions in Organic Synthesis*, Wiley, Chichester/New York (1980).

87. L. L. Murrell: in *Advanced Materials in Catalysis* (eds. J. J. Burton and R. L. Garten), pp. 235–265, Academic Press, New York (1977).
88. D. C. Bailey and S. H. Langer: *Chem. Rev.* **81**, 109 (1981).
89. O. Leal, D. L. Anderson, R. G. Bowman, F. Basolo and R. L. Burwell, Jr.: *J. Am. Chem. Soc.* **97**, 5125 (1975).
90. Z. M. Michalska and D. E. Webster: *Platinum Met. Rev.* **18**, 65 (1974).
91. A. F. Wells: *Structural Inorganic Chemistry*, 4th Ed., Clarendon Press, Oxford (1975).
92. H. P. Boehm: *Adv. Catal.* **16**, 179 (1966).
93. L. H. Little: *Infrared Spectra of Adsorbed Species*, Academic Press, London/New York (1966).
94. M. L. Hair: *Infrared Spectroscopy in Surface Chemistry*, Marcel Dekker, New York (1967).
95. (a) J. J. Fripiat: in *Soluble Silicates* (ed. S. F. James, Jr.), *ACS Symp. Ser. No. 194*, p. 165, Am. Chem. Soc., Washington, DC (1982); (b) D. E. Leyden and W. Collins (eds.): *Silylated Surfaces*, Gordon and Breach, New York/London/Paris (1980).
96. R. K. Iler: *The Chemistry of Silica*, Wiley, New York (1979).
97. K. K. Unger: *Porous Silica*, Elsevier, Amsterdam/Oxford/New York (1979).
98. J. H. de Boer: *Angew. Chem.* **70**, 389 (1958).
99. H. Knözinger and P. Ratnasamy: *Catal. Rev. Sci. Eng.* **17**, 31 (1978).
100. See also references cited in refs. 92–98 and 249.
101. H. A. Bebesi and B. H. C. Winquist: *Adv. Catal.* **27**, 98 (1978).
102. B. A. Morrow and I. A. Cody: *J. Phys. Chem.* **80**, 1995, 1998 (1976).
103. R. S. McDonald: *J. Phys. Chem.* **62**, 1168 (1958).
104. A. V. Kiselev: *Discuss. Faraday Soc.* **52**, 14 (1971).
105. V. Y. Davydov, A. V. Kiselev and L. T. Zhuravlev: *Trans. Faraday Soc.* **60**, 2254 (1964).
106. J. H. de Boer and J. M. Vleeskens: *Proc. K. Ned. Akad. Wet.* **B61**, 2 (1958).
107. J. C. W. Chien: *J. Am. Chem. Soc.* **93**, 4675 (1971).
108. H. P. Boehm: *Discuss. Faraday Soc.* **52**, 264 (1971).
109. B. E. Wagner, J. N. Helbert and E. H. Poindexter: *Surf. Sci.* **67**, 251 (1977).
110. W. K. Hall, H. P. Leftin, F. J. Cheselske and D. E. O'Reilly: *J. Catal.* **2**, 506 (1963).
111. F. J. Cheselske, W. K. Hall and W. E. Wallace: *J. Phys. Chem.* **63**, 505 (1959).
112. R. G. Haldeman and P. H. Emmett: *J. Am. Chem. Soc.* **78**, 2917 (1956).
113. V. Y. Davydov, A. V. Kiselev, S. A. Kiselev and V. O. V. Polotnyuk: *J. Colloid. Interface Sci.* **74**, 378 (1980).
114. H. P. Boehm and R. Sappok: *Z. Anorg. Allg. Chem.* **365**, 152 (1969).
115. V. M. Bermudez: *J. Phys. Chem.* **75**, 3249 (1971).
116. J. J. Fripiat and J. Uytterhoeven: *J. Phys. Chem.* **66**, 800 (1962).
117. A. H. Boonstra and C. A. H. A. Mutsaers: *J. Phys. Chem.* **79**, 1694 (1975).
118. M. L. Hair: *J. Colloid Interface Sci.* **60**, 154 (1977).
119. C. G. Armistead, A. J. Tyler, F. H. Hambleton, S. A. Mitchell and J. A. Hockey: *J. Phys. Chem.* **73**, 3947 (1969).
120. K. H. Rhee and M. R. Basila: *J. Catal.* **10**, 243 (1968).
121. M. L. Hair and W. Hertle: *J. Phys. Chem.* **77**, 2070 (1973).
122. J. K. Lee and S. W. Weller: *Anal. Chem.* **30**, 1057 (1958).
123. R. J. Peglar, F. H. Hambleton and J. A. Hockey: *J. Catal.* **20**, 309 (1971).
124. M. Liefländer and W. Stöber: *Z. Naturforsch.* **156**, 411 (1960).
125. M. Sato, T. Kanbayashi, N. Kobayashi and Y. Shima: *J. Catal.* **7**, 342 (1967).

126. D. J. C. Yates, G. W. Dembinaki, W. R. Kroll and J. J. Elliot: *J. Phys. Chem.* **73**, 911 (1969).
127. H. G. Weiss, J. A. Kneight and J. Shapirs: *J. Am. Chem. Soc.* **81**, 1823 (1959).
128. J. B. Peri and A. L. Hensley, Jr.: *J. Phys. Chem.* **72**, 2926 (1968).
129. W. Hertl: *J. Phys. Chem.* **72**, 1248, 3993 (1968).
130. G. C. Pimenten and A. L. McClellan: *The Hydrogen Bond*, W. F. Freeman, San Francisco (1960).
131. A. J. Van Rosmalen and J. C. Mol: *J. Phys. Chem.* **82**, 2748 (1978).
132. M. L. Hair and W. Hertl: *J. Phys. Chem.* **73**, 2372 (1969).
133. G. Ghiotti, E. Garrone, C. Morterra and F. Boccuzzi: *J. Phys. Chem.* **83**, 2863 (1979).
134. J. B. Peri and A. L. Hensley, Jr.: *J. Phys. Chem.* **70**, 3168 (1966).
135. F. H. Van Cauwelaert, P. A. Jacobs and J. B. Uytterhoeven: *J. Phys. Chem.* **76**, 1434 (1972).
136. J. A. Hockey: *J. Phys. Chem.* **74**, 2570 (1970).
137. B. A. Morrow and I. A. Cody: *J. Phys. Chem.* **77**, 1465 (1973).
138. K. Klier, J. H. Shen and A. C. Zettlemoyer: *J. Phys. Chem.* **77**, 1458 (1973).
139. V. Y. Davydov, L. T. Zhuravlev and A. V. Kiselev: *Zh. Fiz. Khim.* **38**, 2047 (1964).
140. F. H. Van Cauwelaert, F. Vermoortele and J. B. Uytterhoeven: *Discuss. Faraday Soc.* **52**, 66 (1971).
141. W. K. Hall: *Acc. Chem. Res.* **8**, 257 (1975).
142. M. M. Mestdagh, W. E. Stone and J. J. Fripiat: *J. Phys. Chem.* **76**, 1220 (1972).
143. L. B. Schreiber and R. W. Vaughan: *J. Catal.* **40**, 226 (1975).
144. C. G. Armistead and J. A. Hockey: *Trans. Faraday Soc.* **63**, 2549 (1967).
145. W. Hertl and M. L. Hair: *J. Phys. Chem.* **75**, 2181 (1971).
146. L. R. Snyder and J. W. Ward: *J. Phys. Chem.* **70**, 3941 (1966).
147. J. Kunawicz, P. Jones and J. A. Hockey: *Trans. Faraday Soc.* **67**, 848 (1971).
148. B. Stübner, H. Knözinger, J. Conard and J. J. Fripiat: *J. Phys. Chem.* **82**, 1811 (1978).
149. J. B. Peri and R. B. Hannan: *J. Phys. Chem.* **64**, 1526 (1960).
150. J. B. Peri: *J. Phys. Chem.* **69**, 220 (1965).
151. J. B. Peri: *J. Phys. Chem.* **69**, 211 (1965).
152. A. A. Tsyganenko and V. N. Filimonov: *J. Mol. Struct.* **19**, 579 (1973).
153. A. Zecchina: *Discuss. Faraday Soc.* **52**, 89 (1971).
154. C. Morterra, G. Ghiotti, E. Garrone and F. Boccuzzi: *J. Chem. Soc., Faraday Trans. I* **72**, 2722 (1976).
155. E. P. Smirnov and A. A. Tsyganenko: *React. Kinet. Catal. Lett.* **7**, 425 (1977).
156. E. P. Smirnov and V. N. Filimonov: *React. Kinet. Catal. Lett.* **7**, 431 (1977).
157. H. Dunken and H. Hobert: *Z. Phys. Chem.* (*Leipzig*) **260**, 913 (1979).
158. M. R. Basila: *J. Phys. Chem.* **66**, 2223 (1966).
159. P. G. Rouchet and R. E. Sempels: *J. Chem. Soc., Faraday Trans. I* **70**, 2021 (1974).
160. L. B. Schreiber and R. W. Vaughan: *J. Catal.* **40**, 226 (1975).
161. J. B. Peri: *J. Catal.* **41**, 227 (1976).
162. H. G. Weiss, J. A. Knight and I. Shapiro: *J. Am. Chem. Soc.* **81**, 1823 (1959); *ibid.*, **82**, 1262 (1960).
163. E. Lippmaa, M. Mägi, A. Samson, M. Tarmak and G. Engelhardt: *J. Am. Chem. Soc.* **103**, 4992 (1981).
164. J. M. Thomas, L. A. Bursill, E. A. Lodge, A. K. Cheetham and C. A. Fyfe: *J. Chem. Soc., Chem. Commun.* 276 (1981).

165. G. Engelhardt, U. Lohse, A. Samson, M. Mägi, M. Tarmak and E. Lippmaa: *Zeolites* **2**, 59 (1982).
166. M. T. Melchior, D. E. W. Vaughan and A. J. Jacobson: *J. Am. Chem. Soc.* **104**, 4859 (1982).
167. L. A. Bursill, E. A. Lodge, J. M. Thomas and A. K. Cheetham: *J. Phys. Chem.* **85**, 2409 (1981).
168. J. B. Uytterhoeven, L. G. Christner and W. K. Hall: *J. Phys. Chem.* **69**, 2117 (1965).
169. J. W. Ward: *"Infrared Studies of Zeolite Surfaces and Surface Reactions" in Zeolite Chemistry and Catalysis* (ed. J. A. Rabo), *ACS Monog. No. 171*, pp. 118–284, Am. Chem. Soc., Washington, DC (1976).
170. J. W. Ward: *Adv. Chem. Ser.* **101**, 380 (1971).
171. D. Freude, D. Müller and H. Schmiedel: *Surf. Sci.* **25**, 289 (1971).
172. C. M. Hollabaugh and J. J. Chessick: *J. Phys. Chem.* **65**, 109 (1961).
173. C. Morterra, G. Chiotti, E. Garrone and E. Fisicaro: *J. Chem. Soc., Faraday Trans. I* **76**, 2102 (1980).
174. H. Hattori, M. Itoh and K. Tanabe: *J. Catal.* **38**, 172 (1975).
175. M. P. P. Primet and M. V. Mathieu: *J. Phys. Chem.* **75**, 1216 (1971).
176. H. P. Boehm and M. Herrmann: *Z. Anorg. Allg. Chem.* **352**, 156 (1967).
177. M. Primet, P. Pichat and M. V. Mathieu, *C. R. Acad. Sci., Ser. B* **267**, 799 (1968).
178. P. Jackson and G. D. Parfitt: *Trans. Faraday Soc.* **67**, 2469 (1971).
179. G. Munuera and F. S. Stone: *Discuss. Faraday Soc.* **52**, 205 (1971).
180. K. E. Lewis and G. D. Parfitt: *Trans. Faraday Soc.* **62**, 204 (1966).
181. J. M. Criado, J. Domminguez, F. Gonzalez, G. Munuera and J. M. Trillo: *Proc. 4th Int. Congr. Catal., Moscow (1968)*, Vol. 1, p. 504 (1970).
182. P. J. Anderson, R. F. Horlock and J. F. Oliver: *Trans. Faraday Soc.* **61**, 2754 (1965).
183. A. J. Tench: *J. Chem. Soc., Faraday Trans. I* **68**, 187 (1972).
184. D. D. Eley, G. M. Kiwanuka and C. N. Rochester: *J. Chem. Soc., Faraday Trans. I* **72**, 876 (1976).
185. S. Coluccia, M. Deane and A. J. Tench: *Proc. 6th Int. Congr. Catal., London (1976)*, Vol. 1, p. 171, The Chemical Society, London (1977).
186. M. Breysse, B. Claudel, L. Faure and M. Guenin: *Proc. 7th Int. Vac. Congr. & 3rd Int. Conf. Solid Surf., Vienna*, 485–488 (1977).
187. T. Morimoto and H. Naono: *Bull. Chem. Soc. Jpn.* **46**, 2000 (1973).
188. R. K. Webster, T. L. Jones and P. J. Anderson: *Proc. Brit. Ceram. Soc.* **5**, 153 (1965).
189. Y. Iwasawa: unpublished data.
190. T. Morimoto, M. Nagao and F. Tokuda: *Bull. Chem. Soc. Jpn.* **41**, 1533 (1968).
191. A. L. Dent and R. J. Kokes: *J. Phys. Chem.* **73**, 3781 (1969).
192. K. Atherton, G. Newbold and J. A. Hockey: *Discuss. Faraday Soc.* **52**, 33 (1971).
193. G. Mattmann, H. R. Oswald and F. Schweizer: *Helv. Chim. Acta* **55**, 1249 (1972).
194. M. M. Folman: *Trans. Faraday Soc.* **57**, 2000 (1961).
195. R. G. Azrak and C. L. Angell: *J. Phys. Chem.* **77**, 3048 (1973).
196. J. P. Collman and L. S. Hegedus: *Principles and Applications of Organotransition Metal Chemistry*, University Science Books, Mill Valley, California (1980).
197. D. C. Locke, J. T. Schmermund and B. Banner: *Anal. Chem.* **44**, 90 (1972).
198. M. Bartholin, C. H. Graillat, A. Guyot, G. Coudurier, J. Bandiera and C. Naccache: *J. Mol. Catal.* **3**, 17 (1977/78).
199. A. R. Sanger and L. R. Schallig: *J. Mol. Catal.* **3**, 101 (1977/78).
200. K. G. Allum, R. D. Hancock, I. V. Howell, A. McKenzie, R. C. Pitkethly and P. J. Robinson: *J. Organometal. Chem.* **87**, 203 (1975).

201. K. G. Allum, R. D. Hancock, S. McKenzie and P. C. Pitkethly: in *Catalysis*, Vol. 1 (ed. J. W. Hightower), p. 477, North-Holland, Amsterdam (1973).
202. K. G. Allum, R. D. Hancock, I. V. Howell, T. E. Lester, S. McKenzie, R. C. Pitkethly and P. J. Robinson: *J. Organometal. Chem.* **107**, 393 (1976).
203. I. Haller: *J. Am. Chem. Soc.* **100**, 8050 (1978).
204. M. Capka and J. Hetflejs: *Collect. Czech. Chem. Commun.* **39**, 154 (1974).
205. C. R. Hastings, W. A. Aue and F. N. Larsen: *J. Chromatogr.* **60**, 329 (1971).
206. C. Eaborn: *Organosilicon Compounds*, pp. 51–55, Academic Press, New York (1960).
207. E. W. Abel, F. H. Polard, P. C. Uden and G. Nickless: *J. Chromatogr.* **22**, 23 (1966).
208. H. W. Kohlschütter, P. Best and G. Wirzing: *Z. Anorg. Allg. Chem.* **285**, 236 (1956).
209. W. Stöber: *Kolloid Z.* **149**, 39 (1956).
210. L. J. Bellamy: *The Infrared Spectra of Complex Molecules*, Wiley (1975).
211. B. Evans and T. E. White: *J. Catal.* **11**, 336 (1968).
212. R. L. Kaas and J. L. Kardos: *Polym. Eng. Sci.* **11**, 1 (1971).
213. F. Buzek and J. Rathousky: *J. Colloid Interface Sci.* **79**, 47 (1981).
214. H. Engelhardt and D. Mathes: *J. Chromatogr.* **142**, 311 (1977).
215. L. J. Boucher, A. A. Oswald and L. L. Murrell: *Am. Chem. Soc., Div. Pet. Chem. Prepr.* **19**, 155 (1974).
216. J. Conan, M. Bartholin and A. Guyot: *J. Mol. Catal.* **1**, 375 (1975/76).
217. M. Capka and J. Hetflejs: *Collect. Czech. Chem. Commun.* **39**, 154 (1974).
218. B. F. Watkins, J. R. Behling, E. Kariv and L. L. Miller: *J. Am. Chem. Soc.* **97**, 3549 (1975).
219. B. E. Firth, L. L. Miller, M. Mitani, Th. Rogers, J. Lennox and R. W. Murray: *J. Am. Chem. Soc.* **98**, 8271 (1976).
220. J. C. Lennox and R. W. Murray: *J. Am. Chem. Soc.* **100**, 3710 (1978).
221. C. P. Jester, R. D. Rocklin and R. W. Murray: *J. Electrochem. Soc.* **127**, 1979 (1980).
222. C. M. Ellist and R. W. Murray: *Anal. Chem.* **48**, 1247 (1976).
223. K. W. Wiliman, R. D. Rocklin, R. Nowak, K. N. Kuo, F. A. Schultz and R. W. Murray: *J. Am. Chem. Soc.* **102**, 7629 (1980).
224. G. D. Parfitt, J. Ramsbortham and C. H. Rochester: *J. Chem. Soc., Faraday Trans. I* **68**, 17 (1972).
225. P. K. Ghosh and T. G. Spiro: *J. Am. Chem. Soc.* **102**, 5543 (1980).
226. K. G. Allum, R. D. Hancock, I. V. Howell, T. E. Lester, S. McKenzie, R. C. Pitkethly and P. J. Robinson: *J. Catal.* **43**, 331 (1976).
227. R. D. Hancock, I. V. Howell, R. C. Pitkethly and P. J. Robinson: in *Catalysis: Heterogeneous and Homogeneous* (eds. B. Delmon and G. Jannes), pp. 361–371, Elsevier, Amsterdam (1975).
228. A. A. Oswald, L. L. Murrell and L. J. Boucher: *Abstr. 167th Am. Chem. Soc. Meeting, Div. Pet. Chem.* **19**, 162 (1974).
229. K. G. Allum, R. D. Hancock, I. V. Howell, R. C. Pitkethly and P. J. Robinson: *J. Catal.* **43**, 322 (1976).
230. R. Jackson, J. Raddlesden, D. J. Thompson and R. Whelan: *J. Organometal. Chem.* **125**, 57 (1977).
231. T. J. Pinnavaia, J. G.-S. Lee and M. Abedini: in *Silylated Surfaces* (eds. D. E. Leyden and W. Collins), pp. 333–345, Gordon and Breach, New York (1980).
232. M. S. Wrighton, R. G. Austin, A. B. Bocarsly, J. M. Bolts, O. Haas, K. D. Legg, L. Nadjo and M. C. Palazzotto: *J. Am. Chem. Soc.* **100**, 1602 (1978).
233. R. D. Rocklin and R. W. Murray: *J. Phys. Chem.* **85**, 2104 (1981).

234. O. Leal and M. R. Goldwasser, R. Lopez and H. Arzoumanian: *Proc. 7th Int. Congr. Catal. Tokyo (1980), Part B*, p. 1512, Elsevier/Kodansha, Amsterdam/Tokyo (1981).
235. Yu. I. Yermakov: *Proc. 7th Int. Congr. Catal. Tokyo (1980), Part A*, p. 57, Elsevier/Kodansha, Amsterdam/Tokyo (1981).
236. K. G. Allum, S. McKenzie and R. C. Pitkethly: U.S. Pat. 3,726,809 (1973).
237. L. Horner and H. Ziegler: *Justus Liebigs Ann. Chem.* 628 (1976).
238. L. W. Burggraf, D. S. Kendall, D. E. Leyden and F. J. Pern: *Anal. Chim. Acta* **129**, 19 (1981).
239. K. Kochloefl, W. Liebelt and H. Knözinger: *J. Chem. Soc., Chem. Commun.* 510 (1977).
240. L. Horner and F. Schumacher: *Justus Liebigs Ann. Chem.* 633 (1976).
241. A. K. Smith, J. M. Basset and P. M. Maitlis: *J. Mol. Catal.* **2**, 223 (1977).
242. C. U. Pittman Jr., S. K. Wuu and S. E. Jacobson: *J. Catal.* **44**, 87 (1976).
243. M. Capka, P. Svoboda and J. Hetflejs: *Collect. Czech. Chem. Commun.* **38**, 1242 (1973).
244. K. G. Allum, R. D. Hancock, I. V. Howell, R. C. Pitkethly and P. J. Robinson: *J. Catal.* **43**, 322 (1976).
245. Y. M. Zhorov, A. V. Shelkov and G. M. Panchenkov: *Kinet. Katal.* **15**, 1091 (1974).
246. A. J. Moffat: *J. Catal.* **18**, 193 (1970).
247. W. O. Haag and D. D. Whitehurst: in *Catalysis*, Vol. 1 (ed. J. W. Hightower), p. 465, North-Holland, Amsterdam (1973).
248. J. Lahaye and R. Lagarde: *J. Catal.* **60**, 270 (1979).
249. P. W. H. L. Tjan and J. J. F. Scholten: *Proc. 6th Int. Congr. Catal., London (1976)*, Vol. 1, p. 488, The Chemical Society, London (1977).
250. P. R. Rony: *J. Catal.* **14**, 142 (1969).
251. G. J. K. Acres, G. C. Bond, B. J. Cooper and J. A. Dawson: *J. Catal.* **6**, 139 (1966).
252. K. K. Robinson, F. E. Paulik, A. Hershman and J. F. Roth: *J. Catal.* **15**, 245 (1969).
253. G. C. Bond: *Platinum Met. Rev.* **11**, 147 (1967).
254. Monsanto Co.: Belg. Pat. 711,042 (1968).
255. J. Berthoux and J. P. Martinaud: Ger. Pat. 2,261,543 (1973).
256. P. R. Rong and J. F. Roth: *J. Mol. Catal.* **1**, 13 (1975/76).
257. J. Hjortkjaer, M. S. Scurrell and P. Simonsen: *J. Mol. Catal.* **6**, 405 (1979).
258. R. Z. Moravec, W. T. Schelling and C. F. Oldershaw: Br. Pat. 511,556 (1939); Can. Pat. 396,994 (1941).
259. M. Mazzei, W. Marconi and M. Riocci: *J. Mol. Catal.* **9**, 381 (1980).
260. R. L. Burwell Jr., R. G. Pearson, G. L. Haller, P. B. Tjok and S. P. Chock: *Inorg. Chem.* **4**, 1123 (1965).
261. T. J. Pinnavaia and P. K. Welty: *J. Am. Chem. Soc.* **97**, 3819 (1975).
262. R. Raythatha and T. J. Pinnavaia: *J. Catal.* **80**, 47 (1983).
263. J. P. Candlin and H. Thomas: in *Homogeneous Catalysis II* (eds. D. Forster and J. F. Roth), *Adv. Chem. Ser.* **132**, 212 (1974).
264. Yu. I. Yermakov: *Catal. Rev. Sci. Eng.* **13**, 77 (1976).
265. R. L. Burwell Jr.: *Chemtech*, 370 (1974).
266. Y. Iwasawa, Y. Nakano and S. Ogasawara: *Shokubai* **19**, 4 (1977).
267. Y. Iwasawa: *Hyomen* **19**, 160 (1981).
268. Y. Iwasawa: *Shokubai* **26**, 191 (1984).
269. Yu. I. Yermakov, B. N. Kuznetsov, L. G. Karakchiev and S. S. Derbeneva: *Kinet. Katal.* **14**, 709 (1973).

270. B. N. Kuznetsov, Yu. I. Yermakov, V. L. Kuznetsov, Yu. A. Ryndin, L. G. Karachiev, V. G. Shinkarenko, Ye. K. Mamaeva and L. Ya. Startseva: *Kinet. Katal.* **16**, 1356 (1975).
271. (a) Y. Iwasawa and H. Sato: *Chem. Lett.* **507** (1985); unpublished data; (b) S. J. Tauster and S. C. Fung: *J. Catal.* **55**, 29 (1978); S. J. Tauster, S. C. Fung, R. T. K. Baker and J. A. Horsley: *Science* **211**, 1121 (1981).
272. G. Carturan, G. Facchin, G. Cocco, S. Enzo and G. Navazio: *J. Catal.* **76**, 405 (1982).
273. G. Carturan, G. Cocco, L. Schiffini and G. Strukul: *J. Catal.* **65**, 359 (1980).
274. Yu. I. Yermakov, M. S. Ioffe, B. N. Kuznetsov and Yu. A. Ryndin: *Kinet. Katal.* **16**, 807 (1975).
275. Yu. I. Yermakov and B. N. Kuznetsov: *Kinet. Katal.* **18**, 1167 (1977).
276. K. Matsuo and K. J. Klabunde: *J. Org. Chem.* **47**, 843 (1982).
277. B. A. Scott, R. M. Plecenik, G. S. Cargill, T. R. McGuire and S. R. Herd: *Inorg. Chem.* **19**, 1252 (1980).
278. S. C. Davis, S. Severson and K. J. Klabunde: *J. Am. Chem. Soc.* **103**, 3024 (1981).
279. N. G. Maksimov, G. A. Nesterov, V. A. Zakharov, P. V. Stchastnev, V. F. Anufrienko and Yu. I. Yermakov: *J. Mol. Catal.* **4**, 167 (1978).
280. H. E. Evans and W. H. Weinberg: *J. Am. Chem. Soc.* **102**, 872, 2548 (1980).
281. W. L. Carrick, R. J. Turbett, F. J. Karol, G. L. Karapinka, A. S. Fox and R. N. Johnson: *J. Polym. Sci., Part A-1* **10**, 2609 (1972).
282. L. M. Baker and W. L. Carrick: *J. Org. Chem.* **33**, 616 (1968).
283. Y. Iwasawa, Y. Sato and H. Kuroda: *J. Catal.* **82**, 289 (1983).
284. B. N. Kuznetsov, A. N. Startsev and Yu. I. Yermakov: *J. Mol. Catal.* **8**, 135 (1980).
285. Y. Iwasawa, S. Ogasawara and M. Soma: *Chem. Lett.* 1039 (1978).
286. S. A. Best, R. G. Squires and R. A. Walton: *J. Catal.* **60**, 171 (1979).
287. J. Smith, W. Mowat, D. A. Whan and E. A. V. Ebworth: *J. Chem. Soc., Dalton Trans.* 1742 (1974).
288. L. K. Prshevalskaya, V. A. Shvets and V. B. Kazanskii: *Kinet. Katal.* **20**, 1005 (1979).
289. L. K. Prshevalskaya, V. A. Shvets, V. B. Kazanskii and V. T. Kalinnikov: *Kinet. Katal.* **19**, 1277 (1978).
290. M. D. Ward, T. V. Harris and J. Schwartz: *J. Chem. Soc., Chem. Commun.* 357 (1980).
291. M. D. Ward and J. Schwartz: *J. Mol. Catal.* **11**, 397 (1981).
292. M. D. Ward and J. Schwartz: *J. Am. Chem. Soc.* **103**, 5253 (1981).
292a. H. C. Foley, S. J. DeCanio, K. D. Tau, K. J. Chao, J. H. Onuferko, C. Dybowski and B. C. Gates: *J. Am. Chem. Soc.* **105**, 3074 (1983).
293. T.-N. Huang and J. Schwartz: *J. Am. Chem. Soc.* **104**, 5244 (1982).
294. J. Hjortkjaer, M. S. Scurrell and P. Simonsen: *J. Mol. Catal.* **10**, 127 (1981).
295. P. W. H. L. Tjan and J. J. F. Scholten, *Proc. 6th Int. Congr. Catal. London (1976)*, Vol. 1, p. 488, The Chemical Society, London (1977).
296. T. Mole and E. A. Jeffrey: *Organoaluminum Compounds*, pp. 68–138, Elsevier, London (1972).
297. D. Slotfeldt – Ellingsen, I. M. Dahl and O. H. Ellestad: *J. Mol. Catal.* **9**, 423 (1980).
298. K. Soga, K. Izumi, M. Terano and S. Ikeda: *Makromol. Chem.* **181**, 657 (1980).
299. R. H. Grubbs: *Chemtech.* 512 (1977).
300. J. C. W. Chien: *J. Catal.* **23**, 71 (1971).
301. K. Soga, T. Sano and R. Ohnishi: *Polym. Bull.* **4**, 157 (1981).
302. K. Ziegler, E. Holzkampf, H. Breil and H. Martin: *Angew. Chem.* **67**, 541 (1955).

303. K. Weissermel, H. Cherdron, J. Berthold, B. Diedrich, K. D. Keil, K. Rust, H. Strametz and T. Toth: *J. Polym. Sci., Polym. Symp.* **51**, 187 (1975).
304. Solvay: Jpn. Pat. 43 13050; Ger. Pats. 2,000,585 and 2,000, 568; *Chem. Eng. News* 21 (Dec. 5, 1977).
305. M. P. McDaniel, M. B. Welch and M. J. Dreiling: *J. Catal.* **82**, 118 (1983).
306. T. J. Pullukat, R. E. Hoff and M. Shida: *J. Appl. Polym. Sci.* **26**, 2927 (1981).
307. J. C. W. Chien: *J. Am. Chem. Soc.* **93**, 4675 (1971).
308. J. C. W. Chien and J. T. T. Hsiech: in *Coordination Polymerization* (ed. J. C. W. Chien), p. 305, Academic Press, New York (1975).
309. W. Hanke, R. Bienert and H. G. Yerschkewitz: *Z. Anorg. Allg. Chem.* **414**, 109 (1975).
310. G. Öhlmann: *Sitzungsber. Akad. Wiss. DDR, Math-Naturwis-Tech.* **12**, 64 (1978).
311. V. N. Pak: *Zh. Phys. Khim.* **54**, 1879 (1980).
312. O. V. Krylov: *Prepr. 2nd Japan–Soviet Catal. Seminar, Tokyo* 240 (1973).
313. H. Praliaud: *J. Less-Common Met.* **54**, 387 (1971).
314. H. Praliaud and M. Forissier: *React. Kinet. Catal. Lett.* **8**, 451 (1978).
315. V. N. Pak, A. N. Volkova, N. N. Kushakova, S. I. Koltsov and V. B. Aleskovskii: *Zh. Fiz. Khim.* **48**, 2394 (1974).
316. A. D. Shebaldova, V. I. Marin, G. G. Nasledkov, M. P. Khidkel, T. A. Popova and V. A. Akchurin: *Neftekhin* **16**, 533 (1976).
317. M. Che, F. Figueras, M. Forissier, J. McAteer, M. Perrin, J. L. Portefaix and H. Praliaud: *Proc. 6th Int. Congr. Catal. London (1976)*, Vol. 1, p. 261, The Chemical Society, London (1977).
318. A. I. Van Roosmalev, K. Polder and J. C. Mol: *J. Mol. Catal.* **8**, 185 (1980).
319. V. A. Zakharov and Yu. I. Yermakov: *Catal. Rev. Sci. Eng.* **19**, 67 (1979).
320. G. A. Nesterov, V. A. Zakharov, E. A. Paukshtis, E. N. Yurchenko, V. V. Volkov and K. G. Myakishev: *Kinet. Katal.* **21**, 739 (1980).
321. J. Schwartz and M. D. Ward: *J. Mol. Catal.* **8**, 465 (1980).
322. Y. Iwasawa: *J. Mol. Catal.* **17**, 93 (1982).
323. Y. Iwasawa and T. Chiba: to be published.
324. Y. Iwasawa and S. Ogasawara: *Chem. Lett.* 127 (1980).
325. Y. Iwasawa, Y. Sasaki and S. Ogasawara: *J. Chem. Soc., Chem. Commun* 140 (1981).
326. Y. Iwasawa, Y. Sasaki and S. Ogasawara: *J. Mol. Catal.* **16**, 27 (1982).
327. Y. Iwasawa, H. Ichinose and S. Ogasawara: *J. Chem. Soc., Faraday Trans. I* **77**, 1763 (1981).
328. Yu. I. Yermakov and B. N. Kuznetsov: *Prepr. 2nd Japan–Soviet Catal. Seminar. Tokyo* 65 (1973).
329. Y. Iwasawa, M. Yamagishi and S. Ogasawara: *J. Chem. Soc., Chem. Commun.* 871 (1980).
330. Y. Iwasawa, H. Kubo and H. Hamamura: *J. Mol. Catal.* **28**, 191 (1985).
331. Y. Iwasawa, Y. Nakano and S. Ogasawara: *J. Chem. Soc., Faraday Trans I* **74**, 2968 (1978).
332. Y. Iwasawa, T. Nakamura, K. Takamatsu and S. Ogasawara: *J. Chem. Soc., Faraday Trans. I* **76**, 939 (1980).
333. Y. Iwasawa and S. Ogasawara: *J. Chem. Soc., Faraday Trans. I* **75**, 1465 (1979).
334. Y. Iwasawa, H. Kubo, M. Yamagishi and S. Ogasawara: *Chem. Lett.* 1165 (1980).
335. Y. Iwasawa and M. Yamagishi: *J. Catal.* **82**, 373 (1983).
336. Y. Iwasawa, Y. Sato and H. Kuroda: *J. Catal.* **82**, 289 (1983).

337. Y. Iwasawa, M. Yamagishi and S. Ogasawara: *J. Chem. Soc., Chem. Commun.* 246 (1982).
338. Y. Iwasawa and H. Hamamura: *J. Chem. Soc., Chem. Commun.* 130 (1983).
339. Y. Iwasawa and H. Tanaka: *Proc. 8th Int. Congr. Catal. Berlin (1984)*, IV-381 (1984).
340. B. N. Kuznetsov, A. N. Startsev and Yu. I. Yermakov: *J. Mol. Catal.* **8**, 135 (1980).
341. Yu. I. Yarmakov, B. N. Kuznetsov, A. N. Startsev, P. A. Zhdan, A. P. Shepelin, V. I. Zaikovskii, L. M. Plyasova and V. A. Burmistrov: *J. Mol. Catal.* **11**, 205 (1981).
342. A. N. Startsev, B. N. Kuznetsov and Yu. I. Yermakov: *React. Kinet. Catal. Lett.* **3**, 321 (1975).
343. Y. Iwasawa and H. Kobayashi: to be published.
344. B. N. Kuznetsov, Yu. I. Yermakov, A. N. Startsev and V. A. Burmistrov: *Prepr. 6th Japan–Soviet Catal. Seminar Tokyo* 76 (1981).
345. G. V. Lisichkin and S. A. Panichev: in *Supported Metallic Catalysts of Hydrocarbon Reactions* (in Russian), pp. 51–54, Institute of Catalysis, Novosibirsk (1978).
346. M. Wolfsberg and L. Helmholz: *J. Chem. Phys.* **20**, 837 (1952).
347. C. J. Ballhausen and A. D. Liehr: *J. Mol. Spectrosc.* **2**, 342 (1958).
348. S. O. Grim and L. J. Matienzo: *Inorg. Chem.* **14**, 1014 (1975).
349. A. Cimino and B. A. De Angelis: *J. Catal.* **36**, 11 (1975).
350. W. E. Swartz, Jr. and D. M. Mercules: *Anal. Chem.* **43**, 1774 (1971).
351. J. Chatt, C. M. Elson, G. J. Leigh and J. A. Connor: *J. Chem. Soc., Dalton Trans.* 1351 (1976).
352. J. Kress, M. Wesolek and J. A. Osborn: *J. Chem. Soc., Chem. Commun.* 514 (1982).
353. F. N. Tebbe, Q. W. Parshall and D. W. Ovenall: *J. Am. Chem. Soc.* **101**, 5074 (1979).
354. J. L. Herrison and Y. Chauvin: *Makromol. Chem.* **141**, 161 (1971).
355. D. J. Cardin, M. J. Doyle and M. F. Lappert: *Chem. Soc. Rev.* **2**, 99 (1973).
356. R. H. Grubbs: *Prog. Inorg. Chem.* **24**, 1 (1978).
357. T. J. Katz: *Adv. Organometal. Chem.* **16**, 283 (1977).
358. N. Calderon, J. P. Laurence and E. A. Ofstead: *Adv. Organometal. Chem.* **17**, 449 (1979).
359. E. L. Muetterties: *Inorg. Chem.* **14**, 951 (1978).
360. C. P. Casey and T. J. Burkhardt: *J. Am. Chem. Soc.* **96**, 7808 (1974).
361. P. G. Gassman and T. H. Johnson: *J. Am. Chem. Soc.* **98**, 6056 (1976).
362. M. Ephritikine, M. L. H. Green and R. Mackenzie: *J. Chem. Soc., Chem. Commun.* 619 (1976); M. Ephritikine and M. L. H. Green: *ibid.*, 926 (1976).
363. R. H. Grubbs and S. J. Swetnick: *J. Mol. Catal.* **8**, 25 (1980).
364. M. F. Farona and R. L. Tucker: *J. Mol. Catal.* **8**, 85 (1980).
365. D. T. Laverty, J. J. Rooney and A. Stewart: *J. Catal.* **45**, 110 (1976).
366. R. R. Schrock: *J. Am. Chem. Soc.* **97**, 6577 (1975).
367. S. S. M. Ling and R. J. Puddephatt: *J. Chem. Soc., Chem. Commun.* 412 (1982).
368. F. A. Cotton and J. R. Pipal: *J. Am. Chem. Soc.* **93**, 5441 (1971).
369. F. A. Cotton: *Chem. Soc. Rev.* **4**, 27 (1975).
370. Y. Iwasawa, H. Ishii, K. Asakura and H. Kuroda: to be published.
371. Y. Sato, Y. Iwasawa and H. Kuroda: *Chem. Lett.* 1101 (1982).
372. F. A. Cotton, S. Koch, M. Mertis, M. Millar and G. Wilkinson: *J. Am. Chem. Soc.* **99**, 4989 (1977).
373. F. A. Cotton and P. E. Fanwick: *J. Am. Chem. Soc.* **101**, 5252 (1979).
374. A. P. Sattelberger and J. P. Fuckler: *J. Am. Chem. Soc.* **99**, 1258 (1977).
375. J. G. Norman, Jr. and H. J. Kolari: *J. Am. Chem. Soc.* **97**, 33 (1975); *Inorg. Chem.* **16**, 987 (1977).

376. W. H. Böhmer, K. Madeja, E. Kurras and U. Rosenthal: *Z. Chem.* **18**, 453 (1978).
377. M. Benard: *J. Am. Chem. Soc.* **100**, 2354 (1978); *J. Chem. Phys.* **71**, 2546 (1979).
378. M. Colin and A. Herold: *C. R. Acad. Sci., Ser. C* **269**, 1302 (1965).
379. M. A. M. Boersma: *Catal. Rev. Sci. Eng.* **10**, 243 (1974).
380. M. E. Volpin, Yu. N. Novikov, N. D. Lapkina, V. I. Kasatochkin, Yu. T. Struchkov, M. E. Kazakov, R. A. Stukan, V. A. Povitskii, Yu. S. Karimov and A. V. Zvarikina: *J. Am. Chem. Soc.* **97**, 3366 (1975).
381. E. Kikuchi, T. Ino and Y. Morita: *J. Catal.* **57**, 27 (1979).
382. R. C. Croft: *Aust. J. Chem.* **9**, 181 (1956).
383. S. Parkash, S. K. Chakrabartty and J. G. Hooley: *Carbon* **15**, 307 (1977); *ibid.* **16**, 231 (1978).
384. A. W. Syme Johnson: *Acta Crystallogr.* **23**, 770 (1967).
385. J. C. Volta, W. Desquesnes, B. Moraweck and J. M. Tatibouët: *Proc. 7th Int. Congr. Catal. Tokyo (1980), Part B*, p. 1398, Elsevier/Kodansha, Amsterdam/Tokyo (1980).
386. J. C. Volta, M. Forissier, F. Theobald and T. P. Pham: *Faraday Discuss. Chem. Soc.* **72**, 225 (1981).
387. J. C. Volta: *Synth. Mater.* **4**, 319 (1982).
388. J. M. Tatibouët and J. E. Germain: *J. Catal.* **82**, 245 (1983).
389. J. M. Tatibouët, J. E. Germain and J. C. Volta: *J. Catal.* **82**, 240 (1983).
390. R. P. A. Sneeden and H. H. Zeiss: *J. Organometal. Chem.* **28**, 259 (1971).
391. T. Aoki, A. Furusaki, Y. Tomiie, K. Ono and K. Tanaka: *Bull. Chem. Soc. Jpn.* **42**, 545 (1969).
392. E. Kurras and P. Klimsch: *Monatsber. Dtsch. Akad. Wiss. Berlin* **6**, 735, 736 (1964).
393. P. Klimsch and E. Kurras: East Ger. Pat. 45,708 (1964).
394. L. Nodek, D. Mihajlova, A. Andreev, A. Palazov, M. Kraus and D. Shopov: *J. Catal.* **40**, 46 (1975).
395. D. Reinen: *Struct. Bonding (Berlin)* **6**, 30 (1969).
396. Y. Iwasawa, M. Yamada, Y. Sato and H. Kuroda: *J. Mol. Catal.* **23**, 95 (1984).
397. Y. Iwasawa and M. Yamada: *Nikkashi* (in Japanese), 1042 (1984).
398. R. C. Ryan and L. F. Dahl: *J. Am. Chem. Soc.* **97**, 6904 (1975).
399. F. A. Cotton: *Inorg. Chem.* **5**, 1083 (1966).

H. HIRAI AND N. TOSHIMA

POLYMER-ATTACHED CATALYSTS

1. Introduction

Research directed toward the design of catalysts has become an area of intense interest [1]. Catalysis has been one of the most important challenges facing the practicing technologist in the development of new processes or practical reactions. Finding a good catalyst is especially difficult, since catalysis has been thought to be a vast, confusing field, replete with an enormous quantity of empirical facts intermixed with a few general theories. Recently, renewed and increased efforts have been made toward the design of catalysts or the development of tailored catalysts, both in academia and industry. From these points of view, polymer-attached catalysts have received intense interest.

This chapter will describe polymer-anchored transition metal complexes and polymer-protected colloidal dispersions of transition metals as tailored catalysts.

Polymer-anchored transition metal complex catalysts have been investigated from two points of view. One is based on polymer chemistry, and the other on catalysis chemistry. From the viewpoint of polymer chemistry, a polymer-anchored transition metal complex catalyst is basically a functional polymer. Its catalytic function originates from its transition metal complex component, while the polymeric component assists the complex in its work as an active and selective catalyst. The dream of investigators with this viewpoint is the development of an artificial enzyme which can catalyze a highly selective reaction under mild conditions.

From the viewpoint of catalysis chemistry, attention has been focused upon the attachment of a transition metal complex catalyst onto a polymeric support. Transition metal complex catalysts, so-called homogeneous catalysts, often have the advantage of giving high selectivity under mild conditions [2, 3]. Furthermore, their activity and selectivity may be tailored by varying the ligand attached to the metal. Thus, many useful reactions have been found to be catalyzed by homogeneous catalysts and more than 20 processes have been developed commercially [4, 5]. However, heterogeneous catalysts, such as metals or metal oxides, are more often used in practice, since they are readily separated from the reactants and products, and may be used in fixed beds in

Y. Iwasawa (ed.), Tailored Metal Catalysts, 87–140.

gas–solid and liquid–solid contact reactors. Polymers are usually solids and may be bound by any ligand to which the metal can coordinate. In this way homogeneous catalysts can be anchored to the polymer and used in heterogeneous systems as heterogeneous catalysts. In these cases special effects due to th polymer are not expected, although transference of the activity of the original homogeneous catalyst to the solid-supported form is of considerable practical significance.

As far as polymer-protected colloidal dispersions of transition metals are concerned, transition metals dispersed on solid supports such as alumina and silica gel are often used as practical catalysts. The effects of the dispersity of the metal and the character of the supports are often discussed by researchers. Organic polymers may be used as protective colloids or organic supports. Thus 'homogeneous' solutions of colloidal dispersion of metals may be achieved by using water-soluble polymers, which is an example of the application of a polymer as a protective colloid. The effects of dispersity or particle size may become paramount in such systems. In contrast, hydrophobic polymers may be used as supports for metal dispersions. In this case, the organic polymer endows a hydrophobicity on the system which is absent when inorganic supports are employed.

2. Polymer-anchored Catalysts

A number of reviews [6–24] have been published in this field since 1971. In this section some fundamental aspects of polymer-anchored catalysts will be discussed with particular regard to their preparation and characterization. Then, the effects of polymers on the activity and selectivity of the catalyst will be described and illustrated with some interesting reactions. Multifunctional catalysts using organic polymers will also be highlighted.

2.1. PREPARATION AND CHARACTERIZATION

A polymer-anchored transition metal complex catalyst consists of two parts, a transition metal complex and a polymer. Fixation of the complex onto a polymer may usually be accomplished by bonding the ligand(s) to the polymer. The polymer most commonly used for this purpose is polystyrene. A polystyrene-bound complex can be depicted as in Figure 1, where M and L represent the central metal and a ligand group, respectively. The ligand group usually contains coordinative atom(s) such as P, N and O.

The preparation of polymer-anchored catalysts entails three steps: polymerization, bonding of a ligand group(s) and coordination of the metal(s). Polymerization may be performed initially, after bonding of a ligand group, or after

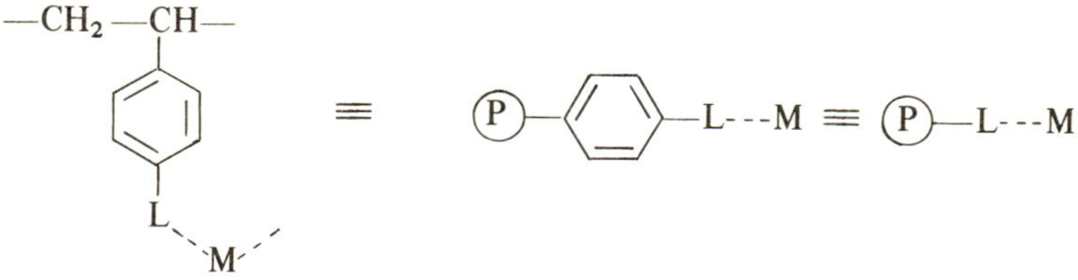

Fig. 1. Representation of polymer-anchored catalyst.

coordination of the metal as shown in Figure 2. If a specific effect of the polymer is desired in the polymer-anchored catalyst, then the design of the polymeric support is very important. Thus a support with specific microstructures and/or specific ligands with a given ratio available for substitution may be prepared by homo- and/or co-polymerization of the monomers with the selected ligand groups. When only immobilization of the homogeneous catalyst is intended, a commercial polymeric support such as polystyrene may be used to bind the ligand group(s).

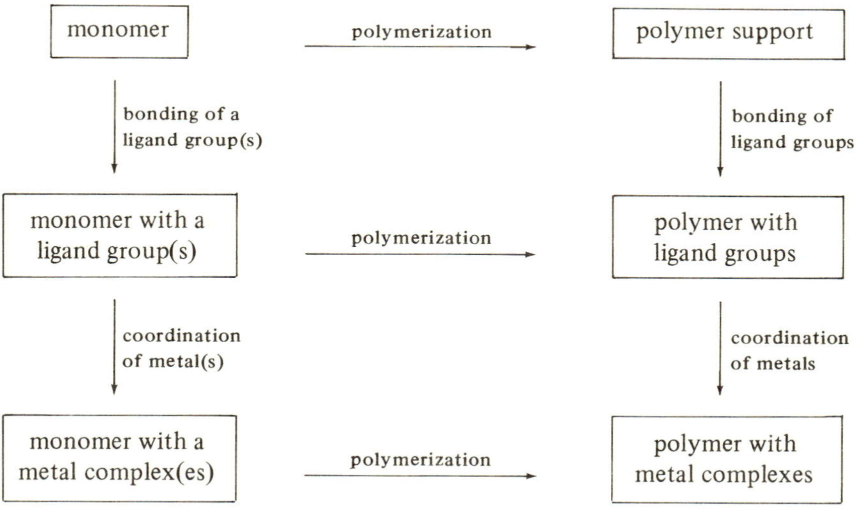

Fig. 2. Fundamental steps in the preparation of polymer-anchored catalysts.

Polystyrene resin, which has been widely used as a polymeric support, is usually a crosslinked copolymer of styrene with a proportion of divinylbenzene (2–20 mol%). The resin is supplied as spherical beads with diameters within the range 1.0 mm to 0.2 mm. In some solvents these may swell to expose micropores. Macroreticular resins [24], which contain many macropores and may be produced by special methods for the copolymerization of styrene and divinylbenzene, are also available commercially.

If such commercially available polystyrene resins are to be used as starting materials, then it is essential that they be pretreated so as to remove impurities [25].

Although unsubstituted styrene copolymers can bind certain transition metal complexes via π-bonding through the benzene ring [26], functionalized polymers containing ligand groups are usually used for the preparation of polymer-anchored complex catalysts.

Normal organic reactions may be used to attach a ligand group(s) onto styrene copolymers via chemical bond(s), although reaction times for polymeric ligands are much longer than those for smaller molecules.

Polymeric phosphine ligands have been most commonly employed for polymer-anchored catalysts. (Some of these are now available commercially from Strem Chemicals Inc., Danvers, MA, U.S.A.) Phosphine ligands bound to polystyrene can be divided into two classes, the first in which a phosphine atom binds directly onto the benzene ring of polystyrene, and the second in which a phosphine atom binds onto the ring via a methylene chain.

Diphenylphosphinopolystyrene is a typical ligand of the former class. Several preparative methods for this ligand have been developed, as illustrated in Figure 3.

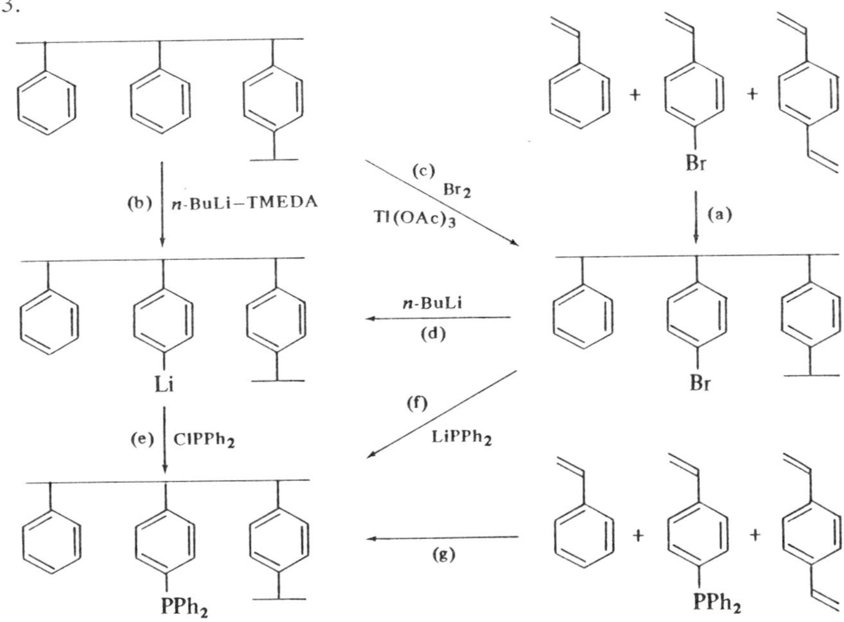

Fig. 3. Preparation of crosslinked diphenylphosphinopolystyrene (from [23]).

Routes (a) and (g) involve copolymerization of a substituted styrene with styrene and divinylbenzene. Random copolymerization of *p*-bromostyrene occurs via route (a), whereas the products obtained from *p*-diphenylphosphino-styrene via route (g) are block copolymers [27], i.e. the diphenylphosphono groups are localized, with the result that the local concentration of the ligand groups may be higher than that in the copolymer obtained from *p*-bromostyrene.

For syntheses employing commercial polystyrene as the starting material, any one of routes (b) to (f) may be chosen. However, the *meta/para* ratio in phenyl substitution depends upon the reaction conditions [28]; for example, thallium acetate is superior to ferrous chloride for bromination at the *para* position [29]. Thus, although it might appear that the same end-product can be obtained by any one of these routes, these end-products are not exactly identical because the various reagents react with the polymer in different ways. It must be emphasized that end-products obtained from the polystyrene resin via several substitution reactions may contain several different functional groups due to incomplete derivatization, with the number and distribution of the phosphine groups differing in each case. Such reactions may be followed by infrared spectroscopy with the phosphorus content of the phosphinated resin finally produced being determined spectrophotometrically as PO_4^{3-}, obtained by the decomposition of the resin with HNO_3/H_2SO_4 and subsequent oxidation with $HClO_4$ [30]. Other kinds of elemental analyses can also be applied.

Such analyses, however, only provide bulk values. The *local* concentration of the ligand in the resin is very important in determining the structure and therefore the catalytic activity of the complex. The distribution of species containing atoms other than C and H may be readily determined in polystyrene copolymer beads by the use of electron microprobe analysis. Cross-sections of the phosphinated polystyrene beads are scanned, and the resulting surface analyses of samples prepared by the differing methods described above reveal that the phosphine distribution is completely different in each case [31]. ^{31}P NMR spectroscopy is also a useful tool for characterizing organophosphorus ligands attached to a polymeric support [18].

Chloromethylated polystyrene, known as Merrifield's resin [32], is often used in the first stage for introducing functional groups. This resin is also available commercially. As shown in Figure 4, the reaction of this resin with diphenyl-phosphine lithium gives diphenylphosphinomethylpolystyrene [33] which has a methylene chain between the phosphine atom and the benzene ring and has been considered to be a polymeric analog of an alkyl phosphine ligand. Characterization of these functional polymers may be performed by procedures similar to those used for diphenylphosphinated polystyrene.

Merrifield's resin has also been used for the preparation of other functionalized polymers allowing the introduction of ligands such as cyclopentadienyl [34], dimethylamino [35] and anthranilic acid [36], as shown in Figure 4.

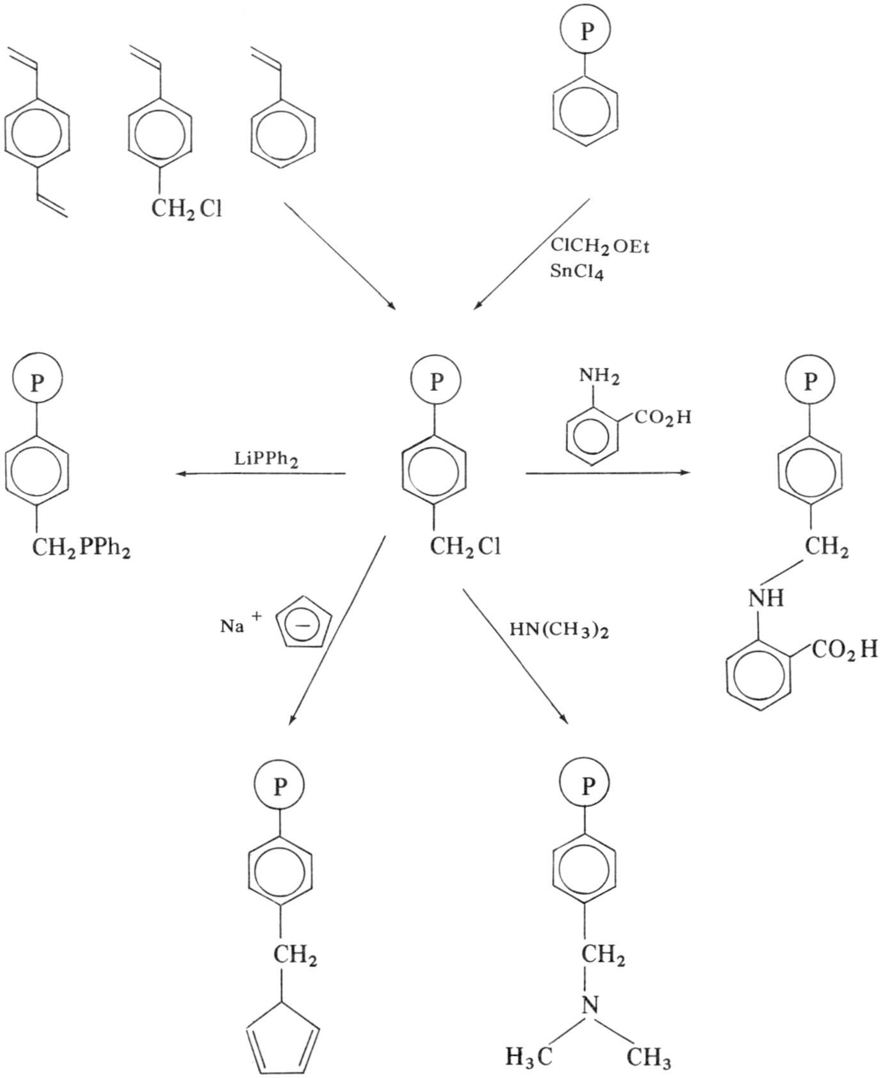

Fig. 4. Preparation of some functionalized polystyrenes.

A detailed knowledge of the molecular structure surrounding the metal atom is very important for a better understanding of the molecular mechanism of a given catalytic reaction. With this in mind, it is possible to prepare structurally ordered polymeric ligands, such as homopolymers of monomers bearing ligand groups of the type *p*-(diphenylphosphino)styrene [37], or alternating copolymers, such as maleic acid with vinyl ethers [38]. Polymerization of a functionalized monomer has also been applied for the preparation of a polymeric ligand containing asymmetic centers as shown in Figure 5 [39].

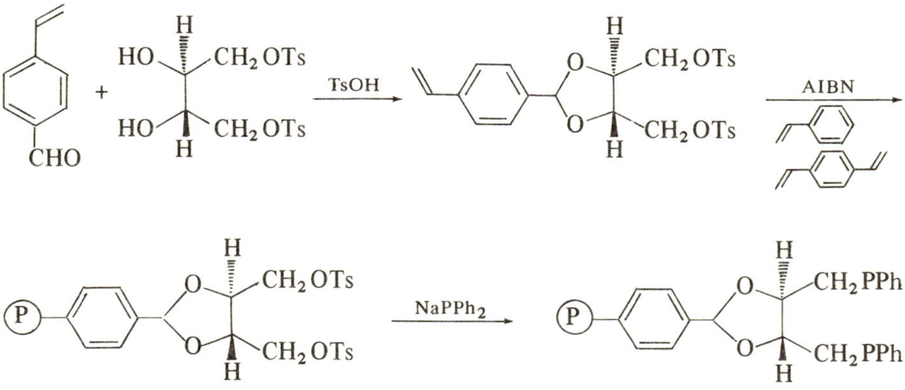

Ts = *p*-toluenesulfonyl

Fig. 5. Preparation of a polymer containing DIOP [2,3-*O*-isopropylidene-2,3-dihydroxy-1,4-bis(diphenylphosphino)butane] (from [23])

For the introduction of a metal onto a polymeric ligand, a number of routes or methods are available, following fundamentally the same types of approach as those used in the preparation of low molecular weight metal complexes. By far the most common route for preparing polymer-anchored catalysts is the displacement of a ligand already coordinated to the metal complex by a polymeric ligand. With phosphinated polymer supports either a metal phosphine complex can be added, so that the metal is attached via phosphine exchange, or a metal complex containing a weaker ligand than phosphine can be added. Under normal reaction conditions, the polymeric ligand will behave quite differently from the low molecular weight ligand. Thus the polymeric ligand will require more time to react, with the rate of reaction being faster on the surface of the resin than inside the resin bead. Another difference with a polymeric ligand is the possibility that two ligand groups may exist close each other, with the result that the metal may bind to both by means of a chelating type of coordination.

As an illustration, a polystyrene-anchored iron complex has been prepared by refluxing a benzene solution of $Fe(CO)_5$ for 20 h with a macroreticular resin consisting of 29% diphenylphosphinated 1% crosslinked polystyrene [40].

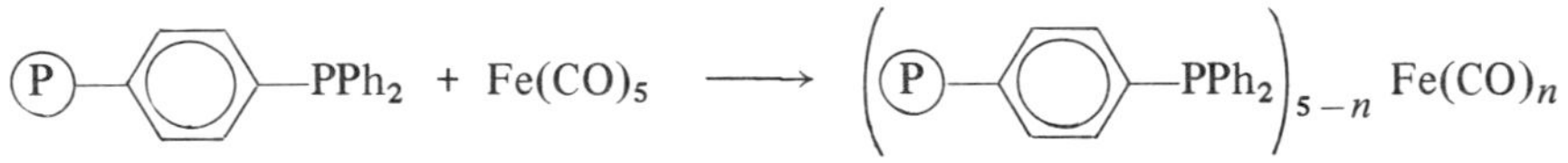

(n = 3, 4)

The polymer-anchored complex thus produced gave a complex IR spectrum corresponding to the two complexes $Fe(CO)_4(PPh_3)$ and *trans*-$Fe(CO)_3(PPh_3)_2$, indicating four CO stretching bands at 2045, 1968, 1932 and 1876 cm^{-1}. The first three bands correspond to the CO stretching of $Fe(CO)_4(PPh_3)$ (2054, 1977.5 and 1942 cm^{-1}), while the last band is close to that for *trans*-$Fe(CO)_3(PPh_3)_2$ which occurs at 1892.5 cm^{-1}. These results indicate that the polymer-anchored complex is a mixture of iron complexes containing one and two polymeric ligand groups. The relative intensities of the respective IR bands indicate that these structures are present in the ratio 1 : 3–5 (for $n = 3/n = 4$).

Instead of displacing a ligand, bridged dimeric complexes can react with polymeric ligands by bridge splitting to produe polymer-anchored complexes. In addition, metals can be introduced into polymeric ligands through the direct reaction of a metal halide with a polymer or by the oxidative addition of a polymer to a metal complex.

Polymer-anchored complexes may be characterized not only by chemical analyses and their IR spectra, but also by NMR, ESR and UV spectroscopy, XPS, electron microscopy, electron-probe microanalysis, etc. For example, scanning electron micrographs have demonstrated that the distribution of metal in polymer beads depends both on the macrostructure and the morphology. Phosphination of a uniformly chloromethylated commercial resin was followed by attachment of (phenanthrene)$Cr(CO)_3$ via irradiation, the large pore sizes (130 nm) in the resin allowing the phosphine and the Cr complex to react with the material throughout the whole of its structure. In contrast, when the pore sizes were small (< 5 nm) only limited penetration occurred into the structure [41].

The polymer-anchored metal complexes thus obtained may be employed as catalysts without any further treatment. However, other anchored complexes of this kind may require activation, since it is essential that the substrate coordinates to the metal complex in order for the whole to function as a catalyst. Such activation can sometimes be achieved under the reaction conditions employed, but in other instances it may be necessary to resort to heat treatment, irradiation by light, addition of a reducing agent or the attachment of a different kind of ligand. In the case of the diphenylphosphinated polystyrene-anchored iron carbonyl complex mentioned above, irradiation allowed further activation of the complex through elimination of CO and enabled the catalyst to be employed for the isomerization and hydroformylation of olefins.

2.2. THE INFLUENCE OF THE POLYMER MATRIX ON CATALYSIS

The polymeric supports employed in the preparation of polymer-anchored catalysts are usually solids forming matrices which are obviously absent in homogeneous metal complex catalysts. Such polymer matrices can affect the reactivity and selectivity of catalytic reactions, as well as the ease of handling of catalysts under practical conditions. The polymer matrix can influence the mobility of the active species or ligand group bound to the polymer, as well as the diffusion of the substrate or other small molecules inside the polymer. Usually such matrix effects lead to a decrease in the catalytic activity relative to that in the corresponding homogeneous systems. However, sometimes matrix effects can have a positive effect on the active species associated with a polymer-anchored catalyst, resulting in a higher activity than that for the unattached complex catalyst. In this section, these effects of the polymer will be compared with the situation where transition metal complex catalysts are employed under homogeneous conditions.

2.2.1. *Isolation of active sites*

As the density of crosslinking increases in a resin, the average mobility of a chain segment or of an attached catalytic site decreases. Thus, the lifetime of a highly reactive site may be increased as a result of such matrix immobilization, particularly if the reactions which destroy the active site involve dimerization or other site–site interactions.

Reduction of α, β-unsaturated carbonyl compounds by carbon monoxide and water has been achieved employing the mixed solvent toluene/dioxan/water at 70 °C under 92 atm pressure of CO over a homogeneous catalyst prepared from $Rh_4(CO)_{12}$ and triethylamine; 8% conversion was achieved after reaction for 5 h.

$$H_2C{=}CHCOOCH_3 + CO + H_2O \longrightarrow CH_3CH_2COOCH_3 + CO_2$$

In contrast, when polystyrene containing an *N,N*-dimethylbenzylamino group was used instead of triethylamine as the catalyst ligand, the conversion increased to 61% under the same reaction conditions [42]. During the reaction, the presence of an Rh_6 complex was detected in the case of the polymer-anchored catalyst, while an Rh_{12} complex was present under homogeneous conditions. The Rh_6 complex has been suggested as the active catalytic species in this reaction, with dimerization which produces the inactive Rh_{12} species being hindered by matrix isolation in the polymer-anchored catalyst.

Wilkinson's complex, $RhCl(PPh_3)_3$, functions as a catalyst for the hydrogenation of olefins. Under homogeneous conditions, the catalytic activity of the complex recovered after hydrogenation of 1-hexene at temperatures up to 45 °C under 1 atm pressure of hydrogen was found to be ca. one-half that of the fresh catalyst. However, when the complex was attached to an aromatic polyamide (Nomex), the activity of the attached complex did not decrease to anything like the same extent despite repeated use of the catalyst at 45 °C [43]. XPS measurements have shown that the polymer-anchored Wilkinson's complex does not contain a bridging chlorine atom in the form Rh—Cl—Rh, but rather the chelating structure shown in Figure 6 where the rhodium atoms are isolated from each other by the polymer matrix.

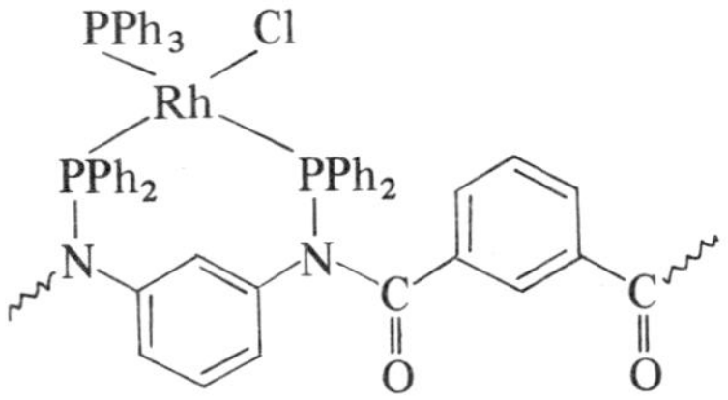

Fig. 6. Structure of aromatic polyamide-anchored Wilkinson's complex (from [23]).

Thus, at temperatures up to 45 °C, homogeneous Wilkinson's complexes produce insoluble dimers or hydride dimers which are inactive as catalysts, while such dimerization is prevented in the polymer-anchored catalyst by the rigid matrix of the aromatic polyamide.

The concept of the isolation of active sites by means of a polymer matrix has also been applied in the carbonylation of allyl chloride catalyzed by palladium. In this case, a cation-exchange resin, poly(styrene sulfonate), has been employed as the counterion of $[Pd(NH_3)_4]^+$ to allow anchorage onto the polymer. Reaction at 100 °C under 1000 psi CO pressure with both homogeneous and polymer-anchored catalysts gave high yields and high selectivity at low catalyst concentrations [44].

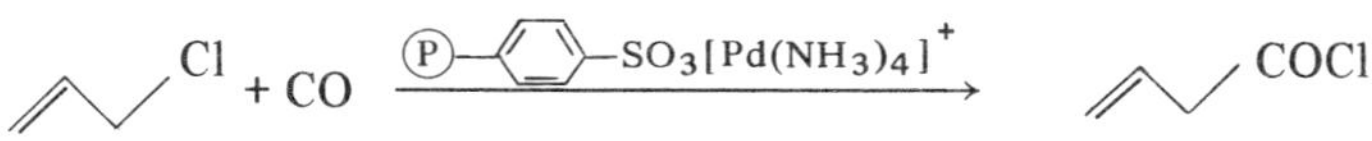

At high concentrations, however, the homogeneous catalyst was less effective, whereas the specific activity of the anchored catalyst remained constant. It was shown that the homogeneous catalyst aggregated to an inactive and uncharacterized complex at higher concentration, while aggregation of the active catalytic sites on the anchored catalyst was prevented by the rigidity of the resin matrix. Thus, the use of a polymer support is effective not only for stabilizing the active site, but also for overcoming any inherent solubility limit as far as the application of the catalyst is concerned.

Clear evidence supporting the view that the polymer matrix maintains the isolation of active catalytic sites has been obtained through studies of a polymer-anchored titanocene catalyst [45]. Titanocene is a reactive intermediate in the reduction of olefins, acetylenes and nitrogen, and as the dichloride may be attached to a 20% crosslinked macroreticular copolymer. On treatment with excess butyllithium or sodium naphthalide, the salmon pink titanocene dichloride polymer was converted to a highly reactive gray polymer which catalyzed the reduction of olefins and acetylenes [46]. However, treatment of unattached titanocene dichloride or benzyltitanocene dichloride with 2 equiv of butyllithium produced a gray catalyst which was only 0.15 times as active per milliequiv of metal in the reduction of cyclohexene as the polymer-anchored metallocene. As shown in Figure 7, titanocene species undergo deactivation by dimerization or polymerization processes, and hence the higher activity of the polymer-anchored catalyst must result from site isolation on the polymer.

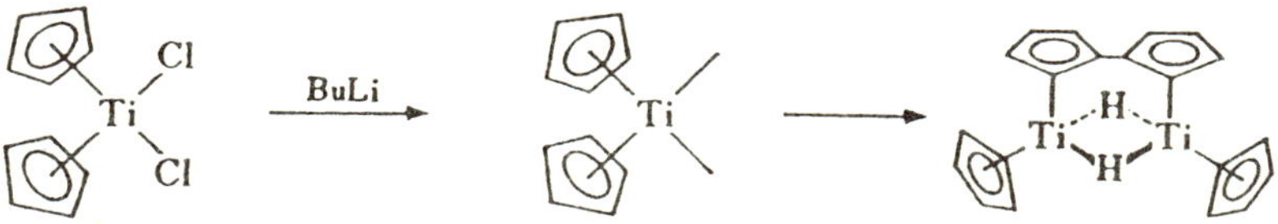

Fig. 7. Deactivation of titanocene by dimerization (from [23]).

Evidence for such a process has been found by measuring the activities of the catalyst per unit mole of Ti as a function of catalyst loading. For the polymer-anchored catalyst where site–site interaction does not occur, the rate of hydrogenation should increase smoothly with increased loading. However, when sites interact the extent of such interaction should increase with loading, and hence the reaction rate should decrease. Hence, for such systems, the rate *versus* loading relationship should exhibit a maximum, as has been found experimentally. However, the same relationship for the unattached catalyst showed no such maximum but rather an incremental decease in the rate of hydrogenation with increasing catalyst concentration, as would be expected if an active catalyst monomer were in equilibrium with an inactive dimer. Recently, polymer-supported titanocene and its derivatives have been applied to other reactions including olefin isomerization, oligomerization, epoxidation and the fixation of dinitrogen [47].

The use of a rigid polymer to prevent dimerization deactivation has been demonstrated in a system comprising an $[RhCl(C_2H_4)_2]_2$ catalyst anchored to poly[1-(4-diphenylphosphinophenyl)ethyliminomethylene] and employed for the hydrogenation of cyclohexene [48].

Complexes of iron or cobalt with porphyrin or phthalocyanine derivatives have been used as oxidation catalysts. In such catalysts, the active species may be a 4-coordinate complex to which the substrate and an oxygen molecule are bound to form the reactive intermediate in the oxidation process. Under homogeneous conditions, however, such complexes readily produce inactive species, which may be direct dimers or μ-dioxo dimers formed by interaction between two complex molecules. If the porphyrin or phthalocyanine ligands are bound to polymer supports, dimerization is inhibited with the result that the catalytic activities of the polymer-anchored catalysts are greater than for the homogeneous catalysts [49]. Site isolation attributable to the polymer matrix has been confirmed by studies of the electronic and ESR spectra for the phthalocyanine complex anchored to a poly(vinylamine) support [50].

2.2.2. *Promotion of coordinative unsaturation*

The first step in any reaction catalyzed by a transition metal complex is coordination of the substrate to the central metal atom. To enable such coordination the central metal should be unsaturated. However, with low molecular weight metal complexes the coordinative site is frequently occupied by other ligands or by aggregation, resulting in deactivation of the complex catalyst even though coordinative unsaturation has been produced initially by activation.

Thus, titanocene dichloride can be activated by butyllithium to yield coordinative unsaturation, but the resulting complex is readily deactivated through dimerization or polymerization. However, if a polymeric ligand is employed instead of the low molecular weight butyl ligand in this complex, the rigid polymer matrix prevents dimerization of the complex so that titanocene associated with a polymeric support is 15–120 times more active than the corresponding low molecular weight complex [46].

Coordination of a palladium complex into a phosphinated polystyrene resin leads to a catalyst which exhibits enhanced activity relative to the corresponding low molecular weight complex $PdCl_2(PPh_3)_2$ for the partial hydrogenation of soy bean fatty acid methyl ester at 25 °C under 1 atm pressure of hydrogen in methanol/benzene as a solvent [51]. The catalytic activity of the low molecular weight complex may be increased by the addition of $SnCl_2$, while that of the polymeric complex is inhibited by $SnCl_2$. Since $SnCl_2$ coordinates strongly to Pd^{II}, the high catalytic activity of the polymeric complex may be attributed to unsaturation of the central palladium atom. Indeed, XPS studies have indicated that the polymeric complex possesses a higher 3d bond energy with a diminished electron density on the Pd atom than the corresponding low molecular weight complex $PdCl_2(PPh_3)_2$ and $PdCl_2$ [52]. On the basis of these results, it has been suggested that the structure of the polymer catalyst is that of a 3-coordinate complex of the type (PSt)—CH_2—PPh_2- - -$PdCl_2$, where PSt depicts the polystyrene resin [53].

The rhodium complex of poly(acrylic acid) serves as a catalyst for the hydrogenation of olefins. The complex of $RhCl_3$ with poly(acrylic acid) was activated through reduction by refluxing the polymeric complex in methanol/water as a mixed solvent [54]. The activated polymeric catalyst (abbreviated as Rh—PAA—H) exhibited extremely high catalytic activity for the hydrogenation of olefins under 1 atm pressure of hydrogen in methanol at 30 °C. The catalytic activity of the polymeric catalyst Rh—PAA—H is compared with that of the known low molecular weight complex of rhodium(II) acetate, $Rh_2(OAc)_4$, in Table I [55].

TABLE I
Hydrogenation of 1-hexene catalyzed by rhodium complexes [55]

Catalyst	Solvent	Catalytic activity[a]	
		Rate of hydrogen uptake[b]	Relative rate
$Rh_2(OAc)_4$	EtOH	0.204	1.0
$Rh_2(OAc)_4$	DMSO	2.68[c]	13.1
$RhCl(PPh_3)_3$	benzene	55.8[d]	273
Rh—PAA—H[e]	MeOH	232	1137
Rh—PAA—Na[f]	MeOH	140	686

[a] [1-hexene] = 0.4 mol dm^{-3}, at 30 °C under 1 atm pressure of hydrogen.
[b] Rate of hydrogen uptake per atom of Rh ($\times 10^{-3}$ mol (Rh atom)$^{-1}$ s^{-1}).
[c] Dodecene may be used as a substrate at 50 °C (see ref. 56).
[d] 1-Heptene may be used as a substrate at 25 °C (see ref. 57).
[e] Rh complex of poly(acrylic acid). Molar ratio of acrylic acid residue per Rh atom = 200.
[f] Sodium derivative of Rh—PAA—H.

From the results quoted in the table, it will be seen that the activity of Rh—PAA—H was > 1000 times the activity of rhodium acetate. When dimethyl sulfoxide was used as a solvent the catalytic activity of rhodium acetate increased, while the polymeric catalyst was deactivated. Since dimethyl sulfoxide is known to function as a ligand which coordinates strongly to Rh^{II}, the high activity of Rh—PAA—H has been attributed to the existence of a highly conentrated Rh^{II} species exhibiting coordination unsaturation. The heterogenized catalyst prepared from $RhCl_3$ and a crosslinked poly(acrylic acid) resin instead of soluble poly(acrylic acid) showed a similar behavior, although the catalytic activity was decreased slightly.

Promotion of coordinative unsaturation has also been demonstrated for the polymer-supported Vaska's complex [16]. Vaska's complex, $Ir(CO)Cl(PPh_3)_2$, can selectively catalyze the reduction of dienes to monoenes, the mechanism of this hydrogenation having been well studied and is as shown in Figure 8 [58, 59].

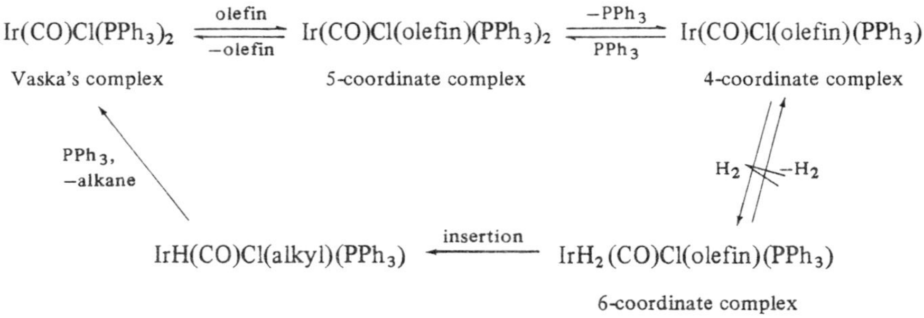

Fig. 8. Schematic diagram for the hydrogenation of olefins catalyzed by Vaska's complex.

The first step in the process is olefin coordination to Vaska's complex, giving the 5-coordinate species. This loses triphenylphosphine reversibly to give the 4-coordinate monophosphine complex, which in turn has hydrogen added to it oxidatively to yield a 6-coordinate complex. This latter step is rate-determining. This rate is, of course, retarded by excess phosphine which reverses the equilibrium from the 4-coordinate complex back toward the 5-coordinate species.

A polymer matrix effect has been investigated in this system by using a variety of phosphinated polystyrene resins containing 1% divinylbenzene [60, 61]. Hydrogenation of 1,5-cyclooctadiene in benzene at 170 °C and under 10 atm pressure of hydrogen was studied over polymer-supported catalysts at different P/Ir ratios and loadings.

As shown in Table II, the selectivity to cyclooctene was quite high in all cases, regardless of whether the catalyst was homogeneous or polymer-supported. However, at low P/Ir ratios the polymer catalyst was very much more active than the corresponding homogeneous catalyst. At P/Ir = 3, the extent of hydrogenation was greater after 15 min with the polymer-supported catalyst than after 33 h with the homogeneous system. On the other hand, when P/Ir = 22, the rates of the homogeneous and polymer-supported systems were approximately equal.

The above observations may be explained in terms of the mechanism of the homogeneous reaction. At low P/Ir ratios, each iridium atom is, to a degree, matrix-isolated from the phosphines ligands with the mobility being ultimately

TABLE II
Hydrogenation of 1,5-cyclooctadiene catalyzed by homogeneous and polymer-supported Vaska's catalysts [16]

Catalyst	P/Ir ratio	Reaction time (h)	Conversion (%)[a]	
			Cyclooctene	Cyclooctane
homogeneous	3	33	37	3
polymer-supported[b]	3	0.25	57	4
homogeneous	22	72	15	0
polymer-supported[b]	22	93	18	2

[a] At 170 °C under 10 atm pressure of hydrogen in benzene.
[b] Phosphinated percentage on polystyrene was 4.02.

restricted by the crosslinking. Thus once a phosphine group has left an iridium atom it cannot readily return to the central metal. Hence, the polymer matrix has the effect of shifting the equilibrium from the 5-coordinate complex toward the 4-coordinate species (see Figure 8). As a result, the concentration of the 4-coordinate species increases relative to that in the corresponding homogeneous system, and consequently the rate of the rate-determining step is also increased. In contrast, at high P/Ir ratios more phosphine is available per iridium atom. In this case, although a phospine group may leave a central iridium atom, its place can be taken quite readily by other phosphines in the vicinity. The consequence of this is that the matrix no longer shifts the equilibrium relative to that existing in the homogeneous system. These various effects of the matrix are depicted in Figure 9.

The influence of phospnine loading in the polymer and of the reaction temperature upon the activity of the polymer-supported iridium catalyst can also be explained in terms of the dissociation equilibrium and immobility of the polymer matrix, which promotes coordinative unsaturation.

Polymer-anchored ruthenium complexes prepared by reacting diphenylphosphinated styrene–1% divinylbenzene resins with $Ru(CO)_3(PPh_3)_2$ under an atmosphere of carbon monoxide function as catalysts for the hydroformylation of 1-pentene, and have been compared to the activity of $Ru(CO)_3(PPh_3)_2$ as a homogeneous catalyst [62]. The normal/branched aldehyde selectivity has also been discussed in this case in terms of the equilibrium between (polymer—$PPh_2)_2RuH_2$(olefin)(CO) and (polymer—$PPh_2)RuH_2$(olefin)$(CO)_2$ within the resin matrix, with the phosphine loading, P/Ru ratio, ligand mobility and swelling again playing contributary roles.

A similar progressive activation has been observed for catalysts prepared from $RhH(CO)(PPh_3)_3$ and maleic acid/vinyl ether alternating copolymers [63].

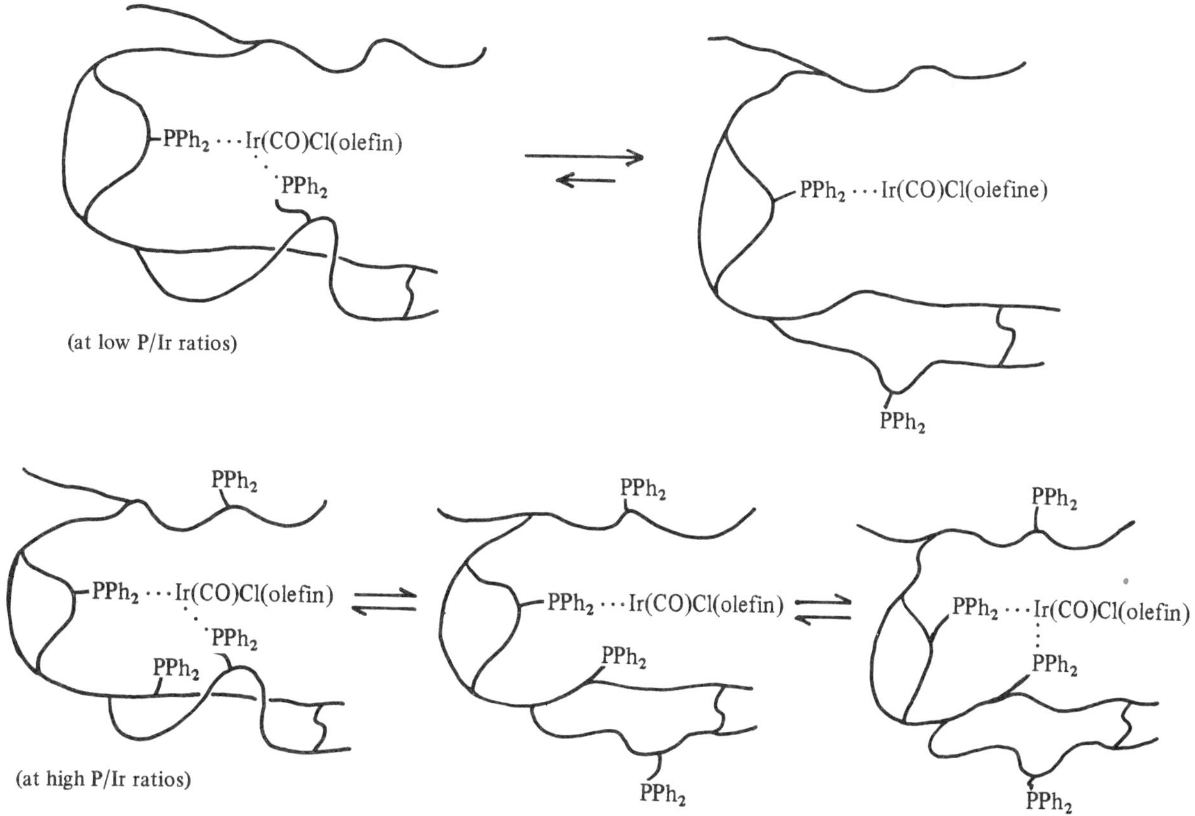

Fig. 9. Effect of P/Ir ratio on the phosphine-iridium dissociation equilibrium of the polymer-supported olefin complex of Vaska's complex [16].

In this case the effect was attributed to the decomposition of the carbonyl species and the progressive release of phosphine, with forming a coordinatively unsaturated species.

An influence of the polymer matrix on catalytic activities has been demonstrated by using a rigid polymer as a support during the hydrogenation of aromatic compounds over rhodium complexes [64]. Poly(diacetylene)-anchored rhodium complexes (**1**) prepared from $[RhCl(norbornadiene)]_2$ and phosphinated poly(diacetylene) have been found to be efficient catalysts for the hydrogenation of arenes under mild conditions. In contrast, the polystyrene-anchored catalyst (**2**) was not active for the same reaction.

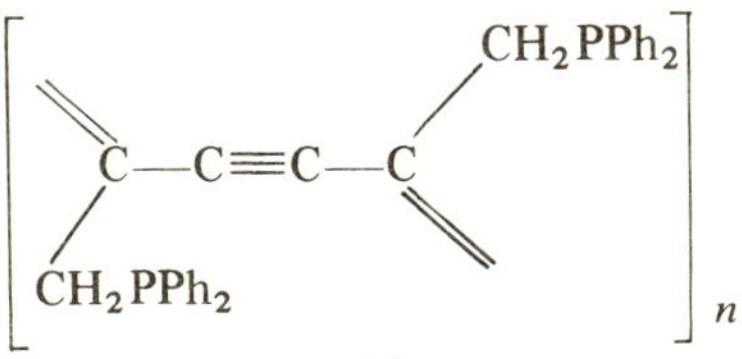

(**1**)

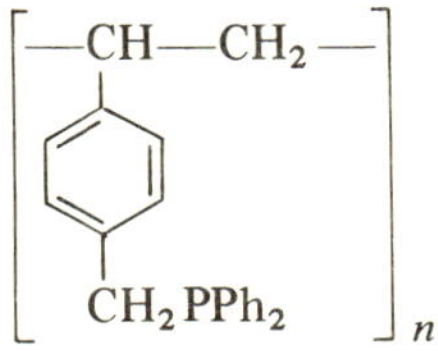

(**2**)

From IR spectroscopic measurements, it has been shown that the polymer-anchored catalysts contain coordinatively unsaturated sites and/or readily removable ligands, and the number of these sites and/or ligands is higher on poly(diacetylene) than on polystyrene. The high activity on poly(diacetylene) is considered to be a reflection of the nature of that particular polymer matrix, i.e., to be due to its rigid polymer backbone.

2.2.3. *Concentration of the ligand*

The ligand loading in any polymer can be made high or low. Furthermore, by controlling the stoichiometry of metal incorporation, polymers with both high or low ligand/metal ratios may be prepared. In contrast, in a homogeneous system, it is sometimes very difficult to prepare a solution with a high ligand/metal ratio containing a high concentration of ligand, although the reverse is easy, especially if there is a heed to recover and re-use the homogeneous catalysts as well as the ligands. This ability to concentrate and retain a ligand in the polymer is one advantage of a polymer-matrix system.

As a result of the concentration of ligands, the central metal atom in a complex may be bound to two or more ligand groups of the polymer, so that the metal–ligand bond in a polymer is stronger than that in a homogeneous system as a result of such chelating type of coordination.

The ligand-concentrating effect can alter the selectivity in reactions catalyzed by polymer-supported metal complexes. An excellent example is provided by hydroformylation reactions catalyzed by rhodium attached to phosphonated styrene–divinylbenzene resins [65, 66]. When hydroformylations of 1-pentene were carried out over $RhH(CO)(PPh_3)_3$ in benzene at 100 °C under 6.8 atm pressure of H_2/CO gas (1 : 1 ratio), the normal/branched (n/b) ratio of the resulting aldehydes was ca. 1.

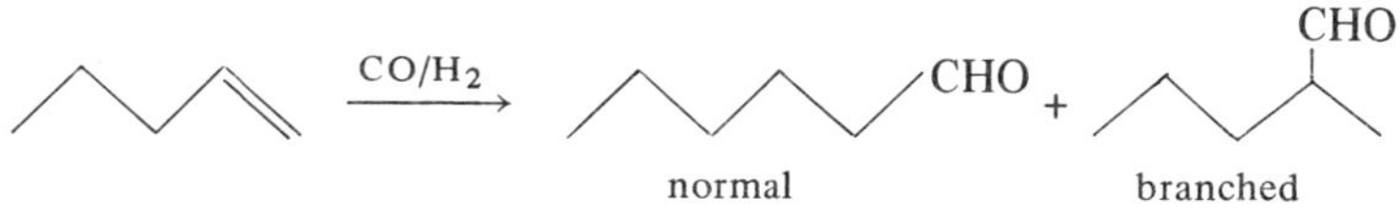

Since the n/b ratio depends on the phosphine/rhodium ratio, addition of a large excess of triphenylphosphine (P/Rh = 19) increased the n/b ratio to 3.3. This change can be explained in terms of the following equilibrium:

$$RhH(CO)_2(PPh_3)_2 \underset{PPh_3}{\overset{-PPh_3}{\rightleftarrows}} RhH(CO)_2(PPh_3)$$

Formation of the normal aldehyde is catalyzed predominantly by the species with two phosphine ligands on the left-hand side of the equation [66]. Over the catalyst supported on 40% phosphinated crosslinked polystyrene the n/b ratio increased to 12.1, even if the same P/Rh ratio of 19 was employed as in the case of the low molecular weight catalyst. Whereas in the homogeneous system the phosphine ligands are distributed throughout the whole solution, in the polymer-supported catalyst the phosphine is localized inside the swollen resin giving up to 10 times the local concentration in the homogeneous solution.

A similar improvement of selectivity by polymer-supported catalysts has been observed in the partial hydrogenation of 1,5-cyclooctadiene. In the absence of excess phosphine, use of the $RuCl_2(CO)_2(PPh_3)_2$ catalyst resulted in the production of cyclooctene and cyclooctane with low selectivity. With excess phosphine (P/Ru = 24), however, the selectivity for cyclooctene was improved up to a value of 94% at 140 °C [67]. Using a 1% phosphinated polystyrene-supported catalyst, enhanced selectivities were observed at high P/Ru ratios if a resin was empoloyed with low phosphine conentration or at low P/Ru ratios if a highly phosphinated resin was employed. Thus, for example, using 47.5% phosphinated polystyrene resin at a P/Ru ratio of 6.6, the selectivity for cyclooctene was 96.9% [68].

High concentrations of a ligand in solution brought about by the addition of excess ligand sometimes improves the selectivity of a reaction catalyzed by complex catalysts. With the hydrogenation of dienes, since the coordinating ability of a diene is stronger than that of a monoene, a diene can coordinate into a complex at relative low concentrations whilst a monoene can only do so when it is present in excess. Unfortunately in a homogeneous system, such excess ligands distribute themselves throughout the bulk of the solution. In contrast, in a polymer-supported catalyst the ligands are concentrated within the polymeric region of the support. Furthermore, in a homogeneous system excess ligands at a high concentration prevent coordination by substrates, resulting in deactivation of catalysts. In a polymer-supported catalyst, however, coordinative unsaturation is promoted by the polymeric matrix as described in the previous section, resulting in the activation of the catalyst.

2.2.4. *Stabilization of catalysts*

Homogeneous complex catalysts are readily deactivated by various reactions. For example, they react with oxygen or water to produce inactive species. Thus, a catalytic solution of Wilkinson's complex, $RhCl(PPh_3)_3$, loses its catalytic activity on exposure to air. However, when the complex is supported on polystyrene resin, it can be employed in air without deactivation [69]. Similarly, a solution of $CuAlCl_4$ in toluene will adsorb carbon monoxide from a mixed gas system. When the mixed gas contains a small amount of moisture, however, the complex decomposes by reaction with the water present in the system and loses its activity for CO adsorption. Addition of polystyrene to this solution results in the formation of a polystyrene complex of $CuAlCl_4$, the activity of which is not affected by moisture, in contrast to the low molecular weight system [70]. Again, most homogeneous complexes corrode metal reaction vessels. In contrast, polymer-supported complex catalysts have no such corrosive effect on metal and retain their activity when the reactions are carried out in metallic reaction vessels.

These protective effects of the polymer can be explained in terms of the polymer matrix. By cross-sectional analysis of polymer beads using an electron probe microanalyzer, it has been shown that the rhodium atoms in a polymer-supported Wilkinson's complex are distributed in a layer 16 μm deep from the surface of the beads [71]. Thus, polystyrene works not only as a polymer matrix to decrease its mobility but also as a ligand for the metal. The stability toward air of a phosphinated polystyrene-supported rhodium complex catalyst has been explained in terms of the coordination of the benzene rings of the polystyrene with the rhodium atom [72]. A similar coordination of the benzene rings in polystyrene with cobalt, as well as a good distribution of the active species within the polymer matrix, have been advanced as reasons for the protective effect of polystyrene on a cyclopentadienyl–cobalt complex against irradiation with light [73].

A polymer-supported oxobis(pentane-2,4-dionato)vanadium(V) complex (**3**) has been shown to catalyze the homolytic decomposition of *t*-butylhydroperoxide, and its reaction with dimethyl sulfoxide, dibutyl thioether and

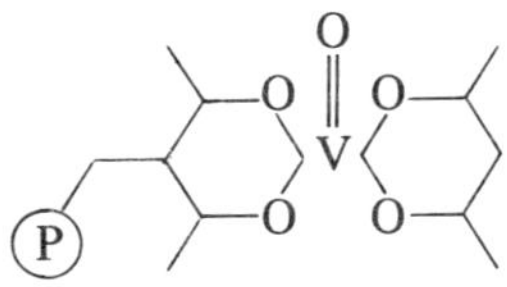

(3)

cyclohexene to give dimethyl sulfate, dibutyl sulfoxide and cyclohexene oxide, respectively [74]. Comparative studies between the polymer-supported and homogeneous systems have demonstrated enhanced stability for the polymer-supported, catalytically active species. The diffusion barrier between the reactants in solution and the polymer influences the rate of a given reaction, and is considered to be responsible for this extra stability.

The polymer matrix can also stabilize the metal in a low valency state, which is frequently the state associated with the active species for catalysis. For example, poly(2-methylethyleneimine) or poly(glutamic acid) forms a water-soluble complex with Ru^{III} or Rh^{I}. These polymeric complexes operate as catalysts for the hydrogenation of methyl acetacetate [75], mesityl oxide [75], 1-hexene [76], etc. in homogeneous aqueous solution. In contrast, when the corresponding low molecular wight complexes were used as catalysts for such hydrogenations, they were reduced to the metallic form resulting in metal precipitation under the conditions employed. Since polymeric ligands form a chelate complex with a metal ion, the formation constant of the complex is larger than that of the corresponding monomeric complex. Furthermore, the conductivity of the microdomain present in the polymeric backbone is smaller than that of an aqueous solution. For these reasons, polymeric ligands can stabilize a central metal ion in a low valency state and prevent its aggregation.

A polymer-supported platinum catalyst is useful for the hydrogenation of benzene at 160 °C in the vapor phase. Such polymer-supported complex catalysts have been prepared by treatment of Nylon-3 with an aqueous solution of H_2PtCl_2 and of poly(acrylonitrile) with an acetone solution of $PtCl_2(\epsilon$-caprolactam$)_2$. Hydrogenation over these polymeric catalysts gave cyclohexene with a selectivity of 40% and 78.6%, respectively. After such hydrogenations, the Cl/Pt atomic ratios remained at the same values as those before reaction [77]. In contrast, hydrogenation of benzene over H_2PtCl_2 or $PtCl_2(\epsilon$-caprolactam$)_2$ supported on alumina produced only cyclohexane. After such hydrogenation it was found that a chlorine atom had been lost from the alumina-supported catalyst. These observations clearly demonstrate that a polymer support can stabilize a platinum atom in a low valency state and prevent its conversion to the metal.

Dechlorination of chlorobenzene by hydrogen has been carried out in the presence of both a polymer-supported and an alumina-supported catalyst. The polymeric catalyst was prepared by hydrogen treatment of a poly-*p*-phenylenephthalamide–H_2PtCl_2 complex, and its use led to the production of benzene in 99.1% selectivity at 97.1% conversion [78]. However, over the corresponding alumina-supported catalyst the main product was cyclohexane with the selectivity for benzene being only 27.6%.

The stability of polymer-attached metal catalysts is an important factor from the viewpoint of practical application. The stability of such catalysts under practical conditions can be examined by their repeated use. With complexes containing phosphine ligands, stabilization of the ligands against attack by air through the use of polymer matrices was insufficient for practical purposes [79, 80]. However, the silica-supported poly(vinylpyrrolidone)–Pd complex was found to catalyze the hydrogenation of nitrobenzene under mild conditions, i.e. at 30 °C under 1 atm pressure of hydrogen. This complex can be re-used several times without any appreciable loss of catalytic activity if small amounts of acetic acid are added to the system [81].

2.2.5. *Effect of substrate size and shape on selectivity*

A metal complex or an active site is sometimes buried inside the polymer matrix in a polymer-supported complex. Under such circumstances, bulky substrate cannot arrive at the active site, resulting in selectivity for a substrate in terms of its size and/or shape.

Thus, during studies of the hydrogenation of olefins of varying sizes in the presence of a Wilkinson's complex catalyst supported on 1.8% crosslinked phosphinated polystyrene beads, the rate of hydrogenation decreased with increasing size of the substrate. On the other hand, with Wilkinson's complex attached to monomeric analogs the difference in rates was quite small. The relevant data are listed in Table III [82].

TABLE III
Substrate selectivity during the hydrogenation of olefins in the presence of Wilkinson's catalyst [82]

Substrate		Relative rate[a]	
Olefin	Size	Polymeric catalyst	Monomeric catalyst
1-hexene	small	2.55	1.4
cyclohexene	↓	1.00	1.00
1-octadecene	↓	0.49	0.71
cyclooctene	↓	0.39	1.00
cyclododecene	↓	0.22	0.67
Δ^2-cholestene	large	0.03	0.71

[a] In benzene at 25 °C under 1 atm pressure of H_2.

Where a polymeric complex is employed, the active catalytic site is located within a pore in the resin bead. The rate of passage of the substrate through the pore varies with the size of the substrate, thus producing the observed selectivity.

The above reaction was carried out in benzene. Addition of ethanol to the benzene solution increased the ratio of the rates of hydrogenation for the olefins. When allyl alcohol or 4-penten-1-ol was used as the substrate, the ratio of the rates of hydrogenation decreased on addition of ethanol. This can be explained in terms of a polarity gradient between the inside and the surface of the resin beads, this gradient being produced by the addition of ethanol to the solution. The gradual change in polarity (nonpolar inside to polar outside) promotes diffusion of a nonpolar substrate toward the inside of beads, and hence to the active site, but retards the diffusion of a polar substrate.

The substitution reaction of esters is catalyzed by zero-valent palladium complexes [83]. However, on a phosphinated polystyrene-anchored palladium complex, the methyl ester of linear dienoic acid was produced more selectively than on the homogeneous analog, $Pd(PPh_3)_4$.

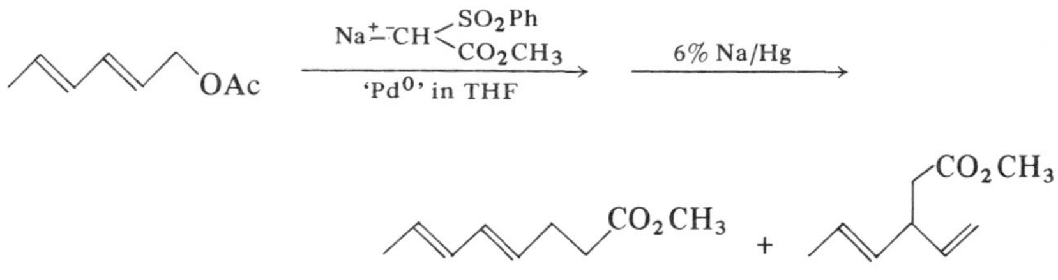

$Pd(PPh_3)_4$	69%	31%
Ⓟ$-CH_2-PPh_2)_nPd(PPh_3)_{4-n}$	80%	20%

This difference was explained in terms of steric hindrance in the polymer matrix. The selectivity difference could be enhanced through substitution with amines. The polymer-anchored catalyst displayed complete stereoselectivity. Thus, employing a *cis* starting compound, a *cis* product was obtained with 100% selectivity; employing a *trans* starting compound, a *trans* product was obtained in 100% yield.

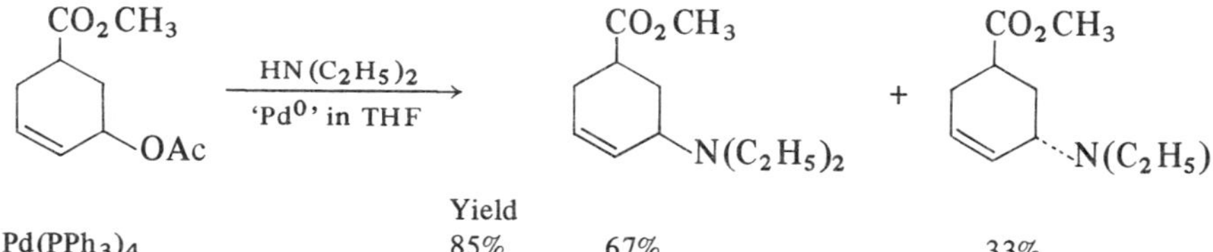

	Yield		
$Pd(PPh_3)_4$	85%	67%	33%
(Ⓟ$-CH_2-PPh_2)_nPd(PPh_3)_{4-n}$	83%	100%	0%

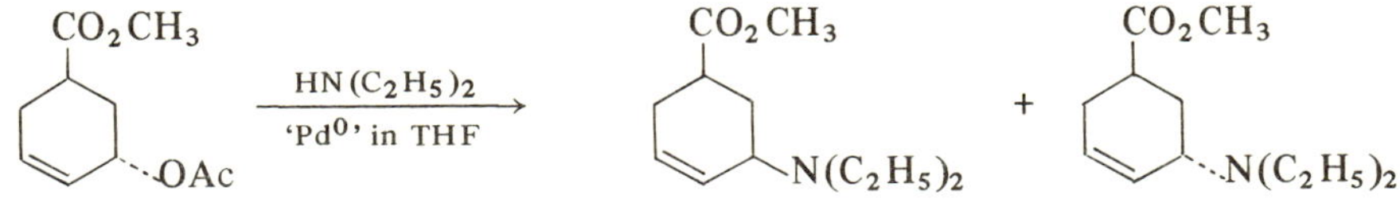

	Yield		
$Pd(PPh_3)_4$	60%	35%	65%
(P)—Pd	86%	0%	100%

A poly(ϵ-benzyloxycarbonyl-L-lysine)—copper(II) complex functions as a catalyst for the dehydrogenation of aliphatic alcohols [84]. For *n*-butyl alcohol, *s*-butyl alcohol and isobutyl alcohol, the rate observed with the polymeric catalyst was much higher than that with copper powder. For di-isobutyl carbinol and di-isopropyl carbinol, however, the polymeric catalyst is inactive. In this case the copper(II) ion coordinates between the amide group of the side-chain and the peptide group of the α-helix main chain, making it impossible for bulky alcohols to approach the copper which is the active center in this case.

2.3. EFFECT OF THE POLYMER 'FIELD'

The active site in polymer-supported catalysts is surrounded by polymer chains, which affect the catalytic reactions. Thus one may define a 'field' of macromolecular groups closely surrounding and influencing the central metal atom. The 'fields' associated with surrounding polymers involve hydrophobic, electrostatic and asymmetric effects. Such 'fields' are not only produced by crosslinked insoluble polymers, but also by soluble polymers. Soluble polymers can, in fact, form microdomains which may be considered as a type of 'field' characteristic of this kind of polymer.

2.3.1. *Hydrophobic 'field'*

The most commonly used polymeric support for polymer-anchored catalysts is polystyrene, which is hydrophobic. Hence, if the catalyst is used in hydrophilic media, the polystyrene support is only slightly swollen forming a hydrophobic 'field' around the central metal atom which is the an active site. Under these circumstances, a hydrophobic reactant may readily approach the central metal atom via hydrophobic interaction. In contrast, a hydrophilic reactant cannot approach the active site because of the hydrophobic–hydrophilic repulsion in the system [82].

For the oxidative cleavage of 4-methylcatechol, the pyridine–Cu complex functions as a catalyst [85, 86]. In the presence of a small amount of methanol, the product consists of the monomethyl ester of *cis,cis*-muconic acid as well as a polymerization product, catechol melanin. However, the addition of methanol to the solvent inhibits cleavage, while addition of benzene brings about an increase in the yield of muconate. A similar increase in yield avoiding the addition of benzene may be effected by using the copolymer of styrene and 4-vinylpyridine as a support for the Cu complex; in this instance the yield was 80% [87].

Poly(vinyl chloride) can be used to protect an $AlCl_3/CuSO_4$ catalyst useful for the cracking and isomerization of hexane [88]. Conversions of 25–60% were achieved after a reaction time of 3–24 h if the catalyst employed had a polymer/$AlCl_3$–$CuSO_4$ ratio of 1:11. The total conversion was greatly affected by the composition of the catalyst. The polymer-protected catalyst showed a much higher catalytic activity than the nonprotected catalyst. This effect may be due to the hydrophobic 'field' generated by poly(vinyl chloride).

2.3.2. *Electrostatic 'field'*

When ionic polymers are used as supports for polymer-anchored catalysts, an electrostatic 'field' is formed by the polymer. In this case, electrostatic interaction or repulsion is of importance.

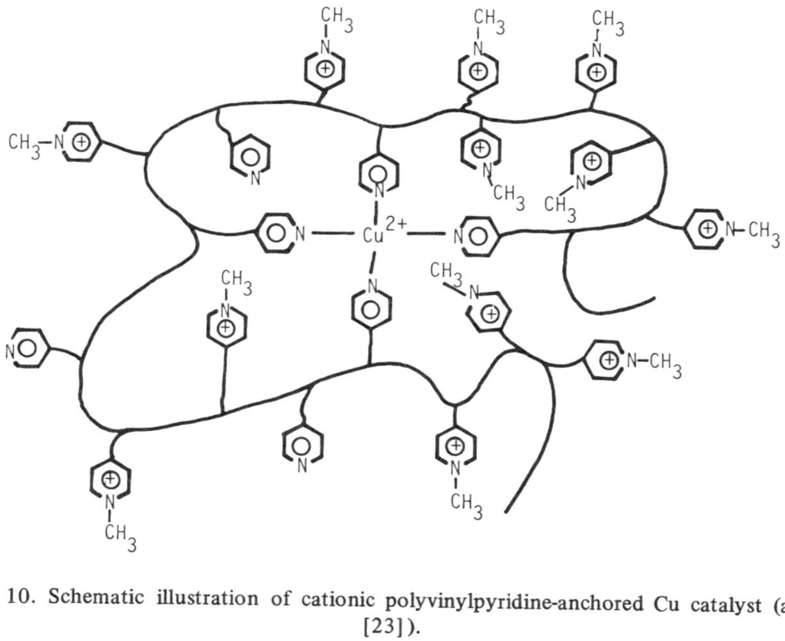

Fig. 10. Schematic illustration of cationic polyvinylpyridine-anchored Cu catalyst (after [23]).

Partially quarternized poly(vinyl pyridine) (PVP) forms a complex with the Cu^{II} ion. This poly(vinyl pyridine)–Cu complex is surrounded by a cationic 'field' generated by the pyridinium residues (as shown in Figure 10) and functions as a catalyst for the oxidation of ascorbic acid (AH_2) to dehydroascorbic acid in the presence of molecular oxygen. The mechanism proposed for this reaction is as follows: The PVP–Cu^{II} complex interacts with the ascorbinate ion, AH^-, to produce a PVP–$CuAH^+$ intermediate which converts to a PVP–Cu^I complex, A^- and a proton. The PVP–Cu^I complex reverts back to the Cu^{II} complex in the presence of molecular oxygen, and A^- forms dehydroascorbic acid. The kinetic parameters for oxidation in the presence of a PVP–Cu catalyst have been investigated in comparison to those with Cu^{II}–bis(2,2′-bipyridyl) and Cu^{II}–aquo complexes [89]. The formation constant, k_c, for PVP–$CuAH^+$ (10 000 mol^{-1} dm^3) is much larger than those for the low molecular weight analogs (280 mol^{-1} dm^3 for a bipyridyl complex and 200 for an aquo complex), while the rate constant for the formation of A^- for the polymeric system (4000 mol^{-1} dm^3 s^{-1}) is ca. 4000 times greater than those for the monomeric system (0.5 mol^{-1} dm^3 s^{-1} for a bipyridyl complex and 1 mol^{-1} dm^3 s^{-1} for an aquo complex). These differences can be explained in terms of the local concentration of the ascorbinate ion (AH^-). Since the cationic 'field' arising from partially quaternized poly(vinylpyridine) interacts with the anion, the local concentration of AH^- around the copper is much higher in the polymeric system than in the monomeric analogs.

Molybdenum(III) complexes with water-soluble polymers containing cysteine residues (copolymer of *S*-benzyl-*N*-benzyloxycarbonyl-4′-vinyl-L-cysteineanilide and acrylic acid or methacrylic acid) have been prepared by the reduction of the corresponding Mo^V complexes with $NaBH_4$. The resulting polymeric Mo^{III} complexes are effective catalysts for the reduction of acetylene in the presence of $NaBH_4$, their catalytic activities being two to three times greater than that of the corresponding low molecular weight Mo–cysteine complex [90]. The higher activity of the polymeric complexes is explained in terms of the stabilization of the active mononuclear species. Thus, the dimerization of the mononuclear Mo^{III} species to an inactive species is considered to be hindered by the spread of polymer chains caused by electrostatic repulsion between the carboxyl groups.

2.3.3. *Asymmetric 'field'*

A catalyst for an asymmetric reaction must have a chiral microenvironment, i.e. an asymmetric 'field'. With polymers such asymmetric 'fields' can arise in two principal ways. In the first type, chirality arises inherently as a result of the secondary structure of the polymer, such as the possession of an α-helix. In the second type, chirality arises from chiral ligand groups bound to the polymeric supports.

The first catalytic reaction employing an asymmetric polymer 'field' was that of the enantioselective hydrogenation of prochiral C=N and C=C bonds using metallic palladium supported on silk fibroin [91, 92]. In this way, for example, (*R*)-phenylalanine was obtained from 4-benzylidene-2-methyloxarol-5-one in 30–70% optical yield.

Asymmetric 'fields' produced by the secondary structure of the polymer can effect asymmetric reactions. Thus, a complex prepared from sodium poly-(*R*)-glutamate and $RuCl_3$ functions as a catalyst for the asymmetric hydrogenation of α-acetamidocinnamic acid to phenylalanine [93]. The optical yield changes with the α-helix content, which can be varied by adjusting the pH of the solution, as shown in Figure 11.

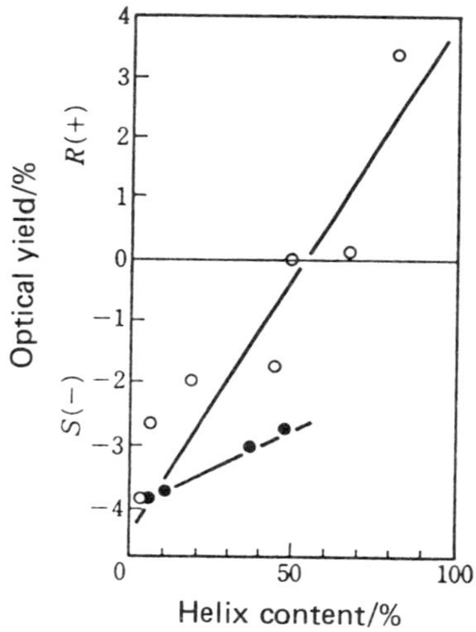

Fig. 11. Asymmetric hydrogenation of α-acetamidocinnamic acid with a poly-(*R*)-glutamic acid–ruthenium complex (molar ratio glutamate/Ru = 5) in an aqueous solution containing 30% ethanol (•) or methanol (○) (after T. Asahara *et al.* (eds.), *Seitai Moderu Kobunshi* (*Models in Biopolymers*), Japan Scientific Soc. Press, Fig. 6, p. 118.)

Apparently the optical yield is controlled by the secondary conformation (helix) rather than the chirality imposed by the asymmetric carbon of the amino acid. Poly-(*S*)-leucine [94, 95] and poly-L-methyleneimine [75] have also been used as supports for palladium and ruthenium complexes, respectively. These polymer-supported complexes also catalyze asymmetric hydrogenation reactions.

Complex **4** prepared from $[Rh(norbornadiene)_2]^+CF_3SO_3^-$ and a chelating phosphine ligand bound to biotin has been explored as a possible catalyst for the asymmetric hydrogenation of α-acetamidoacrylic acid, but the optical yield was less than 2%. However, in the presence of avidin, a protein binding to biotin, the optical yield was improved to more than 40% [96]. Apparently the asymmetric 'field' of the natural protein affects the reaction at the active site, which in this case is the Rh ion.

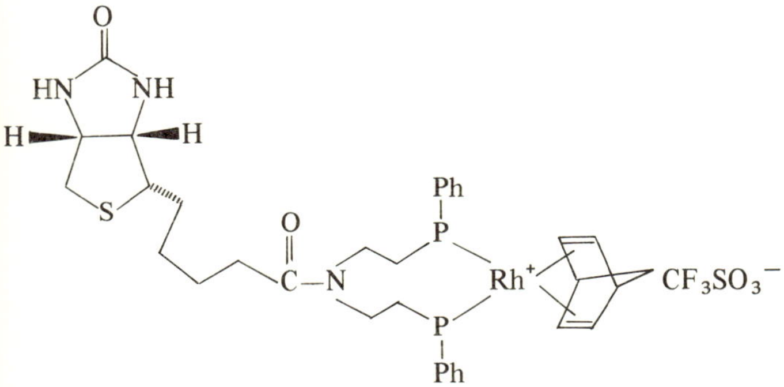

(4)

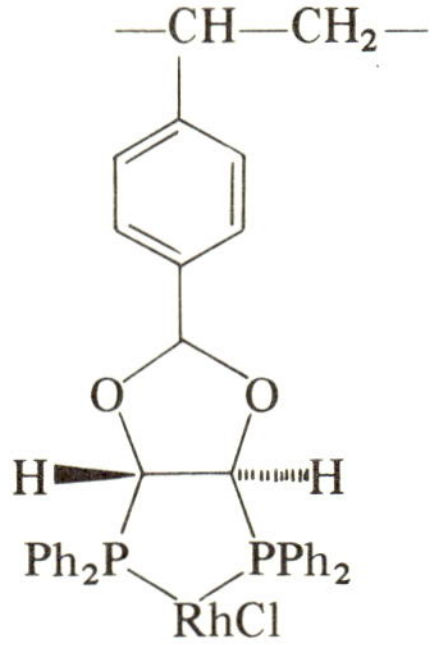

(5)

The most popular polymeric ligand containing a chiral carbon atom is polymer-anchored DIOP [2,3-*O*-isopropylidene-2,3-dihydroxy-1,4-bis(dipheylphosphino) butane]. A rhodium complex catalyst with a (–)DIOP ligand anchored onto poly(styrene–2% divinylbenzene) resins (**5**) has been used in the asymmetric hydrogenation of C=C bonds, such as those in α-acetamidocinnamic acid, using benzene/ethanol as the solvent [72]. However, the catalytic activity was very low due to considerable shrinkage of the resin in the solvent. When a copolymer of styrene and hydroxylethyl methacrylate was used as the polymeric support instead of polystyrene, the polymer complex was swollen by polar solvents and worked as an effective catalyst for the asymmetric hydrogenation of *N*-acetyl-(*R*)-phenylalanine at 25 °C under 1–2.5 atm pressure of hydrogen for 12 h in benzene/ethanol, resulting in 100% conversion and 86% e.e. optical yield [97]. These figures are similar to those for the corresponding homogeneous catalyst. The catalytic activity and asymmetric selectivity did not decrease with repeated use of the recovered catalyst.

An optically active copolymer of styrene and 1-buten-3-ol has been prepared via the asymmetric hydrogenation of a copolymer of styrene and methylvinylketone and used as a support for the complex catalyst mentioned above. However, in the presence of this supportcd catalyst the optical yield of alanine was affected by the confirmation of the pendant alcohol groups [98].

Polystyrene-anchored DIOP has been treated with $[Rh(CO)_2Cl]_2$ or $HRh(CO)(PPh_3)_3$ to produce catalysts for the hydroformylation of olefins. The activity and asymmetric selectivity were lower than those in the presence of the corresponding homogeneous catalyst, but the n/b ratio of the resulting aldehyde was quite different from that in the homogeneous case [99, 100]. Asymmetric reduction of carbonyl groups also occurred over polymer-supported DIOP complex catalysts. While the asymmetric yield was as low as 7.7% for the direct hydrogenation of acetophenone, it was increased to a value as high as 50% by using hydrosilylation [72]. Substrate specificity was also observed in the asymmetric hydrosilylation of ketones using a rhodium complex catalyst supported on polystyrnee containing DIOP. Thus, whereas the optical yields obtained on treatment of acetophenone with dihydrosilanes were very similar to those obtained with the corresponding homogeneous catalyst, with isobutyrophenone it was found that the supported catalyst gave a significantly higher optical yield than the homogeneous catalyst [72].

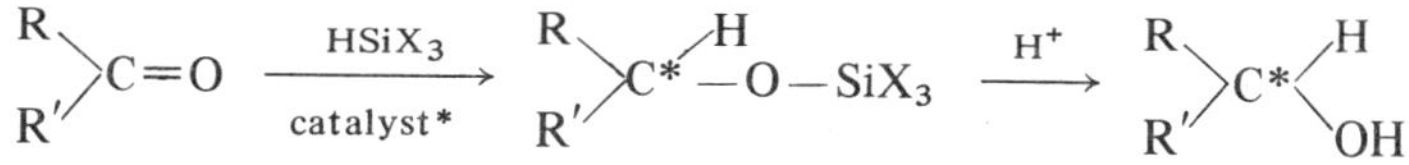

A polymer-supported catalyst containing an asymmetric ligand besides DIOP has also been prepared. Thus, the asymmetric hydrogenation of itaconic acid to methylsuccinic acid has been carried out using a catalyst prepared by treating $\{Rh(COD)(acac) + HClO_4\}$ or $[Rh(COD)Cl]_2$ (COD = cyclooctadiene) with a copolymer of 1-(4-vinylbenzoyl)-(2*S*, 4*S*)-2-dipheylphosphino-4-dipheylphosphinomethylpyrrolidine and hydroxymethylmethacrylate (**6**) [101]. The optical yield using the resulting polymer-supported catalyst was similar to that using the corresponding homogeneous catalyst.

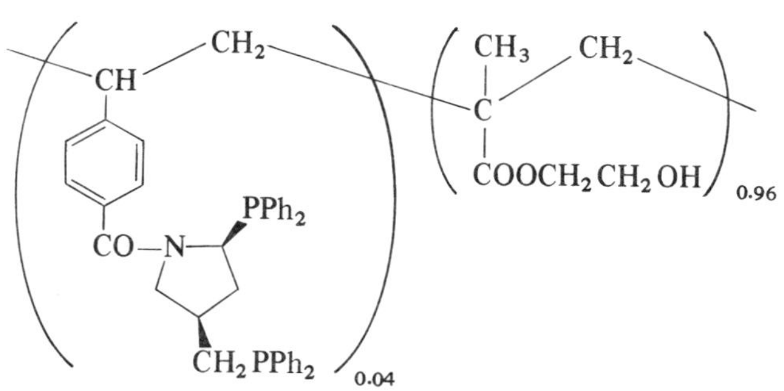

(**6**)

A rhodium(I) complex supported on the polymer with the same asymmetric ligand has been applied in a hydrogenation reaction to obtain an optically active amino acid [102]. The effect of optically active pendant alcohol groups has been discussed in detail in this system [103].

New polymer ligands containing β-dimethylaminoalkylphosphines have been prepared using chloromethylated polystyrene.

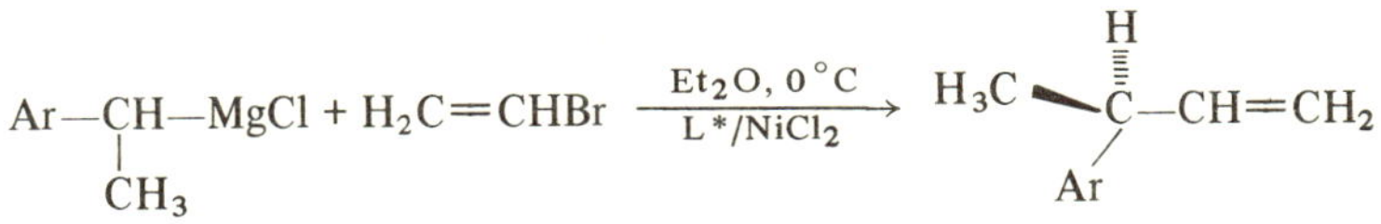

where Ar = phenyl, *p*-tolyl, 2-naphthyl and

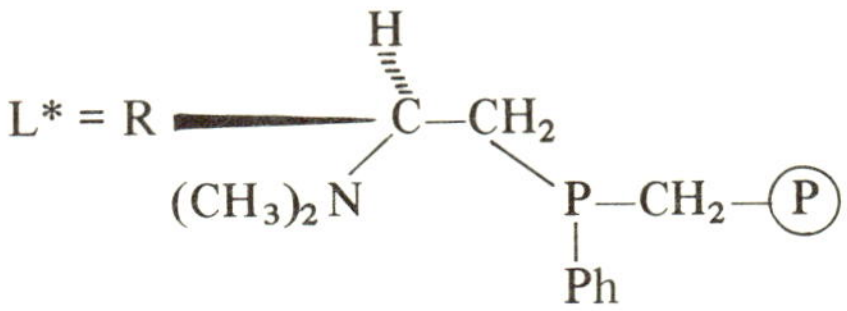

(R = *i*-Pr, $—CH_2Ph$).

These ligands have been used for the nickel-catalyzed asymmetric cross-coupling of *s*-alkyl Grignard reagents with vinyl bromide to yield coupling products with ca. 50% e.e. [104]. The polymer-supported catalyst could be recovered by rapid filtration in air before hydrolysis, and re-used without loss of catalytic activity or stereoselectivity.

2.4. MULTIFUNCTIONAL CATALYSTS

A catalyst which fulfills two or more functions is called a 'multifunctional catalyst'. Strictly speaking, there are two kinds of multifunctional catalyst. In the first case, different functional groups work cooperatively to catalyze a reaction. Thus, one functional group catalyzes the reaction and others assist the catalytic reaction. In the second case, the catalyst contains different active sites, each of which can catalyze the corresponding elementary reaction. Thus, two different complexes may be bound to the same polymeric support, the product of reaction at the first active site being used as the substrate for the second catalytic reaction. In this way sequential multistep reactions may be carried out on one catalyst and the overall reaction process can be simplified.

In this section, the first case will be referred to as cooperative catalysis, and the second case as multistep catalysis. Multifunctional catalysts [105] can be prepared from synthetic polymers, since it is not difficult to bind two kinds of ligand to a synthetic polymer with an appropriate spacial arrangement.

2.4.1. *Cooperative catalysis*

Chemical species which can improve the catalytic activity by cooperation with the catalyst are called 'promoters' or 'cocatalysts'. A polymer-anchored cooperative catalyst incorporates such functional group(s) and a catalytic site within a crosslinked polymer.

To prepare acetic acid from methanol and carbon monoxide, $RhCl(PPh_3)_3$ or $RhCl(CO)(PPh_3)_2$ and iodomethane were initially studied as the catalyst and promoter, respectively, attempts being made to immobilize the catalyst and the promoter on a crosslinked polymer. Subsequently, pentachlorothiophenol was used as the promoter instead of iodiomethane. In this case, the thiophenol was first bound to crosslinked polystyrene, then thiol groups were introduced by treating the resin with NaSH. Rhodium complexes were next coordinated to the thiol groups as shown in Figure 12 [106]. The resulting polymer-anchored cooperative catalyst exhibited the same catalytic activity per mole of Rh as

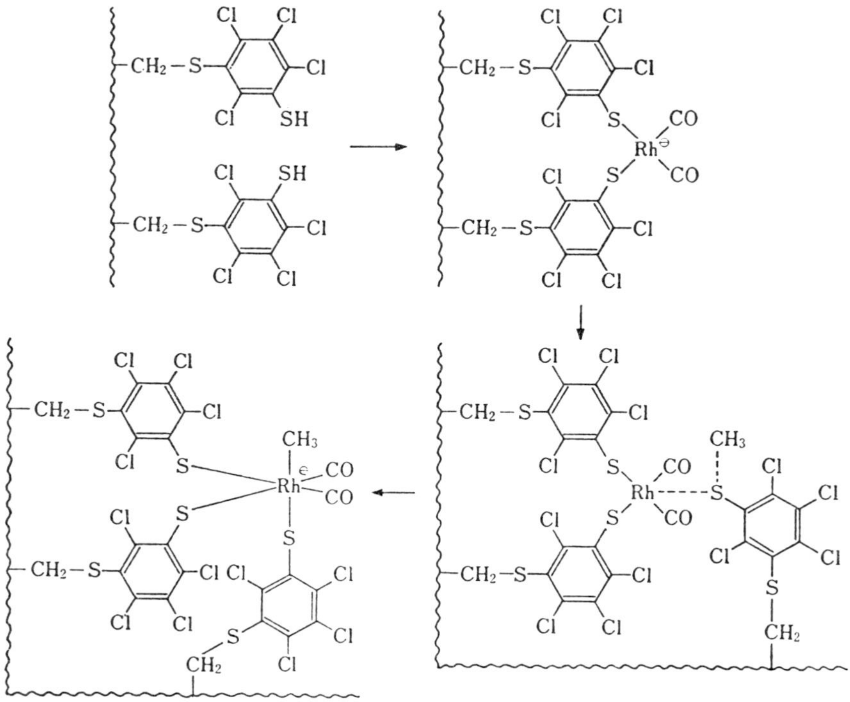

Fig. 12. Preparation and cooperative function of a polymer-anchored catalyst for acetic acid synthesis [23].

the homogeneous catalytic system when the reaction was carried out at 150 °C under 42 atm pressure of CO in methanol/benzene as the solvent. The thiophenol group in the catalyst functioned not only as a ligand to the Rh atom, but also assisted the oxidative addition of methanol.

The homogeneously catalyzed Wacker process for the partial oxidation of ethylene to acetaldehyde includes three chemical reactions:

$$H_2C{=}CH_2 + H_2O + Pd^{2+} \longrightarrow CH_3CHO + Pd^0 + 2\,H^+$$

$$Pd^0 + 2\,Cu^{2+} \longrightarrow Pd^{2+} + 2\,Cu^+$$

$$2\,Cu^+ + 1/2\,O_2 + 2\,H^+ \longrightarrow 2\,Cu^{2+} + H_2O$$

Copper dichloride, when used as a cocatalyst together with $PdCl_2$, re-oxidizes the Pd formed and is then itself re-oxidized by O_2. A polymer-immobilized catalyst for the Wacker process has been designed using a Pd^{II} ion bound to sulfonic groups and benzoquinone instead of the Cu^{II} ion. The homogeneous catalytic system containing $PdCl_2$ and benzoquinone exhibits a high initial reaction rate, but the rate decreases rapidly during the reaction. In contrast, the polymer-immobilized system functions as a catalyst at 80 °C under 1 atm total pressure (ethylene/O_2 = 1 : 1) to produce acetaldehyde as the sole product, the activity being retained for long periods of time [107]. The benzoquinone in the system oxidizes Pd^0 to Pd^{II}. Hydroquinone produced during the oxidation of Pd^0 can be oxidized by oxygen in the air, regenerating the initial benzoquinone.

The copper(II) complex of poly(4-vinylpyridine) catalyzes the oxidative coupling polymerization of 2,6-dimethylphenol, producing poly(2,6-dimethylphenylene oxide). Addition of Mn to this system not only increased the polymerization rate but also the molecular weight of the final polymer by a factor of 1.5 with 85% yield [108]. The formation constants for the attachment of both the Cu and the Mn ion to poly(4-vinylpyridine) are higher than those for the same ions bound separately. Cooperative catalysis by Cu^{II} and Mn^{III} attached to the polymeric chains may be explained in terms of redox cycles. Thus, Cu^{II} oxidizes phenoxide to the phenoxy radical, being itself converted to Cu^{I}. The Cu^{I} is then oxidized by Mn^{III}, producing Cu^{II} and Mn^{II}. Finally, Mn^{II} is oxidized to Mn^{III} by oxygen in the air.

The oxidation of thiol to disulfide is effectively catalyzed by cobalt phthalocyanine in the presence of a base as a promoter. Thus, cobalt(II) 4,4′,4″,4‴-tetrasulfophthalocyanine with a bis(3-aminopropyl)amine base bound to poly(acrylamide) was found to be more active by a factor of 32 than the corresponding homogeneous complex [109].

2.4.2. *Multistep catalysis*

In order to provide for sequential multistep catalytic reactions two kinds of metal complex have been bound to phosphinated crosslinked polystyrene. As a result, sequential butadiene cyclooligomerization and the hydrogenation of the resulting cyclooligomers has been possible. Thus, $Ni(CO)(PPh_3)_2$ and $RhCl(PPh_3)_3$ were immobilized on the same polymer, and a 'single-pot' experimental arrangement devised to allow the use of this polymer-anchored catalyst. The first step in the process was oligomerization of butadiene in the presence of the catalyst in benzene solution at 110–115 °C for 24 h; then the reaction mixture was cooled to 50 °C and maintained under 350 psi pressure of hydrogen for 4 h to enable hydrogenation of the cyclooligomers (see Figure 13) [110]. Neither the nickel nor the rhodium in the catalyst system interferes with the activity of the other, and the system as a whole may be recycled.

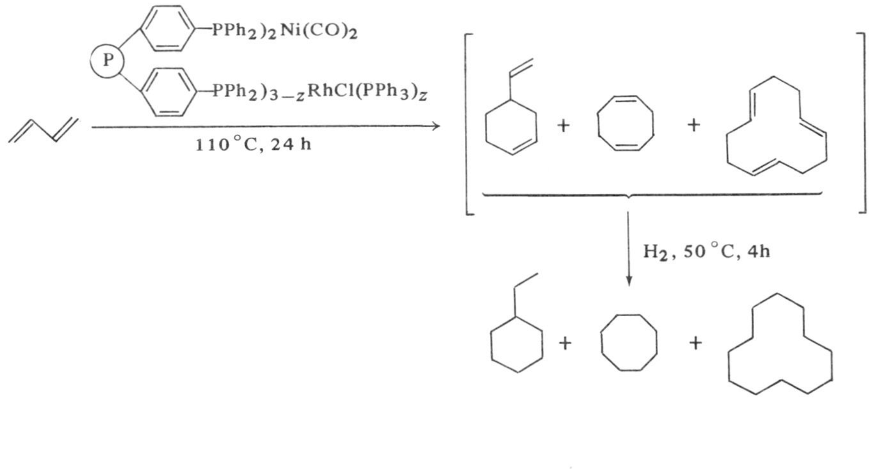

Fig. 13. Sequential multistep 'single-pot' reaction of butadiene using a polymer-anchored Ni and Rh catalyst [23].

Cyclooligomerizations of butadiene have also been coupled in a sequential manner with selective partial hydrogenation of the resulting cyclooligomers to monoenes by using polymer-anchored Ni and Ru complexes as catalysts as shown in Figure 14 [16, 111]. Similarly, the sequential cyclooligomerization of butadiene and hydroformylation of the products could also be catalyzed by Ni and Rh complexes bound onto a single support. However, in this latter case, only the terminal double bond in vinylcyclohexene could be hydroformylated.

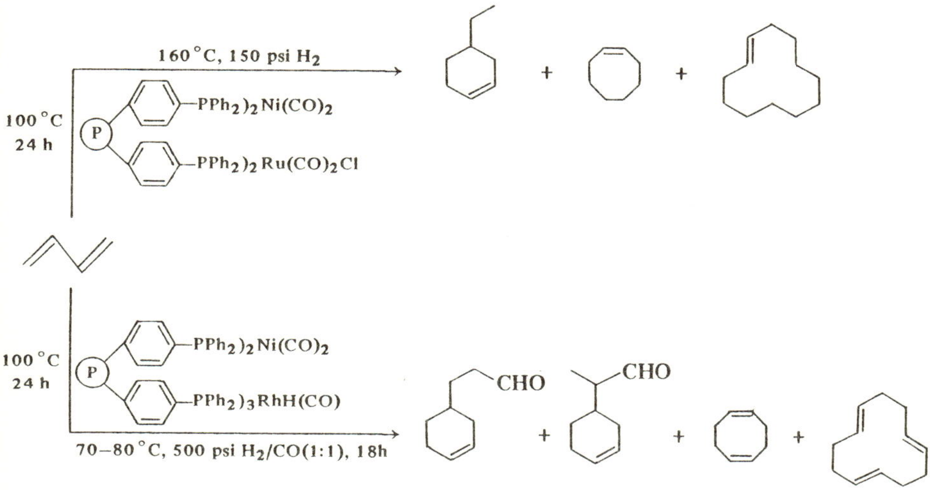

Fig. 14. Sequential multistep 'single-pot' reactions of butadiene using polymer-anchored catalysts.

By using a phosphinated resin-anchored Pd and Rh catalyst, the sequential linear oligomerization–acetoxylation–hydrogenation of butadiene has been achieved [112]. In this case, the oligomerization–acetoxylation was catalyzed by the Pd-containing part of the catalyst while the hydrogenation was promoted by the Rh-containing part (which is similar to a Wilkinson's catalyst).

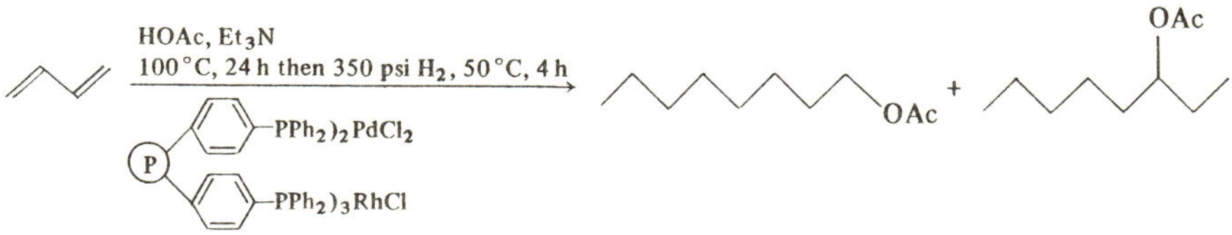

Polymer-anchored catalysts exhibiting a dual function have been prepared by first introducing sulfonic acid groups into crosslinked polystyrene and then depositing crystalline platinum particles onto the support [113]. The sulfonic groups catalyze hydration while the Pt particles promote dehydrogenation. In this way it has been possible to conduct the hydration of propene to 2-propanol, followed by dehydrogenation of the 2-propanol to acetone, as a single process at 118 °C using a gas-phase circulating reactor.

$$H_2C{=}CH{-}CH_3 \xrightarrow[\text{acid}]{H_2O} H_3C{-}\overset{\displaystyle OH}{\overset{|}{C}H}{-}CH_3 \xrightarrow[\text{Pt}]{-H_2} H_3C{-}\overset{\displaystyle O}{\overset{\|}{C}}{-}CH_3$$

A sequential three-step process has also been accomplished under a single set of reaction conditions [114]. This is the Aldox process in which propene is converted to 2-ethylhexanol. In this case, the single, dual-function catalyst employed was prepared by supporting $RhCl(CO)(PPh_3)_2$ on phosphinated polystyrene with subsequent substitution of ethylamino groups onto the polystyrene. The first step in the process is the rhodium-catalyzed hydroformylation reaction which is followed by the amine-catalyzed aldol condensation of the aldehyde produced in the first step. The resulting unsaturated aldol products are hydrogenated in the third step, where the rhodium complex again serve as the catalyst.

$$H_3C{-}CH{=}CH_2 \xrightarrow[CO/H_2]{Rh} H_3C{-}CH_2{-}CH_2{-}CHO$$

$$\xrightarrow[-H_2O]{amine} H_3C{-}CH_2{-}CH_2{-}CH{=}\underset{\underset{CH_3}{|}}{\underset{CH_2}{\underset{|}{C}}}{-}CHO$$

$$\xrightarrow[H_2]{Rh} CH_3{-}CH_2{-}CH_2{-}CH_2{-}\underset{\underset{CH_3}{|}}{\underset{CH_2}{\underset{|}{CH}}}{-}CHO$$

Catalyst employed 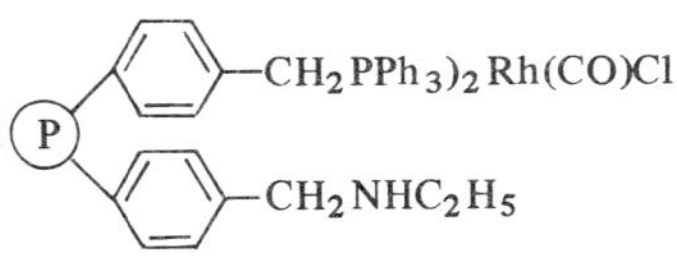

The reactions took place under simple conditions at 80 °C in the presence of a CO/H_2 (1 : 1) gas mixture in benzene as a solvent. The rates of the Aldox process were much higher when the above single, dual-function catalyst was employed than when two sets of resins, one with Rh complexes and the other with amino groups, were used. Thus a 5-fold rate enhancement for hydroformylation, a 15-fold enhancement for aldol condensation, and a 30-fold enhancement for the final hydrogenation step were observed. The amine groups in the catalyst increase the polarity inside the beads employed. Since propene is the most polar compound of those present at the start of the reaction, the enhanced concentration of propene inside the beads accelerates the rate of the first step (hydroformylation). Rate enhancements in the second and third steps in the process result from a diminished resistance to material transfer inside the beads than if the separate operations were conducted on two sets of resins.

In general, interactions between active sites adjacent to each other tend to decrease the overall catalytic activity. Such interaction may be much stronger between different kinds of active sites than between those of the same kind. For this reason, multistep catalysts are usually less active than combinations of the individual monofunctional catalysts, although against this must be set the great advantage of being able to use multistep catalysts in 'single-pot' or one-process reactions. The last example mentioned above is an exception to this rule, in that a higher activity was observed with the multistep catalyst. The precise design of the catalyst is crucial for a multifunctional catalyst on a polymeric support.

3. Polymer-protected Colloidal Catalysts

Transition metals are useful as catalysts for various kinds of chemical reactions, and have often been used in the form of metal particles dispersed onto inorganic supports such as silica gel, alumina and activated charcoal. Dispersed particles on an inorganic support have advantages over metal powders in that (i) the particles can be readily dispersed through the use of the support; (ii) the resulting dispersed particles exhibit a much larger surface area than the bulk metal, thus allowing more effective utilization of the expensive transition metal as a result of dispersion; (iii) the active sites of the metal catalyst can be brought into a homogeneous condition by dispersion; and (iv) the small particles resulting from dispersion may exhibit new catalytic properties.

The science of colloids can be traced back some 140 years to the time when colloidal dispersions of silver chloride and sulfur were first prepared and examined [115]. Although Faraday worked with gold sols in 1856, it was much later that colloidal dispersions of transition metal were used as catalysts for chemical reactions. The protection of colloidal metal dispersions against aggregation is one of the most important reasons for utilizing colloidal metals as catalysts. Faraday, in his work, found that the addition of gelatine to a gold sol protected it from the action of salt, and he suggested that the small particles of gold might be enclosed in 'envelopes of that animal matter'. More recently, synthetic polymers have been used as protective colloids, both in soluble and insoluble form. A polymer-protected colloidal metal catalyst not only has similar advantages to dispersed particles on an inorganic support but also new benefits in that (a) colloidal dispersions can form 'homogeneous' solutions, (b) the protecting polymer can shield a colloidal catalyst from deactivation by catalytic poisons or air, and (c) colloidal dispersions of fine particles transmit light more readily than powders. As a result of this last benefit, colloidal metals have frequently been used as catalysts in recent photochemical investigations [116, 117].

In this section the preparation and characterization of polymer-protected colloidal metal catalysts will be reivewed, together with their characteristic properties and some of their applications [118].

3.1. PREPARATION AND CHARACTERIZATION

Colloidal particles may be generated in two distinct ways: by the subdivision of bulk materials or by growing the particles from molecular precursors. The former method of preparation includes mechanical crushing of the bulk metal, pulverization of the bulk metal by ultrasonic means, deposition of a metallic vapor and arc discharge between metallic electrodes. Polymer-protected colloidal metal particles produced by these methods are usually quite large and have a wide distribution of particle sizes.

In contrast, colloidal particles prepared by growth from molecular precursors are usually small with a narrow size distribution. The chemical reaction most suitable for the preparation of colloidal metals by this method is the reduction of metallic ions. Many kinds of reducing agent have been employed; for example, reductants for the preparation of gold sols include formaldehyde, alcohol, carbon monoxide, phosphorus, hydrogen peroxide, hydrazine, etc. [119].

The colloidal metal particles thus obtained are dispersed in the medium employed. In water, the particles carry an electric charge, the repulsive forces associated with such charges being superimposed on the attractive van der Waals' forces. If the repulsive forces are strong enough an energy barrier is set up against coagulation, the colloidal particles being stabilized by a balance between the repulsive electrostatic forces and the attractive van der Waals' forces. The result is a colloidal dispersion. The presence of an electrolyte destabilizes such colloidal dispersions, with the particles coagulating at some critical concentration of the electrolyte [120].

As observed by Faraday, the addition of polymers such as gelatine to the system protects colloidal metal dispersions against flocculation. Such 'steric stabilization' depends on the nature of the polymers and their adsorption characteristics [121].

The preparation of polymer-protected colloidal metals by chemical methods involves two processes: reduction of the metal ion to the colloidal metal and interaction of the particles in the colloid with the polymer. In practice, reduction can precede or follow such interaction, as shown in Figure 15. If reduction precedes interaction, the colloidal metal particles interact with the polymer mainly as a result of the hydrophobic forces between them. In contrast, if interaction precedes reduction, the interactive forces between the metal ion and the polymer will include electrostatic interaction, hydrogen bonding, coordinate bonding, and so on.

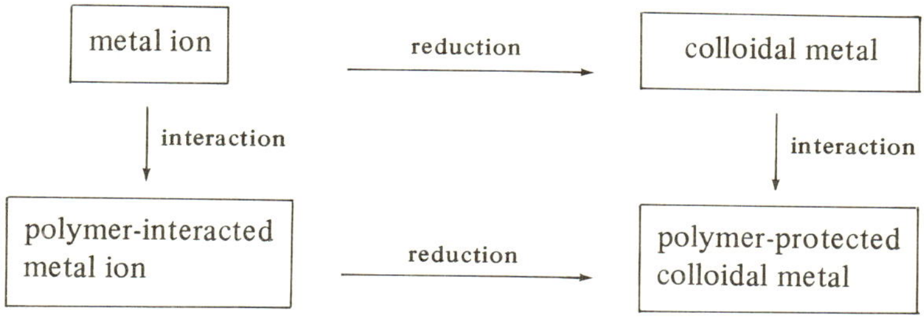

Fig. 15. Schematic representation of the methods available for the preparation of polymer-protected colloidal metals by chemical reaction from the metal ion.

Polymers used for protecting the colloidal metal are usually soluble in the dispersion medium. However, polymeric solid-state resins can also be employed. In this case, small particles of the metal may be dispersed inside the polymer resin or distributed on the surface of the resin bead itself.

Some typical examples will now be presented illustrating the preparation of colloidal dispersions of metal protected by soluble polymers. The classic method of preparation involves the initial production of a colloidal dispersion of the metal, followed by the addition of the protective polymer. In this case, poly(vinyl alcohol) and poly(vinylpyrrolidone) are often used as the protective colloid. The alternative method involves the initial addition of a polymer to the solution of a metal ion followed by reduction of the latter. This method involves the formation of some sort of complex between the metal ion and the polymer. If the interaction between the polymer and the metal ion is not particularly strong, reduction of the metal ion in the presence of the polymer will lead to the same result as in the classic method mentioned above. Nord's colloid is of this type, and involves the reduction of a noble metal ion by molecular hydrogen in systems where NaOH has been added in the presence of poly(vinyl alcohol) [122].

Refluxing alcoholic solutions of noble metal ions in the presence of a protective polymer also gives colloidal dispersions of the corresponding metal [123]. This method has the following advantages: (i) the procedure is very simple and the colloidal dispersions can be readily prepared; (ii) the particle size is small with a narrow size distribution; (iii) the particle size can be controlled by varying the preparative conditions; and (iv) the resulting colloidal dispersions show a high catalytic activity. Variations of the method are listed in Table IV.

In this method, a complex of the noble metal ion is first formed with a polymer followed by reduction of the ion with alcohol under mild conditions. Hence, this is a preparative method involving a polymer–metal complex.

TABLE IV
Preparation of colloidal dispersions of noble metal by reduction of the corresponding metal ion with alcohol [123][a]

Metal salt	Preparative conditions		
	In alcohol[b]	With water[c]	With base[d]
$RuCl_3 \cdot 3H_2O$	–	+	–
$RhCl_3 \cdot 3H_2O$	+	+	+
$PdCl_2 \cdot 2H_2O$	+	+	+
$AgNO_3$	+	–	+
OsO_4	+	+	+
Na_2IrCl_6	+	+	+
$H_2PtCl_6 \cdot 6H_2O$	–	+	+
$HAuCl_4 \cdot 4H_2O$	–	–	+

a The stability of the colloidal dispersions is shown as + (stable) or – (unstable).
b Refluxing in anhydrous ethanol in the presence of poly(vinylpyrrolidone).
c Refluxing in methanol/water in the presence of poly(vinyl alcohol).
d Refluxing in methanol in the presence of poly(vinylpyrrolidone) and then on addition of NaOH.

Polymer-protected metal colloids can be classified into three groups depending on the state of the resulting dispersions, as shown in Figure 16.

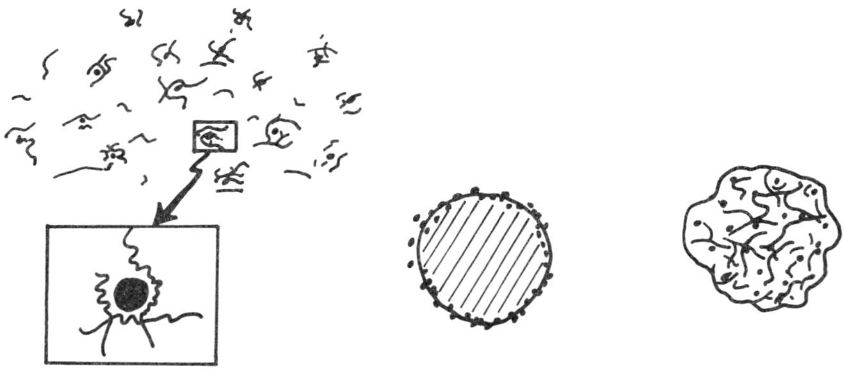

(a) (b) (c)

Fig. 16. Schematic illustration of the three kinds of polymer-protected colloidal metal catalysts.

Figure 16(a) illustrates a typical metal colloid protected in solution by soluble polymers. Figures 16(b) and (c) demonstrate the protection of colloidal dispersions by solid polymers. In (b) the metal particles are distributed on and/or near the surface of the polymer resin, whereas in (c) the particles are distributed uniformly inside the resin.

The usual dispersed solution of a colloidal metal in the presence of a soluble polymer is as illustrated in Figure 16(a). Thus, poly(acrylic acid) when used as a protective polymer in the alcohol reduction of a noble metal as described above yields a dispersion of this kind. The addition of poly(ethyleneimine) to the resulting colloidal solution generates a precipitate which consists of a polyion complex of poly(acrylic acid) and poly(ethyleneimine) containing colloidal metals [124]. In this case the metal particles are distributed within the solid support provided by the polyion complex, as shown in Figure 16(c).

Colloidal metal dispersions of the type illustrated in Figure 16(b) can be produced by using solid polymer beads. Thus colloidal dispersions of a noble metal in an aqueous solution were prepared by the reduction of the corresponding ion with a citrate in the absence of polymer. Addition of polymer beads, such as anion-exchanging polystyrene resin containing quarternary ammonium groups, to this colloidal solution resulted in the deposition of metal particles on the beads. On the other hand, if the polymer beads were added to the solution of the metal ion before reduction, then the metal ions penetrated into the surface of the beads via ion exchange or coordination. Reduction of such metal ions near the surface of the beads can also produce metal-dispersed polymer beads of type (b). For example, Pd^{II} coordinated to the iminodiacetic acid moieties of chelating resin beads may be reduced by refluxing the beads in an alcoholic solution, producing a colloidal palladium catalyst supported on the chelating resin [125]. Colloidal platinum catalysts immobilized on chelating resin may be prepared by a similar method and used to construct a ternary photocatalyst with $[Ru(bpy)_3]^{2+}$ and methyl viologen. Such catalysts assist the photochemical production of hydrogen from EDTA solution [126]. Another example is provided by the photoelectrochemical production of hydrogen by p-type semiconducting photocathodes whose surface had been modified by poly(*N,N'*-dialkyl-4,4'-bipyridinium) $(PQ^{2+})_n$ to which a Pt^0 or Pd^0 colloidal catalyst had been attached [127]. To enable the attachment of Pd^0 to the $(PQ^{2+})_n$ polymer, the Br^- counterions of the latter were replaced by $PdCl_4{}^{2-}$ via an ion-exchange reaction and the Pd^{2+} ions reduced by a reducing agent such as H_2, photoirradiation, or an electrochemical method. A silver colloid has similarly been supported on Nafion by incorporating Ag^I onto Nafion by ion exchange, followed by reduction of Ag^I to Ag^0 with H_2 [128].

Colloidal metal particles may be characterized by electron microscopy, which allows the direct evaluation of the shape and diameter of each particle. The particle size distribution can also be determined by electron microscopy. Particle sizes which include the central metal particle and the surrounding envelope of protective polymers may be measured by light scattering techniques. Auger and ESCA spectroscopy may be employed to determine the position of metal particles in solid polymers, as is possible from electron micrographs of sliced samples.

Interactions between the polymer and the metal particle are very important for the stability of polymer-protected colloidal metal catalysts. Such interactions include physical adsorption by van der Waals' interaction, hydrophobic interaction, electrostatic interaction, hydrogen bonding and coordinate bonding. However, the methods available for the measurement of these interactions are very limited. IR, NMR and UV spectroscopy have used for this purpose with little success. Recent results on the enhanced Raman spectra of poly(vinylpyrrolidone) with attached silver particles [129] seem to provide a better demonstration of this interaction.

3.2. CHARACTERISTIC PROPERTIES OF COLLOIDAL CATALYSTS

The characteristic properties include small particle size and polymer protection. The effect of these on catalytic properties are reviewed in this section.

3.2.1. *Surface area*

Ostwald has defined colloidal dispersions in terms of the size of the particles dispersed in the medium. A colloidal dispersion should have particle sizes within the range from 1 nm to 100 nm. The number of atoms involved and the surface area of a particle in this size range can be calculated. In the case of platinum, if spherical colloidal particles of atomic radius 0.138 nm are close-packed, particles of diameter 1 nm and 10 nm would contain 48 and 48 000 atoms, respectively. In other words, the number of particles obtained from 1 mol of Pt is proportional to r^{-3} where r is the radius of the particle. The surface area of a particle is proportional to r^2. Thus the total surface area of the particles contained in 1 mol of Pt is inversely proportional to r, as illustrated in Figure 17.

3.2.2. *Particle size*

One of the most important characteristics of a polymer-protected colloidal metal catalyst is its small particle size. Generally speaking, the particle size of a metal catalyst can influence both the catalytic activity and selectivity in chemical reactions. Particle size effects can be classified into four groups, as illustrated in Figure 18; these effects are also observed in reaction selectivity.

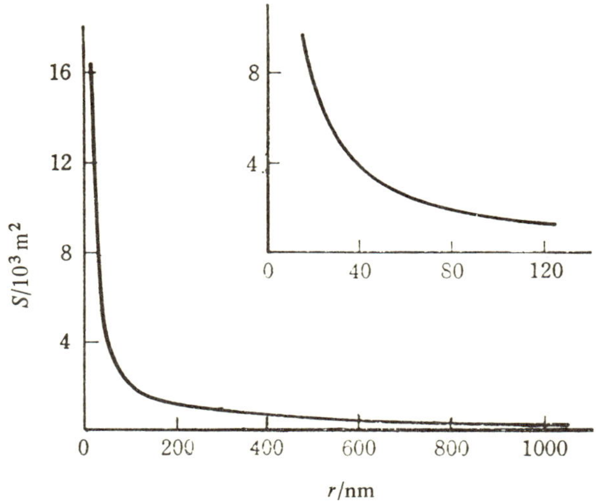

Fig. 17. The relationship between the radius (r) of an individual colloidal platinum particle and total surface area (S) of the particles contained in 1 mol of platinum [116].

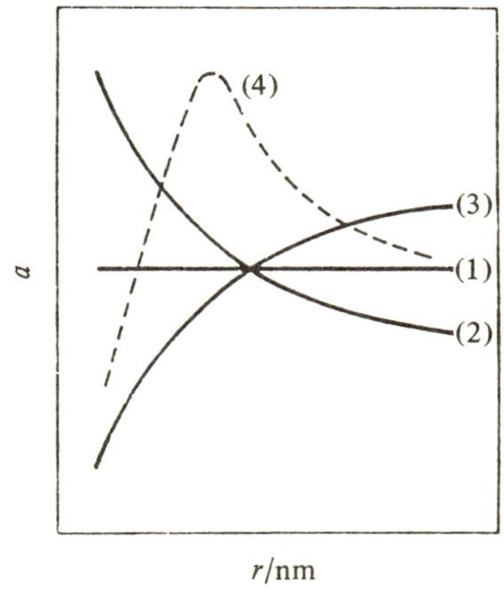

Fig. 18. Effect of particle radius (r) on catalyst activity (a) per atom. In the curves illustrated, the catalytic activity is independent of the particle size in curve 1, it increases (curve 2) or decreases (curve 3) with a decrease in particle size, or exhibits a maximum at a definite size of particle (curve 4) [116].

Poly(vinylpyrrolidone)-protected colloidal rhodium catalysts have been prepared by refluxing solutions of $RhCl_3$ in alcohols [127], the particle size altering with the conditions of preparation. Such colloidal rhodium catalysts influence the hydrogenation of olefins, where the catalytic activity observed increased with a decrease in the particle size of the catalyst. The rate of increase of activity also depended on the nature of the olefin substrate. The results are listed in Table V [130].

TABLE V
Hydrogenation of olefins over a polymer-protected colloidal rhodium catalyst[a]

Substrate	(Type of olefin)	Rate of hydrogenation [mol H_2 (g atom Rh)$^{-1}$ s^{-1}]		
		3.4 nm catalyst	2.2 nm catalyst	0.9 nm catalyst
1-hexene	(terminal)	15.8	14.5	16.9
cyclohexene	(internal)	5.5	10.3	19.2
methylvinylketone	(terminal)	3.7	4.3	7.9
mesityl oxide	(internal)	0.6	4.7	31.5
methylacrylate	(terminal)	11.2	17.7	20.7
methylmethacrylate	(internal)	5.8	15.1	27.6

[a] At 30 °C in methanol under 1 atm pressure of hydrogen; [Rh] = 0.01 mmol dm^{-3}, [substrate] = 25 mmol dm^{-3}.

The most interesting observation arising from these results is that the rate of hydrogenation of an internal olefin is faster than that of the corresponding terminal olefin when a 0.9 nm size polymer-protected Rh catalyst is employed.

Poly(vinylpyrrolidone)-protected palladium colloid catalysts have also been prepared by refluxing solutions of $PdCl_2$ in alcohols in the presence of poly(vinylpyrrolidone) [131]. These catalyzed the selective partial hydrogenation of dienes to monoenes under mild conditions [132, 133], with cyclooctadiene, for example, giving cyclooctene [132]. The particle size of poly(vinylpyrrolidone)-protected palladium catalysts vary with the preparation conditions as shown by the results listed in Table VI.

Polymer-protected colloidal platinum (6.6×10^{-5} mol dm^{-3}) catalyzed the photochemical evolution of hydrogen from Na_2EDTA solutions (5×10^{-2} mol dm^{-3}) in the presence of $Ru(bpy)_3{}^{2+}$ (5×10^{-5} mol dm^{-3}) and methyl viologen (MV) (5×10^{-2} mol dm^{-3}). The rate of hydrogen evolution was dependent on the particle size of the platinum, as illustrated in Figure 19 [134], a maximum rate being observed with a catalyst of ca. 3 nm particle size.

TABLE VI
Partial hydrogenation of dienes to monoenes with palladium catalysts[a]

Pd catalyst		Methyl linoleate		1,5-Cyclooctadiene	
Abbreviation	Average diameter (nm)	Rate of hydrogenation[b]	Selectivity[c]	Rate of hydrogenation[b]	Selectivity[c]
Pd–PVP–MeOH/NaOH[d]	2.5	7.8	97	1.4	97.8
Pd–PVP–EtOH[e]	3.5	0.5	95	—	—
Pd–PVP–MeOH/H_2O[f]	6.5	0.6	92	0.6	94.8
Pd on activated charcoal[g]	—	1.9	71	0.6	93.2

a Hydrogenation at 30 °C under 1 atm pressure of hydrogen in methanol, [diene] = 25 mmol dm^{-3}, [Pd] = 0.1 mmol dm^{-3}.
b Rate of hydrogen uptake in $[H_2]$ $[Pd]^{-1} s^{-1}$.
c Yield of monoene (%) at the point of equimolar hydrogen uptake.
d Pd^{II} was reduced by refluxing in methanol for 30 min and further refluxing for 10 min after addition of NaOH.
e Pd^{II} was reduced by refluxing in ethanol for 4 h.
f Pd^{II} was reduced by refluxing in methanol/water (1 : 1) for 4 h.
g [Pd] = 0.01 mmol dm^{-3}.

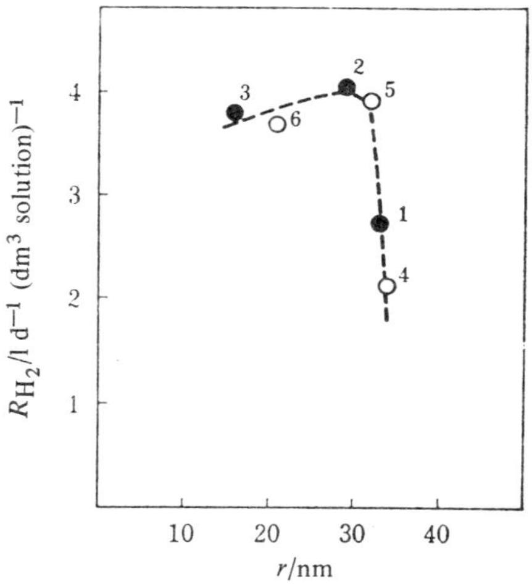

Fig. 19. The relationship between the rate of photochemical evolution of hydrogen (R_{H_2}) and the average particle size (r) of poly(vinylpyrrolidone) (PVP)– or poly(vinyl alcohol) (PVA)-protected colloidal platinum catalysts [134]. Reaction conditions: $[Ru(bpy)_3{}^{2+}]$ = 5×10^{-5} mol dm^{-3}; $[MV^{2+}] = 5 \times 10^{-2}$ mol dm^{-3}; $[Na_2\,EDTA] = 5 \times 10^{-2}$ mol dm^{-3}; $[Pt] = 6.6 \times 10^{-5}$ mol dm^{-3}; 20 ml, $\lambda > 390$ nm. 1, Pt–PVP–MeOH/H_2O; 2, Pt–PVP–EtOH/H_2O; 3, Pt–PVP–EtOH/NaOH; 4, Pt–PVA–EtOH/H_2O; 5, Pt–PVA–EtOH/H_2O, NaOH; 6, Pt–PVA–MeOH/H_2O.

3.2.3. *Effect of protective polymers*

The most important role played by the polymer is that of stabilizing the colloidal metal dispersions. The protective function of the polymer has been expressed quantitatively in terms of its 'gold number' or 'protective value'. The gold number is the amount of the protective colloid in milligrams which just prevents 10 cm^3 of a red gold sol from changing color to violet on addition of 1 cm^3 of a 10% aqueous solution of NaCl. The smaller the gold number, the stronger is the protective function of the polymer [135]. The protective value is the weight of a red gold sol in grams which can just be protected from aggregation by 1 g of the protective colloid on addition of a 1% NaCl solution. Thus, the larger the protective value, the greater is the protective function. The gold number is inversely proportional to the protective value. The gold numbers and protective values of typical protective colloids are summarized in Table VII [136, 137].

The protective value or gold number could be useful as a measure of colloid formation. When platimum colloids were prepared from H_2PtCl_6 by reduction through refluxing in ethanol/water, polymers with a large protective value functioned well as protective colloids and produced a polymer-protected colloidal

TABLE VII
Gold numbers and protective values of typical protective colloids [116].

Protective colloid	Protective value	Gold number
Natural materials		
hemaccel	100	0.004
gelatine	90	0.005, 0.03
bactopeptone		0.03
chondroitin sulfate	4	0.1
gum arabic		0.2
egg ovalbumin		0.2
bovin serum albumin		0.2
pectin		1.0
agar agar		1.1
heparin	0.2	2
algin acid amide	0.2	2
sodium oleate		4.0
ox albumin povite	0.09	5
sodium alginate	0.04	10
pepsin, trypsin	0.04	10
starch		16
Synthetic materials		
poly (acrylic acid hydrazide)	400	0.001
poly (*N*-vinyl-5-methoxazolidon)	70	0.006
poly (vinylpyrrolidone)	50	0.009
N-alkyl polyimine	40	0.02
albatex PO	20	0.02
poly (vinyl alcohol)	5	0.09
N-acetyl polyimine	2	0.2
poly (acrylamide)	1.3	0.3
poly (L-lysin hydrobromide)	1	0.4
poly (acroleine)	0.4	1
zephirol	0.1	4
poly (acrylic acid)	0.07	6
poly (ethyleneimine)	0.04	10
poly (vinylpyridine–methylvinylketone)	0.005	90

platinum catalyst. However, when polymers with a small protective value were employed, precipitates or complexes were formed. These results are summarized in Table VIII [138].

The ratio of the amount of the protective polymer to that of metal (R) is an important factor in determining the stability of colloidal metal dispersions. This has been investigated in the case of a platinum colloid prepared by reduction with sodium citrate, protected with Carbowax 20M (a copolymer of poly

TABLE VIII
Effect of polymers on the preparation of platinum catalysts[a]

Polymer		State of Pt catalyst	Average diameter (nm)
Name (degree of polymerization)	Protective value		
poly(ethyleneimine)	0.04	complex	—
benzalkonium chloride	0.1	precipitate	—
poly(acrylamide)	1.3	precipitate	—
poly(vinyl alcohol) (500)	5	colloid	3.4
poly(vinylpyrrolidone) (3250)	50	colloid	2.9
gelatine	90	colloid	1.6

[a] Prepared by refluxing H_2PtCl_6 in ethanol/water (1 : 1).

(ethylene glycol) and ethylene oxide) and used as a catalyst for the photochemical evolution of hydrogen [139]. The influence of R on the protective function of Carbowax 20 M was studied by adding NaCl to the colloidal solution containing various amounts of the polymer. The results are shown in Figure 20. At least the same weight of Carbowax 20M as that of platinum is necessary to provide a protective function. Hence, the maximum catalytic activity was exhibited by the polymer-protected platinum catalyst when R = 2 during the photochemical evolution of hydrogen.

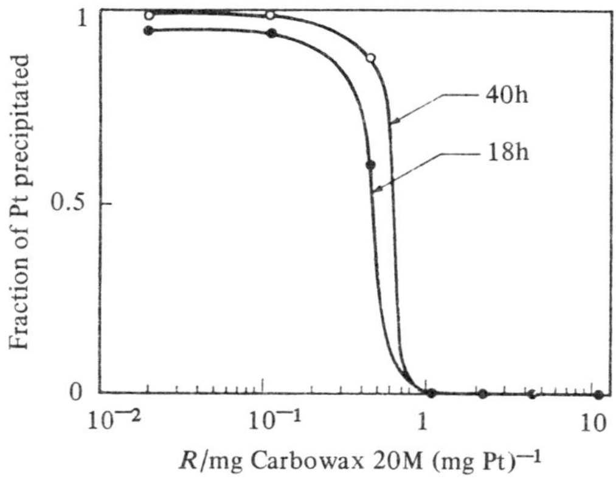

Fig. 20. Effect of the ratio of Carbowax 20M to metal (R) on the protection of a platinum colloid. [NaCl] = 1% (w/v), [Pt] = 45 mg dm^{-3}. Reprinted with permission from *J. Am. Chem. Soc.* **103**, 2924. Copyright 1981, American Chemical Society.

The degree of polymerization of the protective polymer can also affect the activity of polymer-protected colloidal catalysts. Stable colloidal dispersions of copper in water have been prepared via the reduction of copper sulfate with sodium tetrahydroborate in the presence of poly(vinylpyrrolidone) and used as catalysts for the selective hydration of acrylonitrile to acrylamide in water at 80 °C [140]. The acrylamide yield exhibited a maximum when the degree of polymerization of the purified poly(vinylpyrrolidone) used in the preparation of the Cu colloids was ca. 3000, as shown in Figure 21.

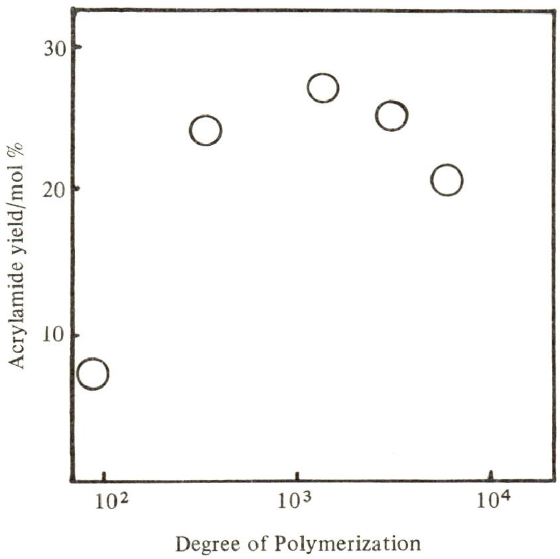

Fig. 21. The influence of the degree of polymerization of poly(vinylpyrrolidone) upon the yield of acrylamide during the hydration of acrylonitrile catalyzed by colloidal copper dispersions protected by poly(vinylpyrrolidone).

Most protective colloids are water-soluble polymers possessing polar groups. It has been proposed that in such polymers the hydrophobic backbones are directed toward the hydrophobic surfaces of the metal particles with the hydrophilic side-chains being directed outwards into the dispersion medium, thus making the colloids hydrophilic [141]. As a result, the polymer-protected metal colloids are stabilized. The protective role of polymers where the overall colloid is hydrophobic has been attributed to repulsion between polymers adsorbed on the colloid surface [142].

The hydrophobicity of organic polymers has been utilized in the concentration of heavy water and tritium [143, 144], a process requiring catalysis of the hydrogen exchange reaction between liquid water and gaseous hydrogen. Small particles of platinum on alumina or activated charcoal are known to catalyze the hydrogen exchange reaction between vaporized water and gaseous hydrogen, but they became deactivated on contact with liquid water since the formation of a film of water on the surface of the platinum causes a substantial reduction in the rate of diffusion of hydrogen molecules to the surface. Substitution of the alumina or charcoal support by one of Teflon or polystyrene resin increases the hydrophobicity of the system, resulting in increased activity for the exchange reaction as shown in Table IX [144, 145].

TABLE IX
Platinum catalysts for the hydrogen exchange reaction

Support	Maker's name	Relative activity k'	Specific surface area S ($m^2\ g^{-1}$)	Specific activity $k'/10^2\ S^{-1}$
Vycor glass	Corning	0.005	154	0.0032
Vycor glass treated with silane	Corning	1.00	87	1.14
activated charcoal	Shimazu	2.59	1000	0.25
Teflon 914J	Mitsui	9.93	0.4	2480
Porapak Q[a]	Waters	170	634	26.8
Shodex 104[a]	Showadenko	500	651	76.8

[a] Macroreticular polystyrene resin.

Although the effects of polymers in polymer-protected colloidal metal catalyst systems have not been investigated precisely to date, polymers added before the reduction of the metal ion might have a different influence on colloidal metal particles from those added after the preparation of colloidal metal dispersions. Vigorous studies of these aspects as well as investigations of new applications of polymer-protected colloidal metal catalysts must be pursued.

3.3. COLLOIDAL CATALYSTS PROTECTED BY FUNCTIONALIZED POLYMERS

The use of polymers is an easy method for introducing multifunctionality into a catalyst. This may be achieved by using functionalized polymers in the preparation of polymer-protected colloidal metal catalysts.

Some examples have been already mentioned. In Section 2.4.2 above, colloidal platinum particles protected by polystyrene resin containing sulfonic groups were described as useful for the catalysis of hydration and dehydrogenation processes, allowing the production of acetone from propene in a single-step process. In this case the sulfonated polystyrene acts as a support of the platinum particles and as an acid catalyst for hydration.

In Section 3.1, colloidal platinum and palladium dispersions in solid poly(*N,N'*-dialkyl-4,4'-bipyridinium), $(PQ^{2+})_n$, were mentioned as surface-immobilized catalysts for the light-induced evolution of hydrogen from p-type semiconducting photocathodes. In this case the $(PQ^{2+})_n$ polymers function as electron carriers and the Pt or Pd metal particles act as catalytic sites for hydrogen formation. A similar system has been developed by using water-soluble polymers. Thus, a polymer of *N*-vinylbenzyl-*N'*-propyl-4,4'-bipyridinium chloride has been prepared and used as a protective colloid during the generation of a polymer-protected colloidal platinum catalyst when platinum ions have been reduced by refluxing in alcohol [146]. This protective polymer also functions as an electron carrier in association with $Ru(bpy)_3{}^{2+}$ as a photosensitizer, as illustrated in Figure 22. The rate of electron transfer from the photosensitizer to a platinum particle in this system was found to be higher than when methyl viologen was used in association with poly(vinylpyrrolidone) as the protective polymer.

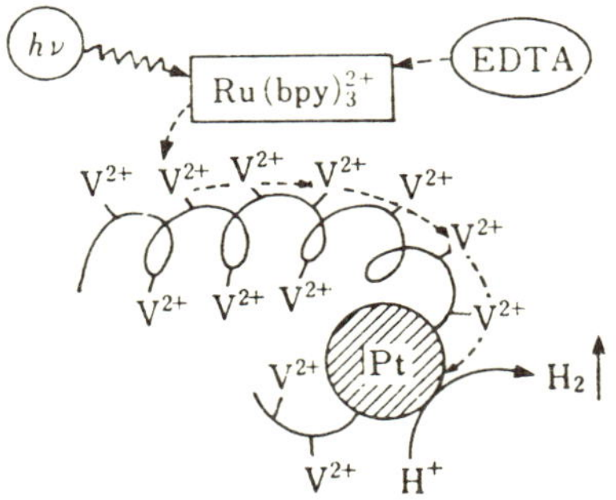

Fig. 22. Schematic illustration of an electron-transport system employing viologen polymer as the protective colloid. V^{2+} depicts a viologen group in the polymer [146].

There is no doubt that polymer-protected colloidal metal catalysts employing functionalized polymers will be developed further in the near future as well as multifunctional catalysts in polymer-anchored forms.

4. Concluding Remarks

The present chapter has described polymer-attached metal and metal complex catalysts as examples of tailored catalysts. As mentioned in the introduction to the chapter, such polymer-attached catalysts have been investigated from two viewpoints that of catalysis chemistry and that of polymer chemistry.

From the field of catalysis chemistry further developments of the practical application of polymer-attached catalysts to industrial processes are to be expected. In this context, the stability of the catalysts will be of great importance. New immobilization methods need to be developed to prevent release of the metals or metal complexes from polymer supports, as well as to retain the activity and selectivity of the catalysts over long periods of use.

The ultimate goal of highly active and selective catalysts must be that of metal enzymes. Polymers should play an important role in such an approach, and multifunctional polymers involving the systematic arrangement of polymers must be investigated further with this in mind. This is a problem in the field of polymer chemistry.

The chemistry of polymer-attached catalysts dates back to the start of the long history of catalysis. Such catalysts have often been used in reactions under reductive conditions such as hydrogenation, hydroformulation, hydrosilylation, etc. However, their usefulness in new processes such as C—C bond formation and photochemical reactions is only just emerging, where the development of new types of polymeric supports is becoming increasingly important. For these various reasons, polymer-attached catalysts are likely to receive far more attention in the future and could grow into an entirely new field of research both in industry and academia.

Department of Industrial Chemistry,
Faculty of Engineering,
The University of Tokyo,
Hongo,
Bunkyo-ku,
Tokyo 113, Japan

References

1. The Chemical Society of Japan: *Shokubai Sekkei* (*Catalyst Design*), (*Kagaku Sosetsu, No. 34*), Japan Scientific Societies Press, Tokyo (1982).
2. J. P. Collman: *Acc. Chem. Res.* **1**, 136 (1968).
3. R. Ugo (ed.): *Aspects of Homogeneous Catalysis*, Vol. 2, Reidel, Dordrecht/Boston (1974).
4. G. W. Parshall: *J. Mol. Catal.* **4**, 248 (1978).
5. F. R. Hartley: *Chemtech.* **10**, 686 (1980).
6. H. Hirai: *Shokubai Kogyo Kagaku* (*Industrial Chemistry for Catalysts*), p. 182, Asakura Shoten, Tokyo (1971).
7. C. Manassen: *Platinum Met. Rev.* **15**, 142 (1971).
8. C. U. Pittman, Jr. and G. O. Evans: *Chemtech.* **3**, 560 (1973).
9. J. C. Bailar, Jr.: *Catal. Rev. Sci. Eng.* **10**, 17 (1974).
10. K. Takemoto, T. Kunitake, Y. Imanishi and T. Shimizu: *Kobunshi Shokubai* (*Polymer Catalyst*), Kodansha, Tokyo (1976).

11. R. H. Grubbs: *Chemtech.* **7**, 512 (1977).
12. Y. Chauvin, D. Commereuc and F. Dawans: *Prog. Polym. Sci.* **5**, 95 (1977).
13. E. Tsuchida and H. Nishide: *Adv. Polym. Sci.* **24**, 1 (1977).
14. F. R. Hartley and P. N. Vezey: *Adv. Organometal. Chem.* **15**, 189 (1977).
15. T. Saegusa, E. Tsuchida and H. Hirai (eds.): *Kobunshi Kinzoku Sakutai – Gosei, Kino, Ohyo* (*Polymer-Metal Complex – Preparation, Function, and Application*), Kagaku Dojin, Kyoto (1978).
16. C. U. Pittman, Jr.: in *Polymer-supported Reactions in Organic Synthesis* (eds. P. Hodge and D. C. Sherrington), p. 249. Wiley, Chichester (1980).
17. N. K. Mathur, C. K. Narang and R. E. Williams: *Polymers as Aids in Organic Chemistry*, p. 220, Academic Press, New York (1980).
18. R. H. Grubbs and S.-C. H. Su: in *Enzymic and Non-enzymic Catalysis* (eds. P. Dunnill, A. Wiseman and N. Blakebrough), p. 223, Ellis Horwood, Chichester (1980).
19. D. D. Whitehurst: *Chemtech.* **10**, 44 (1980).
20. N. L. Holy: *Chemtech.* **10**, 366 (1980).
21. D. C. Bailey and S. H. Langer: *Chem. Rev.* **81**, 109 (1981).
22. F. Ciardelli, G. Braca, C. Carlini, G. Sbrana and G. Valentini, *J. Mol. Catal.* **14**, 1 (1982).
23. H. Hirai and N. Toshima (eds.): *Kobunshi Sakutai Shokubai* (*Macromolecular Complex Catalyst*), Japan Scientific Societies Press, Tokyo (1982).
24. J. Leto, D. Milstein, R. L. Albright, J. V. Minkiewicz and B. C. Gates: *Chemtech.* **13**, 46 (1983).
25. M. J. Farrall and J. M. J. Frechet: *J. Org. Chem.* **41**, 3877 (1976).
26. C. U. Pittman, Jr., P. L. Grube, O. E. Ayers, S. P. McManus, M. D. Rausch and G. A. Moser: *J. Polym. Sci., Part A-1* **10**, 379 (1972).
27. Z. Otero-Schipper, J. Lieto and B. C. Gates: *J. Catal.* **63**, 175 (1980).
28. Ref. 23, p. 25.
29. M. J. Farrall and J. M. J. Frechet: *J. Org. Chem.* **41**, 3877 (1976).
30. G. Innorta, A. Modelli, F. Scagnolari and A. Foffani: *J. Organometal. Chem.* **185**, 403 (1980).
31. R. H. Grubbs and E. M. Sweet: *Macromolecules* 8, 241 (1975).
32. R. B. Merrifield: *J. Am. Chem. Soc.* **85**, 2149 (1963); K. W. Pepper, H. M. Paisley and H. A. Young: *J. Chem. Soc.* 4097 (1953).
33. R. H. Grubbs and L. C. Kroll: *J. Am. Chem. Soc.* **93**, 3062 (1971).
34. E. S. Chandrasekaran, R. H. Grubbs and C. H. Brubaker, Jr.: *J. Organometal Chem.* **120**, 49 (1976).
35. A. J. Schouter, D. Noordegraab, A. P. Jakel and G. Challa: *J. Mol. Catal.* **5**, 331 (1979).
36. N. L. Holy: *J. Chem. Soc. Chem. Comm.* 1074 (1978).
37. C. Carlini, G. Braca, F. Ciardelli and G. Sbrana: *J. Mol. Catal.* **2**, 379 (1977); A. J. Naaktgeboren, R. J. M. Nolte and W. Drenth: *J. Am. Chem. Soc.* **102**, 3350 (1980).
38. G. Sbrana, G. Braca, G. Valentini, G. Pazienza and A. Altomare: *J. Mol. Catal.* **3**, 111 (1977).
39. N. Takaishi, H. Imai, C. A. Bertelo and J. K. Stille: *J. Am. Chem. Soc.* **100**, 264 (1978).
40. R. D. Sanner, R. G. Austin, M. S. Wrighton, W. D. Honnick and C. U. Pittman, Jr.: *Inorg. Chem.* **18**, 928 (1979).
41. D. Tatarsky, D. H. Kohn and M. Cais: *J. Polym. Sci., Polym. Chem. Ed.* **18**, 1387 (1980).

42. T. Kitamura, T. Joh and N. Hagihara: *Chem. Lett.* 203 (1975).
43. T. H. Kim and H. F. Rase: *Ind. Eng. Chem., Prod. Res. Dev.* **15**, 249 (1976).
44. W. O. Haag and D. D. Whitehurst: in *Catalysis* (ed. J. W. Hightower), Vol. 1, p. 465, Am. Chem. Soc., Washington, DC (1973).
45. R. Grubbs, C.-P. Lau, R. Cukier and C. Brubaker, Jr.: *J. Am. Chem. Soc.* **99**, 4517 (1977).
46. R. H. Grubbs, C. Gibbons, L. C. Kroll, W. D. Bonds, Jr. and C. H. Brubaker, Jr.: *J. Am. Chem. Soc.* **95**, 2373 (1973); L. Verdet and J. K. Stille: *Organometallics* **1**, 380 (1982).
47. C.-P. Lau, B.-H. Chang, R. H. Grubbs and C. H. Brubaker, Jr.: *J. Organometal. Chem.* **214**, 325 (1981).
48. A. J. Naaktgeboren, R. J. M. Nalte and W. Drenth: *J. Mol. Catal.* **11**, 343 (1981).
49. T. A. A. Maas, M. Kuijer and J. Zwart: *J. Chem. Soc., Chem. Commun.* 86 (1976).
50. J. Zwart and J. H. M. C. Van Wolput: *J. Mol. Catal.* **5**, 235 (1979).
51. H. Bruner and J. C. Bailar, Jr.: *Inorg. Chem.* **12**, 1465 (1973).
52. K. Kaneda, M. Terasawa, T. Imanaka and S. Teranishi: *Chem. Lett.* 1005 (1975).
53. M. Terasawa, K. Kaneda, T. Imanaka and S. Teranishi: *J. Catal.* **51**, 406 (1978).
54. Y. Nakamura and H. Hirai: *Chem. Lett.* 645 (1974).
55. Y. Nakamura and H. Hirai: *Chem. Lett.* 809 (1974).
56. B. C. Hui, W. K. Teo and G. L. Rempel: *Inorg. Chem.* **12**, 757 (1973).
57. J. A. Osborn, F. H. Jardine, T. F. Young and G. Wilkinson: *J. Chem. Soc. A* 1711 (1966).
58. W. Strohmeier and T. Onada: *Z. Naturforsch. B* **24**, 1493 (1969).
59. M. G. Burnett, R. J. Morrison and C. J. Stregnell: *J. Chem. Soc., Dalton Trans.* 701 (1973).
60. C. U. Pittman, Jr., S. E. Jacobson and H. Hiramoto: *J. Am. Chem. Soc.* **97**, 4774 (1975).
61. S. E. Jacobson, W. Clements, H. Hiramoto and C. U. Pittman, Jr.: *J. Mol. Catal.* **1**, 73 (1975).
62. C. U. Pittman, Jr. and G. M. Wilemon: *J. Org. Chem.* **46**, 1901 (1981).
63. G. Valentini, G. Sbrana and G. Braca: *J. Mol. Catal.* **11**, 383 (1981).
64. T. Okano, K. Tsukiyama, H. Konishi and J. Kiji: *Chem. Lett.* 603 (1982).
65. C. U. Pittman, Jr. and R. M. Hanes: *J. Am. Chem. Soc.* **98**, 5402 (1976).
66. C. U. Pittman, Jr., A. Hirao, C. Jones, R. M. Hanes and Q. Ng: *Ann. N. Y. Acad. Sci.* **295**, 15 (1977).
67. D. R. Fahey: *J. Org. Chem.* **38**, 80, 3343 (1973).
68. C. U. Pittman, Jr. and G. Wilemon: *Ann. N. Y. Acad. Sci.* **333**, 67 (1980).
69. M. Čapka, P. Svoboda, M. Cerny and J. Hetflejs: *Tetrahedron Lett.* 4787 (1971).
70. H. Hirai, M. Komiyama and S. Hara: *Chem. Lett.* 1447 (1981); *Makromol. Chem., Rapid. Commun.* **2**, 495 (1981).
71. R. H. Grubbs and E. M. Sweet: *Macromolecules* **8**, 241 (1975).
72. W. Dumont, J.-C. Poulin, T.-P. Dang and H. B. Kagan: *J. Am. Chem. Soc.* **95**, 8295 (1973).
73. G. Gubitosa, M. Boldt and H. H. Brintzinger: *J. Am. Chem. Soc.* **99**, 5174 (1977).
74. S. Bhaduri, A. Ghosh and H. Khwaja: *J. Chem. Soc., Dalton Trans.* 447 (1981).
75. H. Hirai and T. Furuta: *J. Polym. Sci.* **B9**, 459, 729 (1971).
76. H. Hirai, S. Komatsuzaki and N. Toshima: *Kobunshi Ronbunshu* **37**, 697 (1980).
77. P. Dini, D. Dones, S. Montelatici and N. Giordano: *J. Catal.* **30**, 1 (1973).
78. P. Dini, J. C. J. Bart, and N. Giordano, *J. Chem. Soc., Perkin Trans. 2* 1479 (1975).

79. W. H. Lang, A. T. Jurewiez, W. O. Haag, D. D. Whitehurst and L. D. Rollmann: *J. Organometal. Chem.* **134**, 85 (1977).
80. M. Bartholin, C. Graillat and A. Guyst: *J. Mol. Catal.* **10**, 361 (1981).
81. Y.-J. Li and Y.-Y. Jiang: *J. Mol. Catal.* **19**, 277 (1983).
82. R. H. Grubbs, L. C. Kroll and E. M. Sweet: *J. Macromol. Sci., Chem.* **A7**, 1047 (1973).
83. B. M. Trost and E. Keinan: *J. Am. Chem. Soc.* **100**, 7779 (1978).
84. R. C. W. Welch and H. F. Rose: *Ind. Eng. Chem., Fundam.* **8**, 389 (1969).
85. J. Tsuji and H. Takayanagi: *J. Am. Chem. Soc.* **96**, 7349 (1974).
86. M. M. Rogic and T. R. Demmiu: *J. Am. Chem. Soc.* **100**, 5492 (1978).
87. H. Nishide, T. Minakata and E. Tsuchida: *J. Mol. Catal.* **15**, 327 (1982).
88. J. He, Q. Li and Y. Jiang: *Proc. China–Jpn.–U.S. Trilateral Seminar Organometal. Chem., 1st, 1980*, p. 637, Van Nostrand, New York (1982).
89. Yu. I. Skurlatov, V. Ya. Kovner, V. A. Kavanov, *et al.: Eur. Polym. J.* **15**, 811 (1979).
90. M. Koide, N. Suganuma, E. Tsuchida and Y. Kurimura: *Makromol. Chem.* **182**, 1391 (1981).
91. S. Akabori, Y. Izumi, Y. Fujii and S. Sakurai: *Nippon Kagaku Zasshi* **77**, 1374 (1956).
92. S. Akabori, S. Sakurai, Y. Izumi and Y. Fujii: *Nature (London)* **178**, 323 (1956).
93. H. Hirai: *Shokubai* **15**, 117 (1973).
94. R. L. Beamer, R. H. Belding and C. S. Fickling: *J. Pharm. Sci.* **58**, 1142, 1419 (1969).
95. R. L. Beamer, C. S. Fickling and J. H. Ewing: *J. Pharm. Sci.* **56**, 1029 (1967).
96. M. E. Wilson and G. M. Whitesides: *J. Am. Chem. Soc.* **100**, 306 (1978).
97. N. Takahashi, H. Imai, C. A. Bertels and J. K. Stille: *J. Am. Chem. Soc.* **98**, 5400 (1976); *ibid.* **100**, 264 (1978).
98. T. Masuda and J. K. Stille: *J. Am. Chem. Soc.* **100**, 268 (1978).
99. S. J. Fritschel, J. J. H. Ackerman, T. Keyeser and J. K. Stille: *J. Org. Chem.* **44**, 3152 (1979).
100. J. K. Stille, S. J. Fritschel, N. Takaishi, T. Masuda, H. Imai and C. A. Bertels: *Ann. N. Y. Acad. Sci.* **333**, 35 (1980).
101. K. Achiwa: *Chem. Lett.* 905 (1978).
102. G. L. Baker, S. J. Fritschel, J. R. Stille and J. K. Stille: *J. Org. Chem.* **46**, 2954 (1981).
103. G. L. Baker, S. J. Fritschel and J. K. Stille: *J. Org. Chem.* **46**, 2960 (1981).
104. T. Hayashi, N. Nagashima and M; Kumada: *Tetrahedron Lett.* **21**, 4623 (1980).
105. G. C. A. Schuit and B. C. Gates: *Chemtech.* **13**, 556 (1983).
106. K. M. Webber, B. C. Gates and W. Drenth: *J. Mol. Catal.* **3**, 1 (1977/78).
107. H. Arai and M. Yashiro: *J. Mol. Catal.* **3**, 427 (1977/78).
108. E. Tsuchida and H. Nishide: *ACS Symp. Ser.* **121**, 147 (1980).
109. J. Zwart, H. C. Van der Weide, N. Bröker, C. Rummens, G. C. A. Schuit and A. L. German: *J. Mol. Catal.* **3**, 151 (1977/78).
110. C. U. Pittman, Jr. and L. R. Smith: *J. Am. Chem. Soc.* **97**, 1749 (1975).
111. C. U. Pittman, Jr., L. R. Smith and S. E. Jacobson: in *Catalysis: Heterogeneous and Homogeneous* (eds. B. Delmon and G. Jannes), p. 393, Elsevier, Amsterdam (1975).
112. C. U. Pittman, Jr., S. K. Wuu and S. E. Jacobson: *J. Catal.* **44**, 87 (1976).
113. J. M. Tibbitt, B. C. Gates and J. R. Katzer: *J. Catal.* **38**, 505 (1975).
114. R. F. Batchelder, B. C. Gates and F. P. J. Kuijpers: *Proc. 6th Int. Congress Catal., London*, Paper A40 (1976).
115. D. H. Everett: *Chem. Br.* **17**, 377 (1981).

116. N. Toshima and H. Hirai: *Photolysis of Water with Colloidal Catalysts*, in *Hikari-Enerugi-Henkan* (*Photoenergy Conversion*) (eds. A. Yamada, N. Toshima and M. Kaneko), p. 133, Japan Scientific Societies Press, Tokyo (1983).
117. M. Grätzel: *Ber. Bunsenges. Phys. Chem.* **84**, 981 (1980); *Acc. Chem. Res.* **14**, 376 (1981).
118. H. Hirai and N. Toshima: *Shokubai* **24**, 19 (1982); H. Hirai and M. Komiyama: in *Kobunshi Shokubai no Kogyoka* (*Industrialization of Polymer Catalysts*), p. 151, CMC, Tokyo (1981).
119. The Chemical Society of Japan: *Kaimen to Koroido* (*Surface and Colloid*) (*Shin Jikken Kagaku Koza* **18**), p. 332, Maruzen, Tokyo (1977).
120. B. Jirgensons and M. E. Straumanis: *A Short Textbook of Colloid Chemistry*, 2nd Ed., Pergamon Press, London (1962).
121. T. Sato and R. Ruch: *Stabilization of Colloidal Dispersions by Polymer Adsorption*, Marcel Dekker, New York (1980).
122. W. P. Dunworth and F. F. Nord: *Adv. Catal.* **6**, 125 (1954).
123. H. Hirai, Y. Nakao and N. Toshima: *J. Macromol. Sci., Chem.* **A12**, 1117 (1978); *ibid.* **A13**, 727 (1979).
124. Y. Nakamura and H. Hirai: *Chem. Lett.*, 1197 (1976).
125. H. Hirai, S. Komatsuzaki and N. Toshima: *Bull. Chem. Soc. Jpn.* **57**, 488 (1984).
126. N. Toshima, Y. Yamada, J. Ishiyama and H. Hirai: *J. Chem. Soc. Jpn., Chem. Ind.* 368 (1984).
127. J. A. Bruce, T. Murahashi and M. S. Wrighton: *J. Phys. Chem.* **86**, 1551 (1982).
128. P. C. Lee and D. Meisel: *J. Catal.* **70**, 160 (1981).
129. O. Siiman, A. Lepp and M. Kerker: *Chem. Phys. Lett.* **100**, 163 (1983).
130. H. Hirai, Y. Nakao and N. Toshima: *Chem. Lett.* 545 (1978).
131. H. Hirai, H. Chawanya and N. Toshima: *Reactive Polymers* **3**, 127 (1985).
132. H. Hirai, H. Chawanya and N. Toshima: *Makromol. Chem., Rapid Commun.* **2**, 99 (1981).
133. H. Hirai, H. Chawanya and N. Toshima: *J. Chem. Soc. Jpn., Chem. Ind.* 1027 (1984).
134. N. Toshima, M. Kuriyama, Y. Yamada and H. Hirai: *Chem. Lett.* 793 (1981).
135. R. Zsigmondy: *Z. Anal. Chem.* **40**, 697 (1901); R. Zsigmondy and P. A. Thiessen: *Das Kolloide Gold*, Akad, Verlagsges., Leipzig (1925).
136. H. Thiele and H. S. von Levern: *J. Colloid Sci.* **20**, 679 (1965).
137. H. B. Williams and L. T. Chang: *J. Phys. Colloid Chem.* **55**, 719 (1951).
138. N. Toshima, Y. Yamada and H. Hirai: *Abstr. 4th Int. Conf. Photochem. Conversion and Storage of Solar Energy, Jerusalem (1982)*, p. 273.
139. P.-A. Brugger, P. Cuendet and M. Grätzel: *J. Am. Chem. Soc.* **103**, 2923 (1981).
140. H. Hirai, H. Wakabayashi and M. Komiyama: *Chem. Lett.* 1047 (1983).
141. K. J. Mysels: *Introduction to Colloid Chemistry*, p. 205, Interscience, New York (1959).
142. D. H. Napper: *J. Colloid Interface Sci.* **58**, 390 (1977).
143. I. Iida: *Chem. Ind.* (*Tokyo*) **35**, 391 (1982).
144. N. Toshima: *Kobunshi* **31**, 935 (1982).
145. T. Sato: *Abstr. 4th Course Macromolecular Complexes* (1981), p. 33.
146. T. Nishijima, T. Nagamura and T. Matsuo: *J. Polym. Sci., Polym. Lett. Ed.* **19**, 65 (1981).

R. F. HOWE

THE PREPARATION OF HETEROGENEOUS CATALYSTS FROM MONONUCLEAR CARBONYL COMPLEXES ON INORGANIC SUPPORTS

1. Introduction

The first reported use of a mononuclear transition metal carbonyl complex to prepare a supported metal catalyst was by Baker and Bernstein in 1951 [1], who prepared an active nickel hydrogenation catalyst by decomposing $Ni(CO)_4$ on glass supports. Terenin and Roev [2] subsequently decomposed the carbonyls of chromium, iron and nickel on an alumina support to produce what was considered at the time to be zero-valent metal, although their IR spectra of adsorbed CO and NO show evidence for oxidized species as well. Parkyns [3] demonstrated that the decomposition of $Ni(CO)_4$ on an alumina support produced zero-valent nickel which was however readily oxidized upon exposure to O_2. About the same time, catalysts prepared from the Group VI carbonyl complexes $Mo(CO)_6$ and $W(CO)_6$ on various oxide supports were shown to be highly active for the olefin metathesis reaction [4].

These early reports provided the basis for subsequent experiments in many laboratories aimed at characterizing the species produced by the decomposition of transition metal carbonyl complexes, both mono- and poly-nuclear, on high surface area supports, and investigating the catalytic properties of the materials so produced. More recent efforts to prevent the decomposition of supported carbonyl complexes by anchoring them to functionalized oxide or polymer supports may also be traced back to the early studies.

The general field of immobilized transition metal carbonyl complexes as heterogeneous catalysts was recently reviewed comprehensively by Bailey and Langer [5], who list over 300 references. The specific topics of supported cluster carbonyl complexes and complexes anchored to functionalized supports are considered in the articles by Iwasawa and Ichikawa in this book. The intent of the present chapter is to review what is known after 20 years about the chemistry of mononuclear carbonyl complexes on inorganic supports, the manner in which they decompose and the properties of the resulting catalysts. In this area alone, more than 80 publications have now appeared, with $Mo(CO)_6$ receiving the greatest attention.

Section 2 will review some experimental aspects of catalyst preparation. The

Y. Iwasawa (ed.), Tailored Metal Catalysts, 141–182.

choice of support and pretreatment of the support prior to addition of the carbonyl complex are both important variables which strongly influence the resulting chemistry; the surface properties of the most widely used supports are therefore briefly discussed. Several different methods for preparing and activating the catalysts have been used by the various research groups working in this area. The advantages and disadvantages of each are also considered in Section 2. Section 3 covers the Group VI carbonyl complexes $Cr(CO)_6$, $Mo(CO)_6$ and $W(CO)_6$, Section 4 the Group VIII complexes $Fe(CO)_5$ and $Ni(CO)_4$, and Section 5 the Group V complex $V(CO)_6$ (which to date has received very little attention).

2. General Aspects of Catalyst Preparation

The preparation of a supported transition metal catalyst from a carbonyl precursor involves three steps: pretreatment of the support, contact of the support with the carbonyl complex and decomposition of the complex to the desired extent by an activation treatment. The nature and properties of the species produced will, in general, depend critically on the way in which all three steps are undertaken. These experimental variables are now considered in more detail.

2.1. SUPPORT MATERIALS

2.1.1. *Silica*

The properties of silica as a catalyst support have been reviewed by several authors [6, 7]. Amorphous silica, in the form of a high surface area gel (200–800 $m^2 g^{-1}$) or non-porous spheres (Cabosil or Aerosil, ca. 200 $m^2 g^{-1}$), contains upon exposure to air a monolayer of surface hydroxyl groups (5 OH groups nm^{-2}) plus physical adsorbed molecular water. The physically adsorbed water can be removed by evacuation at 150 °C. The silanol groups remaining are relatively unreactive, having neither basic nor acidic character. It should be noted, however, that the presence of impurities such as aluminium or boron may significantly enhance the Brönsted acidity of the silanol groups.

Dehydration of silica surfaces above 250 °C causes a progressive loss of the silanol groups through a bimolecular elimination of water. The dehydroxylation is initially reversible; after heating to 550 °C, ca. 3 OH groups nm^{-2} removed may be restored by the adsorption of water. At higher dehydration temperatures a relatively hydrophobic surface is produced containing bridging siloxane groups. Extensively dehydroxylated silica surfaces do not show any Lewis acidity, unless significant concentrations of impurities such as aluminium are present.

2.1.2. *Alumina*

The surface chemistry of alumina supports is more complex than that of silica. The two forms of alumina most commonly used, γ and η, have similar surface areas (typically 200 $m^2 g^{-1}$) and the same pseudo-spinel structure, differing only in the packing density of the oxide anions and the crystal planes preferentially exposed [8]. Alumina surfaces exposed to air contain a monolayer of hydroxyl groups (12 OH groups nm^{-2}) plus physically adsorbed water. As with silica, the physically adsorbed water can be removed at 150 °C, and dehydroxylation begins at around 200 °C. Dehydroxylation is virtually complete after heating to 1000 °C although such extreme heat treatment will result in at least partial transformation to the δ-phase of alumina, possibly with some loss in surface area.

At least five different types of hydroxyl group have been identified on the surface of γ-alumina by infrared spectroscopy [9]. Peri [10] suggested that the hydroxyl groups differed in the number of nearest-neighbour oxide ions and oxide ion vacancies surrounding each. Knözinger and Ratnasamy [8] have more recently proposed a model for the alumina surface in which the five different hydroxyl stretching frequencies are accounted for in terms of differing charge densities on hydroxyl groups on the various crystal planes exposed at the surface. The hydroxyl groups on alumina are more basic than those on silica. For example, adsorption of CO_2 on fully hydroxylated γ- or η-alumina surfaces produces bicarbonate species [11, 12]. Upon dehydroxylation, both oxide ions and coordinatively unsaturated aluminium ions are exposed. Partially dehydroxylated alumina surfaces thus contain both acidic and basic sites. The exposed aluminium ions act as Lewis acids, while the oxide ions are residual hydroxyl groups have basic properties. This dual acid–base character of activated alumina surfaces plays a key role in the chemistry of carbonyl complexes on alumina supports.

2.1.3. *Zeolites*

Zeolites are a well-known class of aluminosilicate support characterized by a high degree of crystallinity, regular well-defined pore structures and intense electrostatic fields within the pores [13]. The particular zeolite used to date as a support for carbonyl complexes, zeolite Y, has the largest pores of any known zeolite. The supercages in zeolite Y have a free diameter of ca. 1.2 nm, while the windows to the supercage have a diameter of 0.8 nm. The supercages are thus able to accommodate molecules as large as 1,3,5-triethylbenzene, and the surface chemistry of zeolite Y is dominated by the chemistry of the supercages. The external surface area of the zeolite crystallites is not negligible, however. For example, terminal hydroxyl groups are necessary to terminate the faces of a zeolite crystal where bonding would normally occur with adjacent silicon or aluminium atoms. The number of such external hydroxyl groups has been estimated to vary from about 10^{19} per gram for a 1 μm crystal to 10^{20} per gram for a 0.1 μm crystal [13]. The external surface area of a 1 μm crystal is of the order of 10 $m^2 g^{-1}$ compared with an internal area of about 800 $m^2 g^{-1}$.

The univalent cationic forms of zeolite Y adsorb a maximum of about 260 water molecules per unit cell. The adsorbed water can be removed by evacuation above 200 °C, and such pretreatment is essential if other molecules such as carbonyl complexes are to be adsorbed into the zeolite. In the dehydrated zeolite, univalent cations are distributed between three different sites in the zeolite structure; site I in the centre of the hexagonal prisms or double six-rings, site I′ in the cubooctahedra or β-cages, and site II in the supercages. Only those cations in site II are able to interact directly with molecules adsorbed into the zeolite supercages. The only hydroxyl groups present in a dehydrated univalent zeolite are those on the external surface, which under normal conditions (zeolite crystals tens of microns in size) will make a negligible contribution to the surface chemistry.

Dehydration of zeolite Y containing polyvalent cations produces additional hydroxyl groups within the zeolite due to hydrolysis of the cations and dissociation of water molecules by the electrostatic fields of the cations. Direct introduction of protons via ion exchange from acid solution is not possible due to the instability of the zeolite Y towards strong acids. However suitable heat treatment of the ammonium form of zeolite Y causes decomposition of NH_4^+ to H^+ and ammonia (typically at 350 °C). The protons in this so-called HY zeolite are located near site II in the supercages, and in either the β-cages or the double six-rings, giving rise to two characteristic hydroxyl stretching frequencies in the infrared spectrum. Both types of hydroxyl group show Brönsted acidity, although only one is able to interact directly with molecules adsorbed in the supercages at room temperature. Above room temperature the protons become mobile within the zeolite, and it is then no longer appropriate to distinguish between different sites.

Further thermal treatment of HY zeolite above 500 °C causes gradual dehydroxylation, producing anion vacancies through the elimination of a water molecule from two hydroxyl groups. The decationized zeolite then contains both Brönsted (residual protons) and Lewis acid (exposed aluminium cations) sites. Collapse of the zeolite structure occurs on heating above 750 °C. The decationized zeolite may be stabilized however, by carrying out the heat treatment in the presence of water vapour, which produces a material stable to 1000 °C. The stabilization process probably involves hydrolysis of aluminium from the zeolite framework, forming hydroxyl-aluminium cations.

2.1.4. *Other supports*

Other materials used as supports for carbonyl complexes include magnesium oxide and various forms of carbon. Magnesium oxide is normally prepared by the thermal decomposition of the hydroxide or the carbonate, and can have a surface area as high as 200 $m^2 g^{-1}$ [14]. The fully hydrated surface contains ca. 11 OH groups nm^{-2} plus physically adsorbed water. Dehydroxylation occurs on heating above 350 °C, although some residual hydroxyl groups remain after outgassing to 950 °C [15]. The magnesium oxide surface is considered to be more basic than that of silica or alumina. Adsorption of CO_2 produces carbonate and bicarbonate species [16], while alcohols form surface alkoxy groups [17]. The dehydroxylated surface shows no Lewis acidity.

Carbon supports in the form of charcoal or carbon black may have surface areas as high as 1000 $m^2 g^{-1}$. The surface chemistry of carbon is however complex and poorly understood. Boehm [18] has reviewed the nature of the functional groups present on carbon surfaces under different conditions. A more well-defined form of carbon is 'Grafoil' (Union Carbide), which consists of graphite flakes pressed into self-supporting sheets [19]. The basal planes of the graphite are closely aligned with the macroscopic Grafoil sheets, and the surface area of about 20 $m^2 g^{-1}$ consists almost entirely of basal planes. Since oxidation of graphite occurs preferentially at the edges of the basal planes, the Grafoil surface is largely oxygen-free and thus hydrophobic.

2.2. INTRODUCTION OF THE CARBONYL COMPLEX

Three different methods have been used to load transition metal carbonyl complexes onto supports: dry-mixing, impregnation or adsorption from solution, and adsorption from the vapour phase. The dry-mixing method is the simplest, but undoubtedly the least satisfactory. The carbonyl complex, in the form of a solid, is intimately mixed with the support material in the desired ratio, and the mixture then activated by heating. Of the mononuclear carbonyl complexes, $V(CO)_6$, $Cr(CO)_6$, $Mo(CO)_6$ and $W(CO)_6$ are solids at room temperature. Any molecular interaction between the carbonyl complex and the support will only occur during the activation process. It is essential, therefore, that the complex becomes volatile before it decomposes. As the physical data in Table I indicate, this condition is satisfied for all mononuclear carbonyl complexes, although it may not be for polynuclear complexes. The dry-mixing method has two obvious drawbacks. It is impossible to control exactly the amount and uniformity of loading, since much of the complex originally present may be lost through evaporation during catalyst activation. The method is also not suitable for experiments in which the surface chemistry of the support must be controlled.

TABLE I
Physical properties of mononuclear carbonyl complexes[a]

Complex	Melting point (°C)	Vapour pressure at 30 °C (Torr)
$V(CO)_6$	~70 (dec.)	unknown (can be sublimed *in vacuo*)
$Cr(CO)_6$	130 (dec.)	0.28
$Mo(CO)_6$	150 (dec.)	0.27
$W(CO)_6$	150 (dec.)	0.06
$Fe(CO)_5$	−20 (b.p., 103 °C)	35.4
$Ni(CO)_4$	−19 (b.p., 42 °C)	479

[a] Data from I. Wender and P. Pino (eds.): *Organic Syntheses via Metal Carbonyls*, Vol. 1, Wiley, New York (1967); (dec.) indicates decomposition at the melting point.

Impregnation of the support with a solution of the carbonyl complex has been widely used as a means of achieving a more uniform loading than is possible with dry-mixing. This goal will be obtained, however, only if the complex is molecularly adsorbed from solution onto the support surface and not merely precipitated out when the solvent is evaporated. The mononuclear carbonyl complexes are not particularly soluble in common solvents, but benzene, cyclohexane and pentane have been used successfully. An important factor to consider in the choice of solvent is the possibility of reaction between the solvent and the support, particularly if the support has been pretreated to generate active sites. Appropriate blank experiments must then be undertaken. The purity of the solvent should also be considered.

The solution impregnation method can be adapted to *in situ* preparations with pretreated supports. This may be done in a static experiment in which a solution of the carbonyl complex is added to the pretreated support under vacuum or an inert atmosphere, and the solvent removed by evacuation [20]. Alternatively, the flow method of Burwell and Brenner [21] in which a solution of the carbonyl complex is injected into a stream of carrier gas flowing over the support may be employed. The cautionary remarks in Section 2.3 concerning flow systems should be noted. Solution impregnation cannot be used to introduce carbonyl complexes into the pores of zeolite supports.

Adsorption from the vapour phase is the most direct method for introducing carbonyl complexes onto supports. All of the mononuclear complexes are sufficiently volatile (refer Table I) for adsorption from the vapour phase to be carried out in principle at room temperature. In practice, diffusion of the carbonyl complex into a packed bed of support may be slow at room temperature, and the adsorption is often carried out above room temperature (but below the decomposition temperature of the complex). As with solution impregnation, static or flow methods can be used, and the vapour pressure of the carbonyl complex regulated by adjusting the temperature of the carbonyl source. It is difficult to control the rate of adsorption in a static experiment, and in both static and flow experiments a uniform loading through a packed bed of support may be difficult to achieve.

2.3. CATALYST ACTIVATION

Once loading of a carbonyl complex onto a support has been achieved, decomposition of the supported complex is carried out by heating either *in vacuo* or in a flow of inert gas. Since the decomposition is inhibited by the presence of carbon monoxide, it is essential that the decomposition products be rapidly removed. In the case of catalysts prepared by dry-mixing, the temperature must be raised sufficiently slowly that introduction of the carbonyl complex into the support occurs before the bulk complex decomposes.

Activation in a flow system is particularly suited for temperature programmed desorption measurements. In the so-called temperature programmed decomposition (TPDE) experiments of Brenner *et al.* [21, 22], the gases evolved during decomposition of a supported complex (typically carbon monoxide, hydrogen and some hydrocarbons) are analyzed continuously as the sample temperature is increased linearly. From the resulting TPDE 'spectra', both the stoichiometry of the decomposition reactions and the appropriate activation energies can be obtained, in principle. Since species produced during the decomposition of supported carbonyl complexes are, in general, highly reactive towards oxygen, great care must be taken in flow experiments to exclude all traces of oxygen from the carrier gas. By incorporating Cr^{2+} on a silica trap upstream from their reactor, Brenner *et al.* [22] were able to limit the O_2 content of their helium flow to less than 10^{-6} cm^3 STP min^{-1}. The possible influence of solvent and/or solvent impurities on the activation process should also not be neglected.

An extension of the TPDE experiment for measuring catalyst activities has also been reported by Brenner *et al.* [22]. In this case the helium carrier gas is replaced by a mixture of reactants. The creation and/or destruction of active sites during decomposition of the supported complex can be monitored by following the conversion to products as a function of the linearly increasing activation temperature. The 'activity spectrum' generated can provide information about the nature of the active sites if it is assumed that decomposition in the presence of reactants produces the same species as decomposition in an inert carrier gas (or *in vacuo*).

3. Group VI Carbonyl Complexes

3.1. PREPARATION AND CHARACTERIZATION

3.1.1. *$Mo(CO)_6$–alumina*

The alumina-supported $Mo(CO)_6$ catalysts first used by Banks and Bailey for olefin metathesis [4] were prepared by the solution impregnation method, but no attempt was made to characterize the active catalysts. Kemball *et al.* [23] obtained IR spectra of activated catalysts (as Nujol mulls prepared under nitrogen), and suggested that the active molybdenum species had lost at least some of the original six carbonyl ligands. More detailed IR spectroscopic studies employing an *in situ* technique [24] revealed that activation by heating *in vacuo* caused decomposition of the $Mo(CO)_6$ via a series of intermediate subcarbonyl species. No correlation was found between the activity for olefin metathesis and the concentrations of the various subcarbonyl species, and it was suggested that the active site for metathesis was a fully decarbonylated molybdenum species, possibly oxidized. EPR [25] and XPS [26] measurements provided further evidence for the oxidation of molybdenum during catalyst activation.

The activation of alumina-supported $Mo(CO)_6$ was subsequently investigated by Burwell and Brenner using the TPDE technique [21, 27]. Both carbon monoxide and hydrogen (plus traces of other gases) were evolved during activation in flowing helium. The evolution of carbon monoxide occurred in discrete stages, corresponding to the formation of subcarbonyl species containing five, four and three carbonyl ligands per molybdenum. The decomposition of $Mo(CO)_6$ to $Mo(CO)_3$ could be reversed by the addition of carbon monoxide, whereas loss of the final three ligands was irreversible [28]. The reversibility of the initial decarbonylation has also been demonstrated by IR spectroscopy [29]. The final decarbonylation step was accompanied by the evolution of hydrogen due to oxidation of the zero-valent molybdenum by hydroxyl groups of the support. To explain such an observation, reactions such as the following were suggested to occur:

$$Mo(CO)_3\,(ads) + 2\,OH \longrightarrow \text{‘}(O^-)_2\,Mo^{2+}\text{’} + 3\,CO + H_2$$

Burwell and Brenner also examined the effects of added oxygen, and postulated the formation of species such as '$Mo(CO)_2O_2$' and '$Mo(CO)(O_2)_2$' (although these species have never been observed spectroscopically). Activation in flowing helium to 500 °C caused complete decarbonylation and oxidation to an average valency state between +5 and +6.

Oxidation of supported $Mo(CO)_6$ by support hydroxyl protons may, in principle, be controlled by adjusting the number of hydroxyl groups on the support. For example, activation at 500 °C of $Mo(CO)_6$ on an alumina support pretreated at 950 °C (which removes 99% of the surface hydroxyl protons) produced a catalyst with an average molybdenum valency state of 0.4 [27, 30]. Burwell *et al.* have suggested more recently that such catalysts contain small clusters (< 4 nm) of metallic molybdenum anchored to Mo^{2+} cations [31, 32]. Traces of methane observed during TPDE experiments (e.g. 0.1 CH_4 molecules per Mo atom [30]) are presumably due to a methanation reaction between CO and H_2, since the methane yields were considerably enhanced when TPDE experiments were carried out in flowing hydrogen rather than helium [31]. Brenner and Hucul [33] have argued that this result proves that multinuclear metal centres are not an essential requirement for the catalysis of methanation.

Further IR spectroscopic studies of the alumina-supported $Mo(CO)_6$ catalyst have been undertaken recently by three different research groups, with the object of determining the detailed molecular nature of the various species formed during decomposition of the hexacarbonyl [34–36]. Disagreements between the earlier spectroscopic results [24, 25, 29] and the TPDE results [21, 27] may, for the most part, be attributed to differences in experimental conditions and methods of catalyst preparation. In principle, the combination of IR spectroscopy and the stoichiometry of decomposition determined from TPDE experiments should allow a complete description of the decarbonylation pathway to be obtained. In practice, however, although all three groups obtained similar spectra, their interpretations differ markedly. The different reaction models that were proposed have been reviewed in detail by Brown [37]. Since the alumina-supported $Mo(CO)_6$ catalyst has been more widely studied than any other, current views on the chemistry of the decarbonylation reaction will be reviewed again (more briefly) here. Related chemistry may be anticipated for other supported mononuclear carbonyl catalysts which have not received the same attention.

There is agreement that the initial interaction of $Mo(CO)_6$ with an alumina surface, whether hydroxylated or dehydroxylated, is physical adsorption of the intact complex. The IR spectrum is closely similar to that of crystalline $Mo(CO)_6$ (showing some lowering of symmetry relative to the free molecule). With partially dehydroxylated supports, the ν(OH) bands of the support are broadened on exposure to $Mo(CO)_6$ vapour, indicating a weak hydrogen-bonding interaction of surface hydroxyl groups with the adsorbed complex. Additional bands in the carbonyl regions of the spectrum at frequencies below those of $Mo(CO)_6$ must be attributed to chemisorbed species, and it is the nature of these chemisorbed species which has been the subject of debate [37].

Solution studies have shown that the carbonyl ligands of $Mo(CO)_6$ have increased lability in the presence of base. It thus seem reasonable to postulate that loss of carbonyl ligands from adsorbed $Mo(CO)_6$ is assisted by attack of surface hydroxyl or oxide groups via transient acyl species as follows:

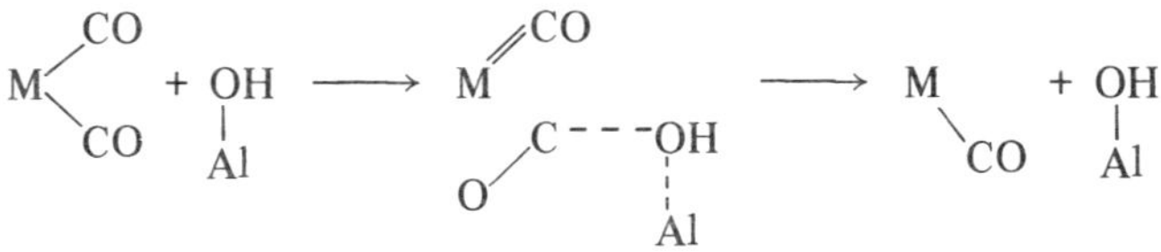

Such labilization of the carbonyl ligand *cis* to that interacting with the support provides a mechanism for the stepwise loss of carbon monoxide seen in TPDE experiments. The coordination sites vacated by carbon monoxide may then be occupied by surface oxide or hydroxyl groups (depending on the state of hydroxylation of the support). The first subcarbonyl speices to be formed, $Mo(CO)_5L$ where L represents a surface ligand, has been identified in the IR spectrum of $Mo(CO)_6$ on a fully hydroxylated alumina [29]. This species appears to have only a transient existence however on dehydroxylated supports.

The carbonyl bands below 1800 cm^{-1} observed when $Mo(CO)_6$ is adsorbed on dehydroxylated supports cannot be attributed to further substituted carbonyl species of the type $Mo(CO)_x L_{6-x}$, since the frequencies are too low to arise from unmodified terminal carbonyl ligands. Laniecki and Burwell [35] have suggested that the low frequency bands are due to carboxylate groups formed by interaction of carbonyl ligands with surface oxide ions:

Kazusaka and Howe [34], on the other hand, attributed the low frequency bands to carbonyl ligands coordinated to Lewis acid sites (exposed Al^{3+} cations) on the alumina surface, either terminal, i.e. $Mo-CO \cdots Al^{3+}$, or bridging, i.e.

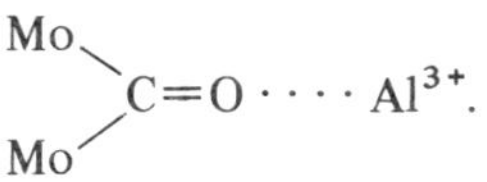

Brown [37] agreed that Lewis acid sites on the support are responsible for the lowering of the carbonyl frequencies, but disputed the assignment to bridging carbonyl ligands, on the grounds that Group VI carbonyl complexes show no propensity for forming bridging structures. As pointed out by Brown [37], it is not necessary to invoke bridging carbonyl ligands to account for carbonyl frequencies as low as those observed by Kazusaka and Howe. On fully hydroxylated supports, the Lewis acid sites are not present and the carbonyl bands are shifted to higher frequency.

The following scheme summarizes the reactions that take place when $Mo(CO)_6$ interacts with an alumina support:

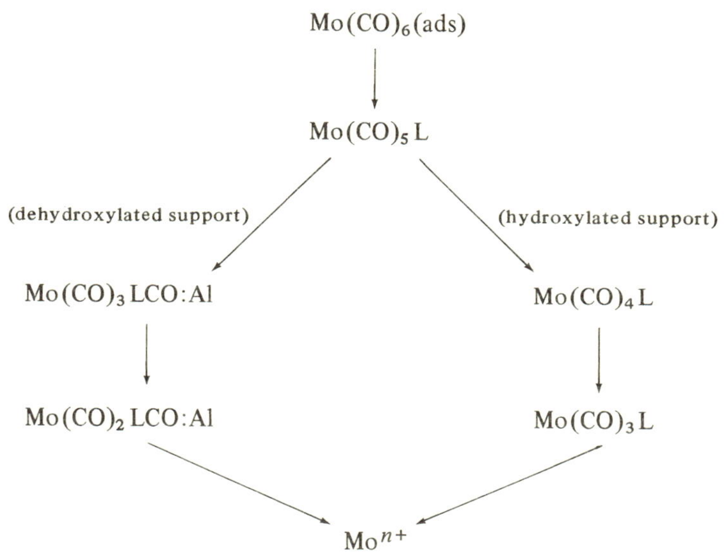

Initial adsorption of $Mo(CO)_6$ produces largely the physically adsorbed complex with minor amounts of the various possible chemisorbed species. On heating above 25 °C decomposition occurs, with up to three carbonyl ligands per molybdenum being lost in a stepwise reversible fashion. The characteristic carbonyl frequencies of each intermediate species can be determined by comparing the IR spectra with the average stoichiometries deduced from TPDE experiments. It should be noted, however, that the IR spectra are, in general, more complicated than the stoichiometric data would indicate. For example, the spectrum reported by Laniecki and Burwell [35] for $Mo(CO)_6$ on a partially dehydroxylated alumina activated at 100 °C for 1 h shows six distinct carbonyl bands, while the stoichiometry was determined to be exactly $Mo(CO)_3$. Evidently several species with the same CO/Mo stoichiometry may be present, such as a mixture of $Mo(CO)_2LCO$:Al and $Mo(CO)_3L$, or $Mo(CO)_2LCO$:Al on two different types of Lewis acid site [37], or a mixture of terminal and bridged carbonyl:Lewis acid adducts [34]. In view of the difficulty of identifying carbonyl complexes from their carbonyl stretching frequencies alone, and the known complexity of alumina surfaces, it is probably unrealistic to expect any better definition of the surface species from $Mo(CO)_6$ on alumina than has already been achieved.

The intermediate species formed on partially or fully dehydroxylated supports are more stable than those on fully hydroxylated supports, but in all cases complete and irreversible decarbonylation can be achieved by heating above 250 °C. Re-adsorption of carbon monoxide on fully decarbonylated catalysts products only weak IR bands around 2200 cm^{-1} typical of carbon monoxide on transition metal oxide surfaces [38].

As mentioned above, both XPS and EPR measurements provided spectroscopic support for the oxidative reaction with surface hydroxyl groups inferred from TPDE experiments. The photoelectron spectra [26] showed a substantial increase in the molybdenum 3d binding energy when alumina-supported $Mo(CO)_6$ was activated; the line widths of the signals implied that a range of oxidation states may be present, and a higher resolution study of this system needs to be undertaken. The EPR evidence for oxidation consists of the observation of signals due to Mo^{5+} upon high-temperature activation *in vacuo*, and of the superoxide ion O_2^- upon subsequent exposure to oxygen [25]. The integrated intensities of the Mo^{5+} signals, however, never exceeded ca. 8% of the total molybdenum content of the catalysts. Comparison of the Mo^{5+} spin concentrations with the amounts of evolved hydrogen measured in TPDE experiments [27] indicates that a considerable fraction of the Mo^{5+} present after activation above 400 °C may not be detected by EPR spectroscopy. A similar descrepancy between EPR and chemical measurements of Mo^{5+} concentrations in the case of alumina-supported MoO_3 catalysts has been shown to be due to spin coupling of adjacent Mo^{5+} ions [39]. Although $Mo(CO)_6$ initially adsorbed on alumina may be molecularly dispersed over the surface, the EPR experiments suggest that during oxidative decarbonylation coupled molybdenum species are formed similar to those found in conventional oxide-derived catalysts. The formation of O_2^- from adsorbed oxygen is also characteristic of molybdenum oxide catalysts (the site for electron transfer being Mo^{5+} [40]), and the concentrations of O_2^- formed on $Mo(CO)_6$ derived catalysts are comparable in magnitude with those found for the conventional catalysts. Further IR spectroscopic evidence for the formation of oxidic molybdenum species upon activation of $Mo(CO)_6$ on alumina has been presented by Kazusaka and Howe [41], who showed that adsorption of nitric oxide produced an identical dinitrosyl species (probably $Mo^{4+}(NO)_2$) to that formed on reduced molybdena–alumina catalysts.

In principle, it should be possible to follow the oxidative reaction of surface hydroxyl groups with zero-valent molybdenum by monitoring the intensities of the ν(OH) bands in the IR spectrum of the alumina support. For example, oxidation of the maximum loading of molybdenum (ca. 2% by weight) on a support pretreated at 450 °C to an average valency state of 5.6 should consume about 35% of the support hydroxyl protons. An overall intensity decrease in the hydroxyl bands of this magnitude was observed by Kazusaka and Howe

[34], although difficulties in measuring IR intensities with quantitative accuracy prevented detailed comparisons. More importantly, the relative intensities of the five alumina hydroxyl bands were unchanged, indicating that the oxidation reaction is not selective for a particular type of surface hydroxyl group. In fact, at temperatures at which oxidation occurs, NMR measurements have shown the alumina protons to be quite mobile, and initiation of the oxidation reaction may be associated with the onset of proton mobility.

3.1.2. *$Cr(CO)_6$–alumina*

The few studies that have been made on this system have revealed significant differences from the corresponding molybdenum catalyst. For example, the TPDE curve for $Cr(CO)_6$ on a partially dehydroxylated alumina support reported by Brenner and Hucul [42] shows that carbon monoxide evolution occurs over a much narrower temperature interval than for $Mo(CO)_6$, and the extent of oxidation of chromium by surface hydroxyl groups, as determined by the amount of hydrogen evolved during activation, was stated to be lower than that for molybdenum. An IR spectroscopic study of similar $Cr(CO)_6$ catalysts has shown that complete decarbonylation may be achieved by heating to 100 °C [43]; no evidence has yet been obtained that $Cr(CO)_6$ can form stable subcarbonyl species of the type described above for $Mo(CO)_6$. Complete decarbonylation at 100 °C could not be reversed by exposure to carbon monoxide. On a fully hydroxylated support, decarbonylation occurred on prolonged evacuation at 45 °C, and could be partially reversed by adding carbon monoxide [29].

The inability of chromium to form stable subcarbonyl species on alumina, either ligand-substituted $Cr(CO)_xL_y$ or Lewis acid adducts, is presumably related to differences in the lability and/or basicity of carbonyl ligands coordinated to chromium compared with the other Group VI transition metals. The lower oxidation states of chromium (Cr^{2+}, Cr^{3+}) are known to be more stable on oxide supports than those of molybdenum, and evidently resist oxidation by protons of the support hydroxyl groups to a greater extent. IR spectra of nitric oxide adsorbed on $Cr(CO)_6–Al_2O_3$ catalysts activated at high temperature indicate the formation of the same Cr^{2+} and Cr^{3+} nitrosyl species found on a conventional reduced chromia–alumina catalyst [41].

3.1.3. *$W(CO)_6$–alumina*

Detailed TPDE [44] and IR spectroscopic studies [43] have been reported for $W(CO)_6$ on partially dehydroxylated alumina supports. According to the TPDE measurements, decarbonylation of $W(CO)_6$ occurs in a stepwise and reversible fashion to $W(CO)_3$; complete and irreversible decarbonylation is achieved at a temperature of ca. 400 °C (significantly higher than for $Mo(CO)_6$). The final decarbonylation is accompanied by evolution of hydrogen due to the redox reaction with surface hydroxyl groups. IR spectra of catalysts similar to those used in the TPDE experiments (containing a maximum possible loading of $W(CO)_6$) showed carbonyl bands which were almost identical (but shifted by 10–40 cm^{-1} to lower frequency) to those observed during activation of the corresponding $Mo(CO)_6$ catalysts. The decarbonylation of $W(CO)_6$ on partially dehydroxylated alumina thus appears to follow the same pathway as that described above for $Mo(CO)_6$.

It was further found, however, that if the amount of $W(CO)_6$ initially adsorbed on the alumina support was limited to less than the saturation coverage (by restricting the time of exposure to the vapour), complete decarbonylation could be achieved at much lower temperatures (typically 100 °C) and stable subcarbonyl species were not observed in the IR spectrum. Similar behaviour has been observed on a fully hydroxylated alumina support [29], and on a partially dehydroxylated η-alumina [45]. The formation of stable subcarbonyl species by $W(CO)_6$ thus appears to depend not only on the pretreatment of the support but also on the surface coverage of $W(CO)_6$. This has been taken as evidence favouring bimolecular structures with bridging carbonyl ligands for the subcarbonyl species [43], although as pointed out above such structures are not known in the solution chemistry of Group VI carbonyl complexes.

Brenner and Hucul [44] have determined the average oxidation state of tungsten as lying between five and six after activation of $W(CO)_6$–Al_2O_3 catalysts at 600 °C. However no EPR signals of W^{5+} could be detected from activated catalysts (at temperatures down to 10 K) [43]. Since isolated W^{5+} cations can be readily observed at room temperature in other systems [46], it was concluded that W^{5+} cations in the $W(CO)_6$–Al_2O_3 catalysts are coupled to an even greater extent than the corresponding molybdenum cations. IR spectra of nitric oxide adsorbed on activated catalysts have revealed the formation of a $W^{4+}(NO)_2$ species similar to that formed on molybdenum catalysts [41].

No spectroscopic studies have been reported of $W(CO)_6$ on fully dehydroxylated alumina supports. The TPDE measurements of Brenner and Hucul [44] have shown however that on an alumina containing less than 1% of a monolayer of hydroxyl groups, tungsten can be oxidized to an average valency state of +1.0. Such catalysts are likely to resemble the corresponding molybdenum systems, but further investigations are necessary.

3.1.4. *$Mo(CO)_6$–silica*

Silica-supported $Mo(CO)_6$ has received less attention than the alumina-supported catalyst, although from the viewpoint of catalytic activity the silica-supported catalyst is more active for olefin metathesis [20]. The early IR spectroscopic studies of Howe *et al.* [24] showed that $Mo(CO)_6$ on a fully hydroxylated silica support could be completely decarbonylated by heating *in vacuo* at 100 °C for 1 h; no intermediate subcarbonyl species could be detected. In later experiments [29], an *in situ* preparation via adsorption from the vapour phase was used to eliminate any possibility of oxidation of molybdenum during preparation, but identical results were obtained. The complete decarbonylation by prolonged evacuation at 45 °C could be partially reversed by exposure to carbon monoxide at this temperature, but not after activation at higher temperatures. In a laser Raman study of $Mo(CO)_6$ on porous Vycor glass, Adams *et al.* [47] were also unable to detect any stable subcarbonyl species.

The spectroscopic results concerning the relative ease of decomposition of $Mo(CO)_6$ on silica supports have been confirmed by TPDE experiments [22]. A single narrow peak in the carbon monoxide desorption spectrum indicates that all six carbonyl ligands are lost in rapid succession, although Brenner *et al.* [22] point out that small amounts of irreversibly formed subcarbonyl species may be present but not detected by IR spectroscopy. Since the capacity of silica to retain $Mo(CO)_6$ is significantly lower than that of alumina, (by a factor of up to 20), there may well be species present which the IR and Raman experiments do not observe. The sensitivity of the IR experiments is sufficient, however, to rule out a major contribution from stable subcarbonyl species.

The oxidation of molybdenum by the support hydroxyl groups occurs on silica to an extent similar to that an alumina (as determined from the amount of hydrogen evolved [22]), although a much lower fraction of the hydroxyl groups is involved. Spectroscopic evidence for the formation of oxidized molybdenum species includes observation of Mo^{5+} and O_2^- EPR signals [22, 25] and the formation of IR-active dinitrosyl species analogous to those formed when nitric oxide is adsorbed on molybdenum oxide catalysts [50]. It should be possible to suppress the oxidation by using fully dehydroxylated silica as the support, but no studies of such systems have as yet been reported.

3.1.5. *Cr(CO)₆–silica*

The decomposition of $Cr(CO)_6$ on a fully hydroxylated silica support was first shown by IR spectroscopy to occur rapidly on evacuation at 100 °C, and over a longer period at 45 °C [29]. The decomposition at the lowest temperature could be partially reversed by adding carbon monoxide. Brenner *et al.* [22] subsequently reported TPDE data for $Cr(CO)_6$ on a partially dehydroxylated silica support which showed a single narrow peak for carbon monoxide evolution. A more detailed IR study by Guglielminotti [48], also with a partially dehydroxylated support, has confirmed the suggestion [29] that significant concentrations of stable subcarbonyl species are not formed in this system. Guglielminotti also observed partial reversal of decarbonylation at 150 °C on subsequent exposure to carbon monoxide.

The oxidation of chromium by silica hydroxyl groups on activation at higher temperatures has been demonstrated by the evolution of hydrogen in TPDE experiments [22] and by the observation of IR spectra characteristic of carbon monoxide adsorbed on Cr^{2+} sites [48]. An EPR spectroscopic study has shown that no paramagnetic chromium species (Cr^{3+} or Cr^{5+}) were produced upon activation *in vacuo*, but that subsequent exposure to oxygen gave Cr^{5+} signals [49]. The chemistry of $Cr(CO)_6$ on silica thus appears to resemble closely that on alumina supports.

3.1.6. *W(CO)₆–silica*

The TPDE curves for the decomposition of $W(CO)_6$ on a partially dehydroxylated silica support have been reported [22] as closely resembling those for silica-supported $Mo(CO)_6$. After activation to 600 °C the average valency state of tungsten was determined to lie between 4 and 5. No EPR signals due to W^{5+} could be detected from activated catalysts, but exposure of such catalysts to oxygen produced the characteristic signal of O_2^- [49, 51]. The O_2^- radical has recently been shown to be formed by electron transfer from oxidized tungsten sites and stabilized on the silica support [52].

3.1.7. *Mo(CO)₆–zeolite Y*

The adsorption and decomposition of $Mo(CO)_6$ within the supercages of zeolite Y was first reported by Coudurier *et al.* [53, 54]. These authors found that a HY zeolite would readily adsorb at least eight $Mo(CO)_6$ molecules per unit cell (an average of one per supercage) when exposed to the vapour at temperatures between 25 °C and 60 °C. X-Ray diffraction located the initially adsorbed $Mo(CO)_6$ at the centres of the 12-membered ring apertures to the supercages, and IR spectra in the $\nu(OH)$ region showed a weak hydrogen-bonding interaction of $Mo(CO)_6$ with zeolite protons in the supercages. Adsorption of the intact $Mo(CO)_6$ molecule has also been demonstrated by ^{13}C NMR spectroscopy [55].

When $Mo(CO)_6$ adsorbed in HY is heated *in vacuo* both carbon monoxide and hydrogen are evolved, and the chemistry of the decomposition resembles that of $Mo(CO)_6$ on alumina quite closely. IR spectra in the $\nu(CO)$ region show that decarbonylation occurs in a stepwise manner, involving at least three distinct subcarbonyl species which can be reversibly interconverted by adding carbon monoxide [54, 56]. The subcarbonyl species do not exhibit the same $\nu(CO)$ frequencies as those on alumina however. Since HY zeolites under the conditions used in this work do not contain significant concentrations of Lewic acid sites, formation of Lewis acid : carbonyl adducts is not possible in the zeolite. Structures of the type $Mo(CO)_4L$, where L represents an oxide ion of the zeolite lattice, have therefore been proposed [54, 56]. Abdo and Howe [56] observed the EPR signal of a radical containing two equivalent molybdenum nuclei from $Mo(CO)_6$–HY samples activated at 100 °C, and suggested that dinuclear anionic carbonyl species analogous to the known complex $[Mo_2(CO)_{10}]^{2-}$ may be formed [56]. Complete decarbonylation, achieved by heating above 200 °C, is irreversible and is accompanied by hydrogen evolution. Spectroscopic evidence for the oxidation of molybdenum by the zeolite protons includes the observation of oxidized molybdenum species in the UV reflectance spectrum [54], Mo^{5+} EPR signals [56, 57] and the disappearance of the zeolite hydroxyl bands in the IR spectrum [54, 56] (all of the zeolite protons are consumed upon prolonged activation at 500 °C).

X-Ray diffraction studies [54] were unable to locate molybdenum cations following decarbonylation and oxidation of $Mo(CO)_6$ in HY zeolite. Recent detailed EPR and IR measurements [56], however, have provided some information about the fate of the molybdenum. For example, the first Mo^{5+} species to appear in the EPR spectrum is located in the supercages, as shown by its ability to interact with adsorbed pyridine, nitric ocide and oxygen. A second Mo^{5+} species, which reaches a maximum concentration only after prolonged activation at 450 °C, is located in hidden sites, either site I in the hexagonal prisms or site I′ in the β-cages. The Mo^{5+} species in hidden sites can interact with ammonia (which can enter the β-cages), but not with pyridine or oxygen. Changes in the low frequency IR spectrum of the zeolite also indicate that molybdenum cations migrate into hidden sites. A complete description of all of the valency states of molybdenum present and their locations within the zeolite after a particular activation treatment is not yet available. Nevertheless, it is clear that following initial decarbonylation of $Mo(CO)_6$ in the supercages, the oxidized molybdenum species produced (either simple cations or possibly oxomolybdenum species such as MoO^{3+}) progressively migrate towards those sites which allow maximum coordination to the zeolite lattice, in the β-cages or hexagonal prisms. After activation at 450–500 °C, most of the molybdenum is in the +5 or +6 valency state and located in sites inaccessible to molecules in the supercages.

When similar experiments were carried out with an NaY zeolite, no oxidizing sites were found to exist, and decomposition of $Mo(CO)_6$ adsorbed in the supercages produced a black material believed to contain zero-valent molybdenum [56, 58]. The decomposition occurred via a single intermediate subcarbonyl species which has been identified from its colour and IR spectrum [56] as the dinuclear anion $[Mo_2(CO)_{10}]^{2-}$. The zero-valent molybdenum produced by the decarbonylation of $Mo(CO)_6$ at 200 °C is quickly oxidized upon exposure to oxygen. If, on the other hand, the zeolite containing $Mo(CO)_6$ is activated at higher temperatures, oxidation upon exposure to oxygen occurs much more slowly. It appears, from the limited information so far available, that the chemistry of $Mo(CO)_6$ in NaY resembles that on dehydroxylated alumina, in that the end results of catalyst activation are clusters of metallic molybdenum. The resistance to oxidation of samples activated at higher temperatures may be arise from sintering of the molybdenum clusters or from migration of molybdenum into sites inaccessible to oxygen. Fruther study of this novel system is needed.

3.1.8. *Cr(CO)₆–zeolite Y*

A brief study of the adsorption and decomposition of $Cr(CO)_6$ in a HY zeolite has been reported by Ben Taarit *et al.* [55]. The initial interaction of $Cr(CO)_6$ with the zeolite was determined to be extremely weak from the ^{13}C NMR spectrum. Upon subsequent activation *in vacuo*, an intense EPR signal characteristic of Cr^{3+} was observed, and subsequent exposure to ^{13}CO gave an NMR signal similar to that found for carbon monoxide adsorbed in transition metal ion-exchanged zeolites. Thus, the decarbonylation and oxidation of $Cr(CO)_6$ in HY zeolite appears to be similar to that of $Mo(CO)_6$, although Ben Taarit *et al.* note that the decomposition of $Cr(CO)_6$ occurs at a lower temperature than that of $Mo(CO)_6$ (as is also the case on alumina supports).

3.1.9. *W(CO)₆–zeolite Y*

An EPR study of the adsorption and decomposition of $W(CO)_6$ in HY and NaY zeolites has been reported by Abdo *et al.* [46]. In HY zeolite, decomposition of $W(CO)_6$ initially adsorbed in the supercages above 200 °C produced three distinct W^{5+} EPR signals due to W^{5+} cations located in the supercages and in two different hidden sites. In NaY zeolite, on the other hand, initial activation at 100 °C produced a new signal which experiments with ^{183}W-enriched $W(CO)_6$ proved was due to a radical containing two equivalent tungsten nuclei. The formation of a dinuclear species at an early stage in the decarbonylation process is consistent with the suggestion made for $Mo(CO)_6$ in NaY zeolite, that in the absence of oxidizing sites aggregation of carbonyl fragments to form clusters of zero-valent metal occurs. As with $Mo(CO)_6$, $W(CO)_6$–NaY samples activated at low temperatures could be readily oxidized with oxygen at room temperature, whereas activation at high temperatures produced oxidation-resistant material.

3.1.10. *$Mo(CO)_6$–magnesia*

An early IR study of $Mo(CO)_6$ on a fully hydroxylated MgO support showed that decomposition of the supported complex occurred in a similar manner to that on alumina [24]. The initial decarbonylation of $Mo(CO)_6$ produced $Mo(CO)_5L$, which subsequently decomposed further via a series of intermediate subcarbonyl species which were not characterized, but which differed from those found on alumina. Complete decarbonylation was achieved at 200 °C.

A more recent IR study has been reported of $Mo(CO)_6$ on a fully dehydroxylated MgO support (pretreated at 800 °C [59]). In this case no physically adsorbed $Mo(CO)_6$ could be detected; the initial interaction of $Mo(CO)_6$ with the dehydroxylated magnesium oxide surface evidently involves chemisorption. The initial reaction was suggested to be:

$$Mo(CO)_6 + Mg^{2+}O^{2-} \longrightarrow [(CO)_5MoCO_2]^{2-}Mg^2$$

This reaction is analogous to that known to occur between carbonyl complexes and alkyl-lithium compounds. Evacuation at room temperature or above caused decomposition via a series of at least three intermediate species. At the same time, an intense absorption band developed below 400 nm in the UV spectrum, which Guglielminotti and Zecchina have attributed to the formation of molybdenum clusters, i.e. the decarbonylation on dehydroxylated MgO produces zero-valent molybdenum species similar to those found on other supports lacking oxidizing sites.

3.1.11. *$Cr(CO)_6$– and $W(CO)_6$–magnesia*

The chemistry of $Cr(CO)_6$ and $W(CO)_6$ on fully dehydroxylated MgO supports is reported [59] to resemble that of $Mo(CO)_6$ closely. Dispersed clusters of chromium or tungsten metal, with an absorption band due to free electron collective transitions extending from the UV to the near-IR, are formed upon complete decarbonylation of the supported complexes.

3.2. CATALYTIC PROPERTIES

Table II summarizes the published work (excluding the patent literature) in which supported Group VI carbonyl complexes have been used as heterogeneous catalysts. Discussion of the literature in this section is organized according to the type of reaction involved.

TABLE II
Catalytic properties of supported Group VI carbonyls

Catalyst and reaction	Comments	Reference
$Mo(CO)_6$–alumina		
metathesis of propylene	first report	[4]
metathesis of propylene	kinetic study (Langmuir–Hinshelwood kinetics)	[61]
metathesis of propylene	reaction rate enhanced by halogenated olefins	[77]
metathesis of propylene	conversion of ethylene to propylene during metathesis	[78]
metathesis of propylene	correlation of activity with infrared spectra	[24]
metathesis of propylene	systematic variation of support pretreatment and activation treatment	[20]
metathesis of propylene	correlation of activity with TPDE	[21]
metathesis of propylene	correlation of activity with TPDE	[64]
metathesis of propylene	activity spectrum	[30]
metathesis of propylene	*in situ* IR study	[36]
metathesis of butenes	kinetics and selectivity	[79]
metathesis of pentene and iso-octene	mechanistic study	[69]
polymerization of ethylene	competition between polymerization, metathesis and exchange of $C_2H_4 + C_2D_4$	[36]
hydrogenation of CO	comparison of activities on fully and partially dehydroxylated supports	[31]
hydrogenation of propylene	activity spectrum	[71]
hydrogenation of propylene		[74]
hydrogenolysis of alkanes	turnover frequencies on partially and fully dehydroxylated supports	[74]
deuterium exchange of alkanes and alkenes	turnover frequencies	[75]
$Mo(CO)_6$–silica		
metathesis of propylene		[4]
metathesis of propylene	correlation of activity with IR spectra	[24]
metathesis of propylene	systematic variation of support pretreatment and activation treatment (also for $Mo(CO)_6$–MgO and $Mo(CO)_6$–silica/alumina)	[20]
metathesis of propylene	NO poisoning experiments	[50]
metathesis of propylene	activity spectrum and correlation with TPDE	[22]
hydrogenation of propylene	activity spectrum and correlation with TPDE	[22]
desulphurization of thiols	liquid-phase reaction	[80]

Table II (continued)

Catalyst and reaction	Comments	Reference
$Mo(CO)_6$–zeolite		
polymerization of ethylene	MoHY: variation of activity with pretreatment and activation conditions	[70]
polymerization of ethylene	MoHY: IR study	[72]
hydrogenation of ethylene		[72]
hydrodesulphurization	comparison of MoHY, MoNaY and CoMoHY with conventional catalysts	[58]
$W(CO)_6$–alumina		
metathesis of propylene		[4]
metathesis of propylene	comparison with other tungsten catalysts	[63]
metathesis of pentene	variation of support pretreatment, activation and added ligands	[45, 81]
hydrogenolysis of alkanes	comparison with molybdenum	[74]

3.2.1. *Olefin metathesis*

The olefin metathesis reaction has been the subject of numerous reviews since its discovery in 1964 (see for example ref. 60). A feature of the heterogeneously catalyzed metathesis reaction is the remarkably high activity shown by the supported carbonyl complexes $Mo(CO)_6$, $W(CO)_6$ and $Re_2(CO)_{10}$ (supported $Cr(CO)_6$ catalysts have no activity) [4]. In reviewing the published data, the following important questions must be addressed. (i) What are the active sites for metathesis over the catalysts prepared from $Mo(CO)_6$ and $W(CO)_6$? (ii) Why are the carbonyl-derived catalysts so much more active than conventional transition metal oxide catalysts?

The nature of the active sites for metathesis on both carbonyl and conventional catalysts has been the subject of many experiments and even more speculation. In the case of catalysts prepared from $Mo(CO)_6$ on alumina, Kemball *et al.* [61] first suggested that the active species may be a coordinatively unsaturated subcarbonyl complex which could coordinate two or more olefin molecules. This is an attractive concept in that it allows direct parallels to be drawn between heterogeneous and homogeneous metathesis catalysts. However, subsequent *in situ* IR experiments reported from the same laboratory revealed that the various subcarbonyl species formed upon mild activation were not perturbed at all on exposure to propylene, and no correlation was found between the metathesis activity and the concentrations of any of the subcarbonyl species [24]. Further evidence against identification of the active site as a subcarbonyl species includes the following: (a) silica-supported $Mo(CO)_6$ catalysts, which do not form stable subcarbonyl species, are more active on a per molybdenum basis than the alumina-supported catalysts [20]; (b) catalysts with comparable activity to that of the $Mo(CO)_6$ systems can be prepared from other organometallic complexes containing no carbonyl ligands, such as $W(CH_3)_6$, $Mo_2[CH_2Si(CH_3)_3]_6$ and $Mo_2(CH_3CO_2)_4$ [63]. This led to the notion that the active site in the $Mo(CO)_6$-derived catalysts is a fully decarbonylated molybdenum species of undefined valency state [24].

Brenner and Burwell [21, 64] subsequently returned to the concept of a subcarbonyl species as the active species by correlating the activity of $Mo(CO)_6$ on alumina catalysts with the various stoichiometric species identified in TPDE experiments. For example, the initial activity was found to increase in the order: '$Mo(CO)_3$' < '$Mo(CO)_3O_2$' < '$Mo(O^-)_2$' < 'oxidized $Mo(O^-)_2$', with the oxidized catalyst being about 36 times more active than the catalyst containing '$Mo(CO)_3$' [64]. Evidence was presented that the active sites in the oxidized catalysts are not the same as in those containing subcarbonyl species. For example, the activity of the '$Mo(CO)_3$' catalyst was enhanced on exposure to successive pulses of propylene, whereas that the oxidized catalyst was not. Exposure to carbon monoxide reduced the activity of the '$Mo(CO)_3$' catalyst but not that of the oxidized catalyst. However, these experimental results do not conclusively eliminate the possibility of small amounts of fully decarbonylated and possibly oxidized species being present in catalysts described as '$Mo(CO)_3$'. In the case of silica-supported $Mo(CO)_6$, Brenner *et al.* [22] concluded that two types of active site exist for metathesis: a subcarbonyl species formed near 100 °C and a fully decarbonylated mildly oxidized molybdenum species formed at higher activation temperatures (the activity passed through a maximum at activation temperatures between 100 °C and 200 °C).

A further IR study of the interaction of pentenes with catalysts prepared from $W(CO)_6$ or substituted $W(CO)_6$ complexes on alumina showed that adsorption of a sufficient quantity of olefin in this case completely removed all carbonyl bands except those due to $W(CO)_6$ [45]. It must be concluded at this point that there is no direct evidence for the existence of a coordinatively unsaturated subcarbonyl species as the active site for metathesis, although such species may well be a precursor of the active site.

If it is accepted that the role of the carbonyl complex precursors is to provide a source of fully decarbonylated low-valent molybdenum or tungsten, the question remains as to which valency state (or states) is active for metathesis. Smith *et al.* [20] first suggested that a common active site, possibly M^{4+}, may be formed in both the carbonyl-derived catalysts (by oxidation of zero-valent molybdenum or tungsten) and in the conventional oxide catalysts (by reduction from the +6 valency state). More recent evidence indicates however that different sites exist on the two different types of catalyst. For example, measurements of the activity of supported carbonyl complexes as a function of activation temperature have shown pronounced maxima between 100 °C and 250 °C [20, 22, 30, 45, 64]. Both EPR studies and hydrogen desorption measurements indicate that the average valency state remains close to zero under these conditions. Conventional supported MoO_3 or WO_3 catalysts, on the other hand, are difficult to reduce below the +4 valency state, yet show activity for metathesis after mild reduction [65–67]. The metathesis activity of a conventional molybdena–alumina catalyst was not poisoned by nitric oxide [68] in contrast to that of catalysts prepared from $Mo(CO)_6$ [50]. Hall *et al.* [68] concluded from their investigations of relative activities and poisoning of metathesis and hydrogenation that the active sites for metathesis on a conventional molybdena–alumina catalyst are rapidly formed at low extents of reduction, and correspond to a higher valency state of molybdenum than the sites for hydrogenation. In this context, a recent theoretical paper is of interest [69a]. Rappe and Goddard have suggested that the active site for metathesis on Group VI oxide catalysts is an oxo-alkylidene complex formed as follows:

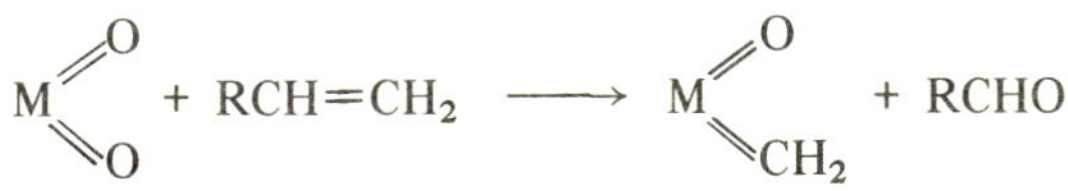

in which the transition metal is formally reduced from the +6 to the +4 valency state by the olefin. On low-valent metal ions, such as those formed from the carbonyl complexes, carbene formation believed to be necessary for metathesis [60], probably occurs via more traditional routes scuch as a 1,2-hydride shift in a coordinated olefin*:

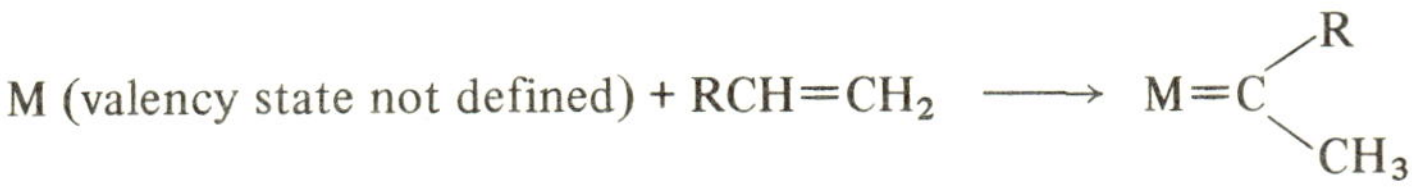

or through allyl and subsequent metallocyclobutane formation [69a]. The differences of orders of magnitude between the activities of organometallic-derived and conventional catalysts are not unexpected if in fact different reaction mechanisms are operating. Clearly, the nature of the active site in the organometallic catalysts needs further experimental and theoretical investigation.

* Very recently, Iwasawa *et al.* [69b] have demonstrated that the initial carbene is formed by methylene-hydrogen elimination followed by 1,2-hydride shift on the Mo^{4+} site derived from the π-allyl complexes (see article by Iwasawa in this volume).

3.2.2. *Olefin polymerization*

Olsthoorn and Moulijn [36] have reported that alumina-supported $Mo(CO)_6$ catalysts are active under certain conditions for the polymerization of ethylene (but not propylene or butene). The maximum activity has been found for $Mo(CO)_6$ on an alumina support pretreated at 250 °C and activated at 100 °C, but insufficient details were given to allow any conclusions to be drawn about the nature of the active sites.

The polymerization activity of $Mo(CO)_6$–HY zeolite catalysts has also been reported [70]. The activity varied with the average valency state of molybdenum (determined by titration with oxygen), reaching a maximum at an average valency state between +1 and +2. The limited data so far available suggest that the active species for polymerization may be the same as that for metathesis.

3.2.3. *Olefin hydrogenation*

The activity spectrum for the hydrogenation of propylene over an $Mo(CO)_6$–alumina catalyst is reported to show a maximum activity on activation at ca. 250 °C [71]. Brenner has suggested that the active site for propylene hydrogenation in this catalyst is approximately zero-valent molybdenum which becomes irreversibly oxidized at higher activation temperatures, causing loss of activity above 300 °C. Since the activity spectrum for hydrogenation is not similar to that for metathesis, different sites are probably involved. Furthermore, the activity of silica-supported $Mo(CO)_6$ catalysts is reported to be orders of magnitude lower than that of the alumina-supported catalysts [22], in contrast to the relative activities for metathesis. Burwell *et al.* [74] have subsequently reported that the hydrogenation activity of catalysts prepared on fully dehydroxylated alumina supports is ca. 10 times higher than that of the catalysts described by Brenner. This appear to support the suggestion that zero-valent molybdenum provides the active sites for hydrogenation. In contrast, Tanaka *et al.* [72] have recently found that catalysts prepared by the decomposition of $Mo(CO)_6$ in HY zeolite show a high activity for ethylene hydrogenation, although extensive oxidation of the molybdenum has occurred in this case. The question of what are the active sites for olefin hydrogenation over $Mo(CO)_6$-derived catalysts (and whether they are the same as the sites on conventional reduced MoO_3 catalysts [68]) cannot be satisfactorily answered at present.

3.2.4. *Hydrogenolysis of alkanes*

The hydrogenolysis of saturated hydrocarbons is usually carried out over supported Group VIII metal catalysts, and little is known about the activity of supported Group VI metals for this reaction, although evaporated molybdenum films do show some activity [73] for the hydrogenolysis of cyclopropane. Burwell *et al.* have reported that catalysts prepared by the decomposition of $Mo(CO)_6$ on fully dehydroxylated alumina supports are active for the hydrogenolysis of cyclopropane at 0 °C, and ethane, propane, butane, pentane and neopentane at 250 °C [74]. In this case, the active sites are almost certainly associated with zero-valent molybdenum. Similar activities were generated by reducing conventional molybdena–alumina and $Mo(CO)_6$ on partially dehydroxylated alumina catalysts in hydrogen at 800 °C; under these drastic reduction conditions both types of catalyst form metallic molybdenum. Zero-valent tungsten catalysts prepared from $W(CO)_6$ on dehydroxylated alumina were also active for hydrogenolysis.

3.2.5. *Deuterium exchange reactions*

Isotopic exchange reactions between hydrocarbons and deuterium have a considerable interest in catalysis because of the information they provide (in principle) about the activation of C—H bonds. The activity of catalysts prepared from $Mo(CO)_6$ on partially or fully dehydroxylated alumina supports for the exchange of deuterium with cyclopentane and neopentane has been investigated by Burwell *et al.* [75]. Maximum activity was shown by $Mo(CO)_6$ on fully dehydroxylated alumina catalysts activated at 300 °C (i.e. containing zero-valent molybdenum). The exchange patterns were primarily of the type resulting from exchange of a single hydrogen at a time, and the suggested reaction mechanism involves adsorption of the alkane at coordinatively unsaturated zero-valent molybdenum sites, followed by C—H bond cleavage:

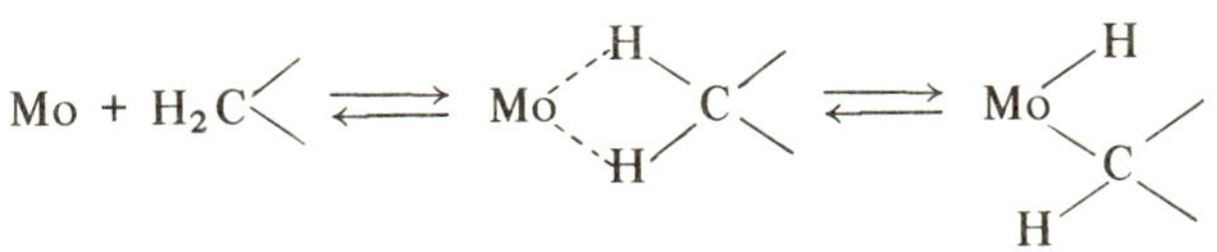

Catalysts containing zero-valent molybdenum also catalyzed the exchange and hydrogenation of propylene with deuterium, with exchange occurring before addition across the double bond.

A related reaction is the self-exchange between ethylene and perdeutero-ethylene. Catalysts prepared from $Mo(CO)_6$ on hydroxylated or partially dehydroxylated alumina have been reported [36] to show some activity for this reaction provided the activation temperature is kept low such that no oxidation occurs. However, the activity was in all cases lower than that for olefin metathesis (which produces CD_2CH_2 as the product in this case). In contrast, catalysts prepared from $Mo(CO)_6$ in both HY and NaY zeolites show a much higher activity for the self-exchange reaction than for metathesis, with a statistical distribution of all possible deutero-isomers being obtained in a few minutes at room temperature [73].

3.2.6. *Hydrogenation of carbon monoxide*

Zero-valent molybdenum catalysts produced by the decomposition of $Mo(CO)_6$ on fully dehydroxylated alumina supports are reported to be active for the hydrogenation of carbon monoxide at 300 °C [31]. The major product is methane, with lesser amounts of ethane and propane, and the turnover frequencies were stated to be about six times lower than those for nickel methanation catalysts under comparable conditions. Bowman and Burwell [31] have suggested that the active sites for methanation are slightly oxidized molybdenum clusters. Similar activity was shown by zero-valent tungsten catalysts similarly prepared. Conventional molybdena–alumina catalysts have a low activity for methanation [76]. The active sites in this case are likely to be the same as those on catalysts prepared from $Mo(CO)_6$ on hydroxylated supports [31, 33], which are an order of magnitude less active than the zero-valent molybdenum catalysts.

3.2.7. *Hydrodesulphurization*

Molybdenum-based catalysts are widely used for the hydrodesulphurization of hydrocarbon feedstocks. Vrinat *et al.* [58] have recently investigated the activity of molybdenum zeolite catalysts prepared from $Mo(CO)_6$ for the hydrodesulphurization of dibenzothiophene. The catalysts studied were MoHY (prepared by decomposition of $Mo(CO)_6$ in HY), CoMoHY (decomposition of $Mo(CO)_6$ in a partially ion-exchanged CoHY) and MoNaY (decomposition of $Mo(CO)_6$ in NaY). The MoNaY zeolite showed a lower activity than the NaY starting material, which was attributed to blockage of the zeolite pores by clusters of metallic molybdenum. MoHY showed a high initial activity which decayed rapidly with time on-stream. The initial activities of CoMoHY catalysts depended strongly on the particular pretreatment used, but in all cases gave the same steady-state activity. Vrinat *et al.* concluded that CoMoHY zeolites were the best hydrodesulphurization catalysts, due to some synergistic interaction between cobalt and molybdenum. The steady-state activities of the CoMoHY catalysts are, however, significantly lower (per gram of catalyst) than those of conventional alumina-supported molybdenum or cobalt–molybdenum catalysts. One factor not considered by these authors is that diffusion of the test molcule dibenzothiophene into the zeolite pores may well limit the apparent activity of the zeolite catalysts. A smaller test molecule such as thiophene may be more appropriate for investigating the hydrodesulphurization activity of molybdenum zeolites.

4. Group VIII Carbonyl Complexes

4.1. PREPARATION AND CHARACTERIZATION

4.1.1. *Fe(CO)₅–alumina*

The first detailed studies of the decomposition of $Fe(CO)_5$ on alumina supports were those of Brenner *et al.* [82, 83], who employed the TPDE technique to examine $Fe(CO)_5$, $Fe_2(CO)_9$ and $Fe_3(CO)_{12}$ on a partially dehydroxylated alumina support. Activation of $Fe(CO)_5$ on alumina at 120 °C caused the loss of approximately three CO ligands per iron atom, without significant evolution of hydrogen. The formation of the $Fe(CO)_2$ subcarbonyl species at this point in the activation sequence was proposed, the decarbonylation being capable of partial reversal by adding carbon monoxide. Activation at higher temperatures (above 300 °C) caused complete and irreversible decarbonylation, and the evolution of one hydrogen molecule per complex species, indicating that the zero-valent iron is oxidized to the +2 valency state by the hydroxyl groups on the support. The analogous decomposition of $Fe_2(CO)_9$ and $Fe_3(CO)_{12}$ will not be considered further here, except to note that the subcarbonyl species formed from the cluster complexes were more stable than those formed from $Fe(CO)_5$ (i.e. complete decarbonylation required higher activation temperatures). The major conclusion reached by Brenner *et al.* is that decomposition of the carbonyl complexes of iron on alumina at low temperatures (to prevent oxidation) produces much higher dispersions of zero-valent iron than can be achieved by reduction of the conventional supported oxide catalysts.

An IR spectroscopic study [84] has shown that the initial interaction of $Fe(CO)_5$ with a hydroxylated alumina support involves nucleophilic attack of the support hydroxyl groups to give a hydrido cluster anion:

$$3\,Fe(CO)_5 + AlOH^- \longrightarrow Al^+[HFe_3(CO)_{11}]^- + 3\,CO + CO_2$$

Evidence for cluster formation included the fact that subsequent addition of HCl produced $Fe_3(CO)_{12}$ (a known reaction of $[HFe_3(CO)_{11}]^-$ in solution), and that an identical surface species was formed from adsorbed $Fe_3(CO)_{12}$. It should be noted, howver, that the $[HFe_3(CO)_{11}]^-$ species does not correspond to the $Fe(CO)_2$ stoichiometry determined from TPDE experiments. Furthermore, EPR measurements by Basset *et al.* [86] have revealed that the end-products of decarbonylation on both hydroxylated and partially dehydroxylated supports are largely metallic iron (giving a characteristic ferromagnetic resonance signal), with partial oxidation to Fe^{3+} observed on fully hydroxylated supports.

4.1.2. *Fe(CO)$_5$–magnesia*

The initial interaction of $Fe(CO)_5$ with hydroxylated MgO supports is reported [84] to resemble closely that with alumina, forming the same $[HFe_3(CO)_{11}]^-$ species. Cluster formation also occurs [85] on fully dehydroxylated MgO supports, but at a later stage in the activation sequence. In this case, the initial interaction is believed to involve formation of a mononuclear dianion analogous to that formed by the Group VI carbonyl complexes on fully dehydroxylated MgO.

$$Mg^{2+}O^{2-} + Fe(CO)_5 \longrightarrow [(CO)_4FeCO_2]^{2-}Mg^{2+}$$

The end-product of activation on fully dehydroxylated MgO is zero-valent iron. Ferromagnetic resonance spectra [86] have revealed the presence of superparamagnetic (and thus very small) clusters of zero-valent iron on the hydroxylated or partially dehydroxylated supports, although as in the case of alumina, some oxidation to Fe^{3+} also apparently occurs.

4.1.3. *Fe(CO)$_5$–zeolite Y*

A detailed IR spectroscopic study of the adsorption and decomposition of $Fe(CO)_5$ in the hydrogen form of zeolite Y has been described by Ballivet-Tkatchenko and Coudurier [87]. Initial adsorption of $Fe(CO)_5$ from the vapour phase at 25 °C gave bands assigned to $Fe(CO)_5$ and a monosubstituted $Fe(CO)_4L$ species, where L represents an oxide or hydroxyl group of the zeolite. On heating to 60 °C *in vacuo*, the IR spectrum became identical to that of the cluster complex $Fe_3(CO)_{12}$ adsorbed in the zeolite, but the original spectrum could be restored by subsequent exposure to CO. Cluster information from $Fe(CO)_5$ via $Fe(CO)_4$-type intermediates is known to occur in solution and in low-temperature matrices; similar reactions were postulated to take place within the supercages of zeolite HY

$$2\,Fe(CO)_5 + Fe(CO)_4 \underset{CO}{\overset{-CO}{\rightleftarrows}} Fe_3(CO)_{12}$$

On heating to higher temperatures, gradual decarbonylation occurred via a series of undefined intermediates, such decarbonylation being complete at 200 °C. At the same time hydrogen plus traces of hydrocarbons and water were evolved, due to oxidative reactions with zeolite protons. The final products of activation were believed to be Fe^{2+} cations (identified from their interaction with adsorbed

NO and CO) which migrate into hidden sites as the activation temperature is raised. Ballivet-Tkatchenko and Coudurier point out that the preparation of Fe^{II} Y zeolites via this indirect method of oxidative decomposition of adsorbed carbonyl complexes offers several advantages over direct ion exchange: the crystal structure is not damaged, the Fe^{2+} ions are located in more accessible sites (at least for low-activation temperatures), and counteranions (which may be trapped as impurities in solution ion exchange) are not involved.

Conflicting conclusions about the end-products of the activation of $Fe(CO)_5$ in HY have been reached by Derouane *et al.* [88, 89]. These authors studied both thermal and photochemical decompositions of $Fe(CO)_5$ in HY, and determined that metallic iron was produced in both cases. The ^{13}C NMR spectrum of $Fe(CO)_5$ initially adsorbed in HY at room temperature was similar to that of the pure liquid, indicating the existence of a weak interaction only with the zeolite (a heat of adsorption of about 8 kJ mol^{-1} was estimated from the temperature dependence of the NMR line width). Subsequent evacuation up to 70 °C caused a gradual decrease in intensity, increase in line width and shift in frequency (to higher chemical shift) of the ^{13}C resonance. These changes, which could be partially reversed by exposure to CO, were attributed to stepwise loss of carbonyl ligands from adsorbed $Fe(CO)_5$:

$$Fe(CO)_5 \underset{CO}{\overset{-CO}{\rightleftharpoons}} Fe(CO)_4 \underset{CO}{\overset{-CO}{\rightleftharpoons}} Fe(CO)_3, \text{etc.}$$

The increase in chemical shift was attributed to an increase in the extent of back-bonding from iron to carbonyl ligands as the ligands are progressively removed. Complete decarbonylation (above 70 °C in the NMR experiment) could not be reversed by adding CO. Fully decarbonylated samples gave a ferromagnetic resonance signal characteristic of metallic iron, and electron microscopic examination revealed the presence of iron crystallites between 1 nm and 5 nm in size.

The photochemical decomposition of $Fe(CO)_5$ in HY at room temperature caused, in contrast with the thermal decomposition, a gradual decrease in the ^{13}C chemical shift as decarbonylation proceeded. This was attributed to a stronger interaction with the zeolite in the case of photochemical decomposition, resulting in net electron transfer from iron to the zeolite and hence reduced back-bonding to the carbonyl ligands. No metallic iron crystallites (larger than 1 nm) could be detected by electron microscopy following photochemical decarbonylation. Derouane *et al.* concluded that photochemical decomposition produces very small clusters of metallic iron within the zeolite, too small to be detected by electron microscopy.

Production of zero-valent iron clusters on photochemical decompositon of adsorbed $Fe(CO)_5$ is not unexpected, since oxidation by the zeolite protons is unlikely to occur at room temperature. Further consideration must be given, however, to the apparent disagreements between the two different spectroscopic studies of the thermal decomposition. One relevant factor is that the NMR experiment will not necessarily detect all carbonyl species present. It is well known that NMR spectra of adsorbed species can only be observed in the usual single-pulse experiments if dipolar coupling and chemical shift anisotropies are averaged to zero, either by rapid motion of the adsorbed species or by exchange between strongly held and weakly interacting species [90]. Thus, the changes observed by Derouane *et al.* in the ^{13}C NMR spectra as adsorbed $Fe(CO)_5$ is gradually heated *in vacuo* might be attributed to the clustering reaction observed in the IR spectroscopic experiments, particularly if the $Fe_3(CO)_{12}$ formed is interacting strongly with the zeolite support and as a result does not contribute to the measured ^{13}C NMR spectrum.

The formation of metallic iron in the one case, and Fe^{2+} in the other, is not so readily accounted for. However, if a competition exists between sintering of the $Fe_3(CO)_{12}$ species produced initially from $Fe(CO)_5$ and oxidative decarbonylation to produced dispersed Fe^{2+} ions, the relative concentrations of metallic and oxidized species produced on activation may depend on several experimental variables, such as the amount of $Fe(CO)_5$ initially adsorbed in the zeolite and the heating rate employed. In NaY zeolites, where no oxidizing sites exist, there is no doubt that the end-product of activation will be metallic iron [91, 92].

4.1.4. *$Fe(CO)_5$–graphite*

Detailed Mössbauer and volumetric studies have been reported of the thermal decomposition of $Fe(CO)_5$ on graphite supports [93, 94]. Only two iron-containing species were detected by Mössbauer spectroscopy: physically adsorbed $Fe(CO)_5$ and the end-product of decomposition, metallic iron. It is noteworthy that no intermediate subcarbonyl species, either mononuclear or clustered, were detected on graphite. The activation treatment necessary to fully decompose the adsorbed $Fe(CO)_5$ was found to depend strongly on the pretreatment that the support had received. Quantitatively, the rate of decomposition as a function of support pretreatment temperature passed through a maximum at ca. 500 °C. Dumesic *et al.* suggested from this and other evidence that decomposition of $Fe(CO)_5$ occurs at oxygen free edge or defect sites. The increase in rate of decomposition as the support pretreatment temperature is raised to 500 °C is then attributed to either an increase in the reactivity of the decomposition sites (due to removal of adsorbed oxygen) or to an increase in the rate of diffusion of adsorbed $Fe(CO)_5$ to the decomposition sites as adsorbed water is removed. The decline in decomposition rate at high pretreatment temperatures is due to annealing of dislocations in the graphite basal planes.

The nature of the metallic iron crystallites produced from $Fe(CO)_5$ was also investigated by Dumesic and co-workers. The Mössbauer parameters for the metallic phase were not identical to those of bulk iron. From computer simulations of the Mössbauer spectra obtained after partial and complete decomposition of $Fe(CO)_5$ on graphite supports pretreated at different temperatures, it was concluded that in all cases metallic 'rafts' of iron of the order of 1 nm thick and 10 nm in lateral dimension are formed. Carbon monoxide chemisorption measurements supported this conclusion (one CO molecule was adsorbed for every two Fe atoms). The unusual (but not unique) morphology of iron crystallites prepared from $Fe(CO)_5$ may be contrasted with that of graphite-supported iron catalysts prepared by conventional (impregnation and reduction) techniques.

4.1.5. *$Ni(CO)_4$–alumina*

Parkyns [3] reported the first IR spectroscopic study of the decomposition of a carbonyl complex on alumina, viz. $Ni(CO)_4$. He found that $Ni(CO)_4$ was both physically adsorbed and dissociatively chemisorbed on a partially dehydroxylated alumina support. The dissociatively chemisorbed species were identified as linear and bridged forms of CO chemisorbed on highly dispersed nickel crystallites, possibly as small as one or two atoms. On subsequent exposure to oxygen, the nickel became oxidized, and CO desorbed and either oxidized to CO_2 or re-adsorbed on nickel oxide sites. The dissociatively adsorbed $Ni(CO)_4$ was thus considered to be a direct intermediate in the alumina-catalyzed oxidation of $Ni(CO)_4$ vapour. It should be noted that a similar chemistry did not occur on a silica support; physically adsorbed $Ni(CO)_4$ was the only species observed in this case.

A ^{13}C NMR spectroscopic study of $Ni(CO)_4$ adsorbed on partially dehydroxylated alumina supports was reported by Derouane *et al.* [95]. The physically adsorbed complex showed a ^{13}C chemical shift which was slightly lower than that of the free molecule, an observation attributed to weak interaction with the support. Decomposition at 200 °C produced metallic nickel crystallites 5–15 μm in size. Closer examination revealed that the crystallites consisted of ensembles of relatively uniform spherical clusters 50–200 nm in size.

TPDE measurements on the $Ni(CO)_4$–alumina system have been described by Bjorklund and Burwell [96]. These authors found, like Parkyns, that $Ni(CO)_4$ is both physically adsorbed and dissociatively chemisorbed on alumina supports, the chemisorbed species having a characteristic bright yellow colour. Exposure to carbon monoxide reversed the chemisorption reaction, and all of the initially adsorbed $Ni(CO)_4$ could be recovered from the support in flowing carbon monoxide. Decomposition of the chemisorbed species to metallic nickel was observed to begin at temperatures as low as 50 °C, and on heating to 100 °C a total of ca. 3.6 CO molecules per nickel atom were evolved. At higher activation temperatures hydrogen evolution occurred, to an extent depending on the degree of hydroxylation of the support. For example, activation of $Ni(CO)_4$ on a fully hydroxylated support at 500 °C produced 0.9 hydrogen molecules per nickel atom, indicating that almost all of the metallic nickel formed initially reacts with surface hydroxyl groups:

$$Ni + 2\,OH \longrightarrow Ni^{2+} + 2\,O^- + H_2$$

On partially dehydroxylated aluminas, 0.3–0.4 H_2 molecules per Ni atom were evolved, while no hydrogen evolution at all was detected on fully dehydroxylated supports.

Bjorklund and Burwell compared the chemistry of $Ni(CO)_4$ on alumina with that of $Mo(CO)_6$. In both cases the initial interaction with the support involves subcarbonyl species [$Mo(CO)_5$ or $Ni(CO)_3$] which are reconverted to the parent complexes on addition of carbon monoxide. The further decomposition of the subcarbonyl species differs however; relatively stable lower subcarbonyl intermediates of stoichiometry $M(CO)_4$ and $M(CO)_3$ are formed from $Mo(CO)_5$, whereas $Ni(CO)_3$ species agglomerate to form metallic nickel covered with carbon monoxide. The metallic nickel is subsequently oxidized to Ni^{2+} if the support contains sufficient hydroxyl groups, i.e. metallic nickel is a definite intermediate in the activation sequence. With $Mo(CO)_6$, on the other hand, final decarbonylation on hydroxylated supports is accompanied by oxidation. Only on fully dehydroxylated alumina supports is a similar chemistry found for $Mo(CO)_6$ and $Ni(CO)_4$. The same chemistry for the two complexes is also reported to exist on fully dehydroxylated MgO supports [59].

4.1.6. *Ni(CO)₄–zeolite*

The ^{13}C NMR spectra of $Ni(CO)_4$ adsorbed in NaY and HY zeolites are reported to differ significantly [95]. In NaY zeolites, the ^{13}C chemical shift is higher than that of the free molecule, which was attributed by Derouane *et al.* to interaction of the $Ni(CO)_4$ with electron donor sites (lattice oxide ions) in the zeolite. In HY zeolites, on the other hand, the chemical shift is lower than that of the free molecule, and the ^{13}C signal is a doublet due to dipolar coupling with a zeolite proton. The adsorbed $Ni(CO)_4$ must be interacting directly in this case with a zeolite hydroxyl group in the supercage. A detailed investigation of the decomposition of $Ni(CO)_4$ in NaY and HY zeolites has not been reported as yet.

The decomposition of $Ni(CO)_4$ in NaX zeolite to produce metallic nickel has been described by Galinski *et al.* [97]. Infrared spectra indicated that the decomposition occurs via a complex series of intermediate species and is only complete at 300 °C. The dispersion of the metallic nickel produced was measured by oxygen chemisorption, O/Ni ratios of between 0.5 and 0.6 were typical. Subsequent reduction in hydrogen caused further sintering.

4.2. CATALYTIC PROPERTIES

With the exception of a report of benzene hydrogenation over nickel zeolite catalysts prepared from $Ni(CO)_4$ [97], all catalytic work so far undertaken with supported mononuclear Group VIII carbonyl complexes has been concerned with the Fischer–Tropsch synthesis. The importance of the Fischer–Tropsch reaction needs no further emphasis here [98]. Several research groups have investigated the possibility of improving the activity and/or selectivity of supported iron Fischer–Tropsch catalysts by preparing them from carbonyl precursors.

Basset *et al.* [86] reported the activity and selectivity of catalysts prepared from $Fe(CO)_5$ on a variety of oxide supports. The $Fe(CO)_5$ was adsorbed from pentane solution onto the fully hydroxylated supports in an inert atmosphere and the catalysts then activated by heating to reaction temperature (typically 270 °C) in a 1 : 1 mixture of CO and H_2. Table III summarizes some of the data obtained. The initial activities of the $Fe(CO)_5$-derived catalysts were all comparable in magnitude (taking into account the different iron loadings)

TABLE III
Fischer–Tropsch activities and selectivities of supported $Fe(CO)_5$ catalysts[a]

Catalyst (wt.%)	Conversion at 270 °C (%)	Selectivity to hydrocarbons (%)	Olefin fraction (%)	Fraction $> C_5$ (%)
$Fe(CO)_5-Al_2O_3$ (1.85)	3.3	100	42.8	9
$Fe(CO)_5-La_2O_3$ (0.30)	3.4	44.2	58.1	11.5
$Fe(CO)_5-La_2O_3$ + pyridine	6.1	44.1	62.8	13.5
$Fe(CO)_5-SiO_2$ (2.80)	5.1	42.6	68.9	7
$Fe(CO)_5-MgO$ (0.50)	1.4	100	60.3	9
$Fe-Al_2O_3$ (conventional catalyst, 8.1%)	19.4	62.6	38.4	35.9

[a] Data from ref. 86.

with that of a conventional alumina-supported iron catalyst. However, there are significant differences in initial selectivity. The $Fe(CO)_5$-derived catalysts were, in general, more selective for lower molecular weight olefins than the conventional catalyst. This high initial selectivity was reported to decrease after several hours on-stream, while at the same time the specific activity increased.

Basset *et al.* concluded that the high selectivity for low molecular weight olefins is associated with the small crystallite sizes of metallic iron produced from $Fe(CO)_5$. The gradual loss of selectivity is then attributed to progressive sintering of the iron crystallites under the reaction conditions. The fact that the activity increases at the same time undoubtedly has mechanistic implications which should be explored further. Variation of the support (Table III) appears to affect only the total selectivity to hydrocarbons and not the olefin fraction.

A useful comparison can be made with the activity and selectivity data reported by Ballivet-Tkatchenko *et al.* [91] for iron zeolite Fischer–Tropsch catalysts prepared from $Fe_3(CO)_{12}$. The comparison is valid, since initial adsorption of $Fe(CO)_5$ in zeolite Y produces $Fe_3(CO)_{12}$ [87]. Catalysts prepared from $Fe_3(CO)_{12}$ in HY zeolite showed no Fischer–Tropsch activity at all, presumably due to the fact that the iron is completely oxidized by zeolite protons during activation. In the case of NaY zeolite supports, Fischer–Tropsch activity was exhibited by catalysts fully decarbonylated *in vacuo* prior to exposure to CO and H_2, and by catalysts containing the parent complex exposed directly to the reactants at the reaction temperature (250 °C). The selectivities of the two types of catalyst were however significantly different. Catalysts which had not been previously decarbonylated gave an approximately Schulz–Flory distribution of hydrocarbons up to C_9, with negligible quantities of hydrocarbons higher than C_9 being formed. The decarbonylated catalysts gave products deviating from a Schulz–Flory distribution, with much higher yields of C_6–C_{10} hydrocarbons.

Ballivet-Tkatchenko *et al.* considered the selectivity of their iron zeolite catalysts to be determined by the location of the iron crystallites within or on the external surface of the zeolite. In catalysts which had not been decarbonyllated, the zero-valent iron clusters presumably remain within the zeolite supercages, and the absence of products beyond C_9 can be attributed to steric restraints imposed by the zeolite framework (similar shape-selective effects have been reported for ruthenium–zeolite Fischer–Tropsch catalysts [98]). In the decarbonylated catalysts, on the other hand, iron crystallites several tens of nanometres in diameter are formed on the external surface of the zeolite, and higher molecular weight products may then be formed. It would appear, however, that the possibility of selectivity being determined by iron crystallite size, as suggested for the oxide-supported $Fe(CO)_5$ catalysts, cannot be ruled out as an alternative to the steric explanation offered by Ballivet-Tkatchenko and co-workers.

Mention should also be made of the preparation of alumina-supported Fischer–Tropsch catalysts by McVicker and Vannice [99] from the carbonyl complex $K_2Fe(CO)_4$. Conventional iron catalysts can be promoted by the addition of potassium. McVicker and Vannice argued that more efficient promotion might be achieved by incorporating potassium into the catalyst precursor. Catalysts prepared from $K_2Fe(CO)_4$ on alumina (decomposed in hydrogen at 500 °C) were in fact found to be both more active than a conventional catalyst containing the same amount of iron and potassium, and more selective for C_3 and C_4 olefins. However, similar effects were not found for silica-supported $K_2Fe(CO)_4$, nor for the alumina-supported iridium and ruthenium catalysts prepared from the corresponding potassium carbonylate salts.

5. Group V Carbonyl Complexes

Although $V(CO)_6$ is less stable than the Group VI carbonyl complexes (Table I), it is volatile and stable at room temperature. There is no reason, therefore, why supported vanadium catalysts should not be prepared from $V(CO)_6$ via vapour-phase adsorption onto oxide supports or into zeolites in a similar manner to the catalysts described above. In the only study reported to date, Brenner and Hucul [33] have briefly described a TPDE experiment with alumina-supported $V(CO)_6$. Decarbonylation of the supported complex produced a small amount of methane (0.1 CH_4 per vanadium) comparable in magnitude to that formed from alumina-supported Group VI carbonyl complexes. The chemistry of supported $V(CO)_6$ may thus resemble that of the Group VI complexes, but further investigation of this system is clearly necessary.

6. Concluding Remarks

This review has attempted to summarize the wide range of chemistry associated with supported mononuclear carbonyl complexes. Only a few systems [$Mo(CO)_6$ in particular] have been studied in detail to date, but several general comments can be made at this stage:

1. The formation of stable carbonyl species as intermediates in the decomposition of supported carbonyl complexes depends strongly on the nature of the support. Intermediate species appear to be stabilized on alumina and magnesium oxide supports, and in zeolites, but not on silica or graphite.
2. Despite earlier claims to the contrary, there exists no firm evidence at this time that subcarbonyls are themselves active sites for any catalytic reaction. The various carbonyl species appear to be functioning as precursors of the catalytically active species.

3. There appear to be two alternative pathways available to supported mononuclear carbonyl complexes upon decarbonylation: oxidation by the support protons to produce transition metal cations and ultimately sites similar to those found on supported transition metal oxide catalysts, or aggregation of zero-valent fragments to produce metallic clusters, often more highly dispersed than in conventional supported metal catalysts. The pathway followed by a particular complex appears to depend on the availability and reactivity of the support protons, and the tendency of the transition metal concerned to undergo cluster formation. In some cases both oxidized and zero-valent species may be formed, depending on experimental parameters such as the amount of complex adsorbed on the support and the rate of activation.

Future research in this area of heterogeneous catalysis must involve further characterization of the various supported complexes with as many different techniques as possible, and wider investigation of the catalytic properties of the unusual materials produced from carbonyl precursors. A major objective is the production of more active and selective catalysts for reactions such as the Fischer–Tropsch synthesis. A second objective to which research in this area may also be directed is a better understanding of the phenomenon of catalyst sintering. The production of mononuclear fragments from the supported carbonyl complexes provides a way of studying the initial stages of cluster formation and growth under mild conditions, particularly in well-defined environments such as zeolite supports.

7. Literature Update

Since this chapter was written in 1982, a number of further papers have appeared on the subject of mononuclear carbonyl complexes on inorganic supports. A brief account of the most recent work is given here.

$Fe(CO)_5$ continues to be a popular precursor for the preparation of supported iron catalysts. Kazusaka *et al.* [100] have reported that the oxidation of zero-valent iron by hydroxyl groups on an alumina support during decomposition of adsorbed $Fe(CO)_5$ could be inhibited by pretreatment of the alumina with KOH. Nitric oxide was used as a spectroscopic probe of the valency state of iron, an IR band at 1720 cm^{-1} being characteristic of NO chemisorbed on zero-valent iron. The authors suggest that KOH treatment results in the replacement of the hydroxyl protons by K^+, thus removing the oxidizing sites. As expected, catalysts prepared by the decomposition of $Fe(CO)_5$ on the treated alumina were much more active for the hydrogenation of ethylene than those prepared from untreated alumina (which contained predominantly Fe^{2+}).

The adsorption of $Fe(CO)_5$ vapour in a variety of different zeolite structures has been investigated in detail by Bein and Jacobs [101]. For the large pore zeolites Y, L and ω, exposure to excess $Fe(CO)_5$ vapour resulted in completed filling of the zeolite cavities (ca. three molecules per cavity for the sodium, caesium and hydrogen forms of zeolite Y, and 1.5 per unit cell or per channel for zeolites ω and L, which have a one-dimensional pore geometry). The adsorption was largely irreversible; ca. 70% of the adsorbed $Fe(CO)_5$ remained after prolonged outgassing at room temperature. Infrared spectra suggested that $Fe(CO)_5$ is adsorbed largely intact. Only 0.25 molecules per channel were adsorbed into sodium mordenite, and the authors suggest that this may be due to blockage of the entrances of the one-dimensional pores in this particular zeolite by the first $Fe(CO)_5$ molecules to be adsorbed. The aluminium-free zeolite silicalite adsorbed only 0.05 molecules per channel; in this case the adsorption was believed to occur only on the external surfaces of the silicalite crystallites.

A detailed study of the low-temperature photolysis of $Fe(CO)_5$ adsorbed onto silica has been reported by Trusheim and Jackson [102, 103]. $Fe(CO)_5$ is adsorbed reversibly on fully hydroxylated silica surfaces, and exposure to higher pressures of $Fe(CO)_5$ vapour causes multilayer condensation in the silica pores. Irradiation with UV light from N_2 or Nd/YAG lasers caused rapid conversion of the adsorbed $Fe(CO)_5$ to $Fe_3(CO)_{12}$, as shown by changes in the UV–vis and IR spectra, and evolution of CO. By carrying out the photolysis below room temperature, an intermediate species identified as $Fe(CO)_4$ (silica) could be observed. The adsorbed $Fe_3(CO)_{12}$ could subsequently be removed from the silica surface by prolonged evacuation at room temperature. In this particular system, the adsorbed carbonyl complexes must interact only weakly with the silica support, in contrast to the strong interactions of $Fe(CO)_5$ with alumina or MgO (Section 4.1). The use of laser photolysis to induce a selective cluster synthesis without decomposition or sintering clearly deserves wider application.

TiO_2-supported iron catalysts have been prepared by Dumesic *et al.* [104] by the decomposition of adsorbed $Fe(CO)_5$. The TiO_2 support is of particular interest in view of the current debate over 'strong metal–support interactions' (SMSI) that exist between Group VIII metals and TiO_2 [105]. Mössbauer spectroscopy was employed to show that the decomposition of $Fe(CO)_5$ on TiO_2 by heating *in vacuo* at 380 K produced a mixture of Fe^{2+} (66%) and zero-valent iron. The zero-valent iron may have been an iron subcarbonyl species, but the Mössbauer experiment was incapable of detecting the presence of carbonyl ligands. At higher temperatures, all of the iron present appeared as Fe^{2+}, i.e. oxidation by surface hydroxyl groups attached to the TiO_2 support becomes more complete as the decomposition temperature is increased. Subsequent reduction in H_2 at 700 K produced metallic iron particles with Mössbauer parameters identical to those of bulk iron, although the dispersion of the iron on TiO_2 was significantly higher than that achieved with conventional Fe–TiO_2 catalysts (as measured by CO adsorption). The kinetic parameters for ammonia synthesis over Fe–TiO_2 catalysts prepared from $Fe(CO)_5$ by reduction at 700 K were found to be similar to those for conventional iron catalysts supported on MgO or SiO_2.

Two pieces of evidence are presented by Dumesic *et al.* for the onset of SMSI when $Fe(CO)_5$-derived catalysts are reduced at higher temperatures (above 770 K). The activation energy and reaction order with respect to ammonia in the ammonia synthesis reaction both increase, and the chemisorption of CO decreases by an order of magnitude [106]. The reduction temperature at which this occurs corresponds to that needed to produce SMSI in conventional Fe–TiO_2 catalysts, i.e. such SMSI does not depend on the natureof the iron precursor. The only major difference between $Fe(CO)_5$-derived catalysts and those prepared conventionally is the smaller particle sizes of iron achieved from the carbonyl.

Further catalytic studies of the alumina-supported Group VIB carbonyls $Cr(CO)_6$, $Mo(CO)_6$ and $W(CO)_6$ have been reported by Thomas and Brenner [107]. The activity of these materials for the hydrogenation of olefins has been measured as a function of the activation treatment. Optimum activation (at 200 °C in flowing helium) produces hydrogenation activities significantly higher than those of conventional hydrogenation catalysts (such as the noble metals). Thomas and Brenner suggest that this is due to the facility with which high dispersions of low valency states may be prepared from carbonyl precursors.

An interesting development in the preparation of molybdenum zeolites from $Mo(CO)_6$ is the characterization of bimetallic Pt–Mo zeolite catalysts by Gallezot *et al.* [108]. These authors first prepared a PtHNaY zeolite by ion exchange of NaY with an ammonia solution of $PtCl_2$, calcination and subsequent reduction in H_2 to produce PtHNaY containing 1 nm platinum particles trapped in the zeolite cages. This material was then exposed to $Mo(CO)_6$ vapour at 340 K for 12 h and reduced in H_2 at 600 K to decompose the adsorbed carbonyl complex. Characterization of the resulting catalysts was undertaken by means of electron microscopy, chemisorption of CO and H_2, IR spectroscopy, X-ray adsorption edge spectroscopy and XPS. Part of the added molybdenum interacts directly with the platinum particles, producing a small increase in the average particle size and a progressive decrease in the amounts of CO and H_2 chemisorbed on platinum. The initially electron-deficient platinum particles become closer to bulk platinum as molybdenum atoms are added, presumably due to shielding from the electron-deficient sites of the zeolite support. A part of the molybdenum which does not interact with platinum becomes oxidized by zeolite protons in the usual way (see Section 3.1).

The direct interaction of the two metallic components in the PtMoY zeolites may resemble that in the grafted catalysts prepared by Yermakov and Kuznetsov by sequential reaction of the complexes $Pt(\pi\text{-}C_4H_7)_2$ and $Mo(\pi\text{-}C_3H_5)_4$ with silica supports [109]. In any event, interesting catalytic properties may be anticipated for the modified platinum clusters.

Finally, a preliminary study has been made of the adsorption and decomposition of the hitherto unpopular Group V complex $V(CO)_6$ (see Section 5 above) in zeolites X and Y [110]. Adsorption of $V(CO)_6$ in dehydrated NaX and NaY zeolites from the gas phase results in formation of the stable anion $V(CO)_6^-$, as demonstrated by IR and UV–vis spectroscopy. Decomposition of the $V(CO)_6^-$ on heating *in vacuo* is believed to produce zero-valent vanadium, although the state of metallic dispersion within the zeolite is not yet known. The corresponding adsorption of $V(CO)_6$ in HY zeolite produces a complex mixture of carbonyl species which may include the intact neutral molecule, although an EPR spectrum of the paramagnetic $V(CO)_6$ could not be measured at 77 K. In this case, decomposition produced oxidized vanadium species through reaction with zeolite protons, particularly the paramagnetic VO^{2+} which was observed by EPR spectroscopy.

Chemistry Department,
University of Auckland,
Private Bay,
Auckland, New Zealand

References

1. L. J. Baker and R. B. Bernstein: *J. Am. Chem. Soc.* **73**, 4434 (1951).
2. A. Terenin and L. Roev: *Actes du Deuxieme Congres International de Catalyse*, Vol. 2, p. 2183, Editions Technip, Paris (1961).
3. N. D. Parkyns: *Proc. 3rd Int. Congr. Catal. Amsterdam (1964)*, Vol. 2, p. 914, North-Holland, Amsterdam (1965).
4. R. L. Banks and G. C. Bailey: *Ind. Eng. Chem., Prod. Res. Dev.* **3**, 170 (1964).
5. D. C. Bailey and S. H. Langer: *Chem. Rev.* **81**, 110 (1981).
6. J. R. Anderson: *Structure of Metallic Catalysts*, Chap. 2, Academic Press, London (1975).
7. R. K. Iler: *The Chemistry of Silica*, Wiley, New York (1979).
8. H. Knözinger and P. Ratnasamy: *Catal. Rev.* **17**, 31 (1978).
9. J. B. Peri: *J. Phys. Chem.* **69**, 211 (1965).
10. J. B. Peri: *J. Phys. Chem.* **69**, 220 (1965).
11. N. D. Parkyns: *J. Phys. Chem.* **75**, 526 (1971).
12. C. Morterra, A. Zecchina, S. Coluccia and A. Chiorino: *J. Chem. Soc., Faraday Trans. I* **73**, 1544 (1977).

13. See, for example, D. W. Breck: *Zeolite Molecular Sieves*, Wiley-Interscience, New York (1974).
14. See, for example, B. Linsen (ed.): *Physical and Chemical Aspects of Adsorbents and Catalysts*, Academic Press, New York (1970).
15. P. Anderson, R. Horlock and J. Oliver: *Trans. Faraday Soc.* **61**, 2754 (1965).
16. J. V. Evans and T. L. Whateley: *Trans. Faraday Soc.* **63**, 2769 (1967).
17. R. J. Tench, D. Giles and J. F. Kibblewhite: *Trans. Faraday Soc.* **67**, 854 (1971).
18. H. P. Boehm: *Adv. Catal.* **16**, 179 (1966).
19. M. Bretz: *Phys. Rev.* **A8**, 1589 (1973).
20. J. Smith, R. F. Howe and D. A. Whan: *J. Catal.* **34**, 191 (1974).
21. R. L. Burwell and A. Brenner: *J. Mol. Catal.* **1**, 77 (1975).
22. A. Brenner, D. A. Hucul and S. J. Hardwick: *Inorg. Chem.* **18**, 1478 (1979).
23. E. S. Davie, D. A. Whan and C. Kemball: *J. Chem. Soc., Chem. Commun.* 1430 (1969).
24. R. F. Howe, D. E. Davidson and D. A. Whan: *J. Chem. Soc., Faraday Trans. I* **68**, 2266 (1972).
25. R. F. Howe and I. R. Leith: *J. Chem. Soc., Faraday Trans. I* **69**, 1967 (1973).
26. D. A. Whan, M. Barber and P. Swift: *J. Chem. Soc., Chem. Commun.* 198 (1972).
27. A. Brenner and R. L. Burwell: *J. Catal.* **52**, 353 (1978).
28. A. Brenner and R. L. Burwell: *J. Am. Chem. Soc.* **97**, 2565 (1975).
29. R. F. Howe: *Inorg. Chem.* **15**, 486 (1976).
30. A. Brenner and D. A. Hucul: *Proc. 3rd Int. Conf. Chem. Uses of Molybdenum*, Climax Molybdenum Co., London (1979).
31. R. G. Bowman and R. L. Burwell: *J. Catal.* **63**, 643 (1980).
32. C. DeFosse, M. Laniecki and R. L. Burwell: *Proc. 7th Int. Congr. Catal. Tokyo (1980)*, p. 1331, Elsevier/Kodansha (1981).
33. A. Brenner and D. A. Hucul: *J. Am. Chem. Soc.* **102**, 2484 (1980).
34. A. Kazusaka and R. F. Howe: *J. Mol. Catal.* **9**, 183 (1980).
35. M. Laniecki and R. L. Burwell: *J. Colloid Interface Sci.* **75**, 95 (1980).
36. A. Olsthoorn and J. A. Moulijn: *J. Mol. Catal.* **8**, 147 (1980).
37. T. L. Brown: *J. Mol. Catal.* **12**, 41 (1981).
38. A. Kazusaka and R. F. Howe: unpublished results.
39. S. Abdo, A. Kazusaka and R. F. Howe: *J. Phys. Chem.* **85**, 1380 (1981).
40. S. Abdo, R. F. Howe and W. K. Hall: *J. Phys. Chem.* **82**, 969 (1978).
41. A. Kazusaka and R. F. Howe: *J. Catal.* **63**, 447 (1980).
42. A. Brenner and D. A. Hucul: *Prepr. Am. Chem. Soc., Div. Petr. Chem* **22**, 1221 (1977).
43. A. Kazusaka and R. F. Howe: *J. Mol. Catal.* **9**, 200 (1980).
44. A. Brenner and D. A. Hucul: *J. Catal.* **61**, 216 (1980).
45. J. Bilhou, A. Theolier, A. K. Smith and J. M. Basset: *J. Mol. Catal.* **3**, 245 (1977).
46. S. Abdo, J. Gosbee and R. F. Howe: *J. Chim. Phys.* **78**, 885 (1981).
47. D. M. Adams, I. R. Gardner and N. D. Parkyns: *J. Catal.* **45**, 145 (1976).
48. E. Guglielminotti: *J. Mol. Catal.* **13**, 207 (1981).
49. R. F. Howe: *J. Chem. Soc., Faraday Trans. I* **71**, 1689 (1975).
50. R. F. Howe and C. Kemball: *J. Chem. Soc., Faraday Trans. I* **70**, 153 (1974).
51. A. Kazusaka, L. K. Yong and R. F. Howe: *Chem. Phys. Lett.* **57**, 592 (1978).
52. W. Timmer and R. F. Howe: to be published.
53. G. Coudurier, P. Gallezot, H. Praliaud, M. Primet and B. Imelik: *C. R. Acad. Sci. Paris* **282**, 311 (1976).

54. P. Gallezot, G. Coudurier, M. Primet and B. Imelik: in *Molecular Sciences II, ACS Symp. Ser. No. 40*, p. 144 Am. Chem. Soc., Washington, DC (1977).
55. Y. Ben Taarit, G. Wicker and C. Naccache: 'Magnetic Resonance in Colloid and Interface Science', *NATO Advanced Study Inst. Ser. C* **61**, 497 (1980).
56. S. Abdo and R. F. Howe: *J. Phys. Chem.* **87**, 1713 (1983).
57. P. Rommelfaenger and R. F. Howe: *J. Chem. Soc., Chem. Commun.* 123 (1979).
58. M. L. Vrinat, C. G. Gachet and L. de Mourgues: *Catalysis by Zeolites*, p. 219, Elsevier, Amsterdam (1980).
59. E. Guglielminotti and Z. Zecchina: *J. Chim. Phys.* **78**, 891 (1981).
60. R. L. Banks: *Catalysis – Specialist Periodical Reports*, Vol. 4, p. 100, Chemical Society, London (1981).
61. E. S. Davie, D. A. Whan and C. Kemball: *J. Chem. Soc., Chem. Commun.* 1202 (1971).
62. W. Mowat, J. Smith and D. A. Whan: *J. Chem. Soc., Chem. Commun.* 34 (1974).
63. J. Smith, W. Mowat, D. A. Whan and E. A. V. Ebsworth: *J. Chem. Soc., Dalton Trans.* 1742 (1974).
64. A. Brenner and R. L. Burwell: *J. Catal.* **52**, 364 (1978).
65. R. Westhoff and J. A. Moulijn: *J. Catal.* **46**, 414 (1977).
66. O. V. Krylov *et al.: React. Kinet. Catal. Lett.* **8**, 347 (1978).
67. N. Giordano *et al.: J. Catal.* **42**, 381 (1976).
68. E. A. Lombardo, M. Lo Jacono and W. K. Hall: *J. Catal.* **64**, 150 (1980).
69. (a) M. F. Farona and R. L. Tucker: *J. Mol. Catal.* **8**, 85 (1980); (b) Y. Iwasawa and H. Hamamura: *J. Chem. Soc., Chem. Commun.* 130 (1983).
70. S. Namba, T. Komatsu and T. Yashima: *Chem. Lett.* 115 (1982).
71. A. Brenner: *J. Mol. Catal.* **5**, 157 (1979).
72. K. Tanaka, P. Ravines and R. F. Howe: to be published.
73. R. Merta and V. Ponec: *Proc. 4th Int. Congr. Catal. Moscow (1968)*, p. 896.
74. R. Nakamura, R. G. Bowman and R. L. Burwell: *J. Am. Chem. Soc.* **103**, 673 (1981).
75. C. Defosse, M. Laniecki and R. L. Burwell: *Proc. 7th Int. Congr. Catal Tokyo (1980)*, p. 1331, Elsevier/Kodansha (1981).
76. G. A. Mills and F. W. Steffgen: *Catal. Rev. Sci. Eng.* **8**, 159 (1973).
77. E. S. Davie, D. A. Whan and C. Kemball: *J. Chem. Soc., Chem. Commun.* 1202 (1971).
78. P. P. O'Neill and J. J. Rooney: *J. Am. Chem. Soc.* **94**, 4383 (1972).
79. E. S. Davie, D. A. Whan and C. Kemball: *Proc. 5th Int. Congr. Catal. Miami* (1972), Vol. 1, p. 1205, North-Holland, Amsterdam (1973).
80. H. Alper and C. Blais: *J. Chem. Soc., Chem. Commun.* 169 (1980).
81. J. M. Basset *et al.: Proc. 6th Int. Congr. Catal. London (1976)*, Vol. 1, p. 570, The Chemical Society, London (1977).
82. A. Brenner: *J. Chem. Soc., Chem. Commun.* 251 (1979).
83. A. Brenner and D. A. Hucul: *Inorg. Chem.* **18**, 2836 (1979).
84. F. Hugues, A. K. Smith, Y. Ben Taarit, J. M. Basset, D. Commereuc and Y. Chauvin: *J. Chem. Soc., Chem. Commun.* 68 (1980).
85. E. Guglelminotti, A. Zecchina, F. Boccuzzi and E. Borello: *Growth and Properties of Metal Clusters*, p. 165, Elsevier, Amsterdam (1980).
86. F. Hugues, P. Bussiere, J. M. Basset, D. Commereuc, Y. Chauvin, L. Bonneviot and D. Oliviet: *Proc. 7th Int. Congr. Catal. Tokyo, 1980*, p. 418, Elsevier/Kodansha (1981).
87. D. Ballivet-Tkatchenko and G. Coudurier: *Inorg. Chem.* **18**, 558 (1979).

88. J. B. Nagy, M. Van Eenoo and E. G. Derouane: *J. Catal.* **58**, 230 (1979).
89. J. B. Nagy, M. Van Eenoo, E. G. Derouane and J. C. Vedrine: 'Magnetic Resonance in Colloid and Interface Science', *NATO Advanced Study Inst. Ser. C* **61**, 591 (1980).
90. See, for example, W. N. Delgass, G. Haller, R. Kellerman and J. H. Lunsford: *Spectroscopy in Heterogeneous Catalysis*, Academic Press, New York (1979).
91. D. Ballivet-Tkatchenko, G. Coudurier and H. Mozzanega: *Catalysis by Zeolites* (eds. B. Imelik *et al.*), p. 309, Elsevier, Amsterdam (1980).
92. D. Ballivet-Tkatchenko, G. Coudurier, H. Mozzanega, I. Tkatchenko and A. Kiennemann: *J. Mol. Catal.* **6**, 293 (1979).
93. J. Phillips, B. Clausen and J. Dumesic: *J. Phys. Chem.* **84**, 1814 (1980).
94. J. Phillips and J. Dumesic: *Appl. Surf. Sci.* **7**, 215 (1981).
95. E. G. Derouane, J. B. Nagy and J. C. Vedrine: *J. Catal.* **46**, 434 (1977).
96. R. B. Bjorklund and R. L. Burwell: *J. Colloid Interface Sci.* **70**, 383 (1979).
97. P. N. Galich, V. S. Gutyrya and A. A. Galinski: in *Proc. 5th Int. Conf. Zeolites, Naples (1980)* (ed. L. V. Rees), p. 661, Heyden, London (1980).
98. See, for example, P. A. Jacobs: in *Catalysis by Zeolites* (eds. B. Imelik *et al.*), p. 293, Elsevier, Amsterdam (1980).
99. G. B. McVicker and M. A. Vannice: *J. Catal.* **63**, 25 (1980).
100. A. Kazusaka, H. Suzuki and I. Toyoshima: *J. Chem. Soc., Chem. Commun.* 150 (1983).
101. T. Bein and P. A. Jacobs: *J. Chem. Soc., Faraday Trans. I* **79**, 1819 (1983).
102. R. L. Jackson and M. R. Trusheim: *J. Am. Chem. Soc.* **104**, 6590 (1982).
103. M. R. Trusheim and R. L. Jackson: *J. Phys. Chem.* **87**, 1910 (1983).
104. J. Santos, J. Phillips and J. A. Dumesic: *J. Catal.* **81**, 147 (1983).
105. B. Imelik *et al.* (eds.): *Studies in Surface Science and Catalysis*, Vol. 11, Elsevier, Amsterdam (1982).
106. J. Santos and J. Dumesic: in *Studies in Surface Science and Catalysis* (eds. B. Imelik *et al.*), Vol. 11, p. 43, Elsevier, Amsterdam (1982).
107. T. J. Thomas and A. Brenner: *J. Mol. Catal.* **18**, 197 (1983).
108. J. M. Tri, J. P. Candy, P. Gallezot, J. Massardier, M. Primet, J. C. Vedrine and B. Imelik: *J. Catal.* **79**, 396 (1983).
109. Y. I. Yermakov and B. N. Kuznetsov: *J. Mol. Catal.* **9**, 13 (1980).
110. T. Forrester, A. Kucernak, L. Young and R. F. Howe: to be published.

M. ICHIKAWA

SURFACE-SUPPORTED METAL CLUSTERS AND THEIR CATALYSIS IN CO-BASED REACTIONS

1. Introduction

Metal clusters are characterized by their unique framework consisting of three or more (up to 35), metal atoms which are stabilized by appropriate ligands such as carbonyls, phosphines and halides, and which can exist as electroneutral species or ions (see Figure 1). Among the more interesting clusters is a rhodium carbonyl species having an icosahedral Rh_{13} unit in its structure and a platinum species containing 19 metal atoms whose dimensions (0.8 × 1.1 nm) are close to those of the smallest crystallites in petroleum-reforming supported catalysts. We might expect that such metal clusters with or without ancillary ligands will exhibit structural and catalytic properties intermediate between those of mononuclear complexes and bulk metal surfaces, including conventionally supported metal crystallites which possess a spectrum of different particle sizes.

In spite of the inherent chemical interest in cluster complexes, catalysis by metal clusters is often difficult to interpret because the clusters can fragment in solution to give mono- and/or sub-nuclear complexes (which may also be catalytically active species), or they can agglomerate to give metallic colloids having particle sizes in the range 2–5 nm which may also exhibit catalytic activity. Most reported examples of metal cluster catalysis need to be re-investigated since often characterization of the catalyst has been insufficient to establish that the active species involved are discrete molecular clusters in which the metal frame work is retained during reaction [1].

Two approaches are feasible in considering the potential application of metal clusters in heterogeneous catalysis. Firstly, particular cluster complexes which have been shown to be active species in homogeneous catalysis may be anchored directly by reaction with functional groups such as phosphines and amines attached onto organic and/or inorganic supports, such attachment often being achieved by simple ligand exchange processes. Collman *et al.* [2] reported that $Rh_6(CO)_{16}$ attached onto a phosphino polystyrene support catalyzed the hydrogenation of arenes with activities similar to those achieved with a commercial Rh/Al_2O_3 catalyst. Gates *et al.* [3] extended this work to the preparation of well-defined substituted metal clusters such as $Rh_6(CO)_{13}[PPh_2\text{–}Ⓟ]_3$ and

Y. Iwasawa (ed.), Tailored Metal Catalysts, 183–263.

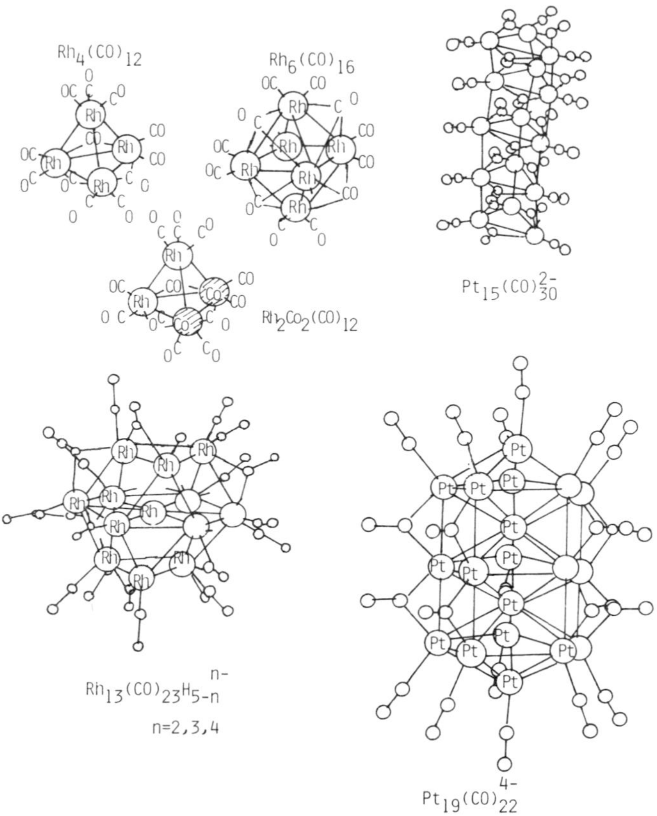

Fig. 1. Typical examples of carbonyl metal clusters and their metal frameworks.

$Ir_4(CO)_{11}$ [PPh_2–Ⓟ] which are active for the hydrogenation of ethylene and cyclohexene under mild conditions. Secondly, metal clusters have been used as the starting materials for the preparation of quite small metal aggregates (crystallites) of homo- or hetero-metallic constituents dispersed on inorganic supports, usually metal oxides having a higher surface area. Thus, preparations involving metal clusters as the discrete metal units may result in supported metals exhibiting noble metal/support interactions and allow favorable control of metal particle sizes down to less than 1 nm. Unusual activities and selectivities could well be expected for such catalysts.

However, both methods pose difficulties in terms of characterizing on a molecular scale the immobilized or impregnated metal species derived from the precursor clusters. A number of papers have been published recently which report the improved characterization of the cluster-derived catalysts using a variety of different techniques under ambient conditions, e.g. IR, UV, Ramen and Mössbauer spectroscopies, XPS and EXAFS techniques, and electron microscopy. Such methods allow a better understanding of metal dispersions, their surface compositions, and their chemical and catalytic behavior, as summarized in Figure 2.

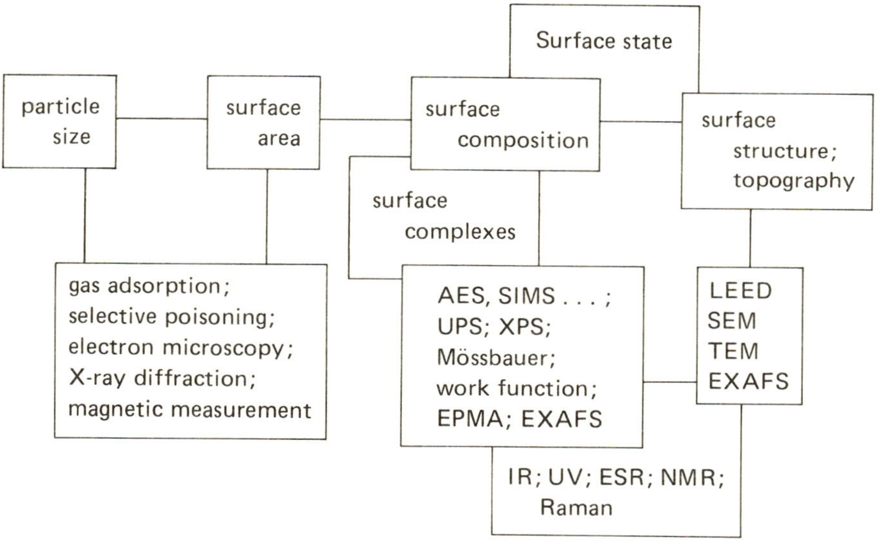

Fig. 2. Methods available for the characterization of metallic and bimetallic cluster catalysts.

In considering the potential applications of metal clusters in heterogeneous catalysis, it is appropriate to take account of some generalizations regarding conventional metal catalysts, comprising metals and/or metal oxides. Metals are usually dispersed on inorganic oxide supports both to increase the effective surface areas responsible for catalysis and to dilute the active metal species and thus prevent metal agglomeration. Industrially, heterogeneous catalysts are preferred principally on the grounds of their considerably higher activity, but also because of process engineering requirements, such as the ease of separation of products from the catalysts and the ability of the catalysts to withstand vigorous working conditions, e.g. high pressures and high temperatures.

The activities and selectivities of supported metal catalysts are generally influenced by the states of metal dispersion (particle sizes), chemical properties and structural circumstances of the supporting materials. In particular, if the metal is bi- or multi-metallic, its electronic and conformational properties may be associated with its composition and the geometric factors involved at the metal surface. The dispersion of the metal depends on the system and the preparative conditions employed. Various methods have been suggested for the preparation of, highly dispersed metals and bimetallic alloys, e.g. conventional impregnation methods or ion-exchange methods involving impregnation of the support (which may be a refractory inorganic oxide such as alumina or silica, or a molecular sieve) by an aqueous solution of the metal halide or nitrate. The resulting metal ion-impregnated materials are dried and calcined to form metal oxide aggregates, which are further reduced by treatment with H_2 (at 300–700 °C). The difficulty with this method however, is that sufficiently narrow size distributions are impossible to achieve with metal dispersions in conventional catalyst preparations and that reduction of impregnated ions and oxides of early transition metals such as Mo, Cr, Mn, W. Fe and Co to metal particles by means of hydrogen is only partially successful.

The alternative of using transition metal carbonyl compounds for better control in the preparation of dispersed metal catalysts was first suggested by Parkyns [4], Bjorklund and Burwell [5] and Howe *et al.* [6] who showed that $Ni(CO)_4$, $Mo(CO)_6$ and $W(CO)_6$ could be decomposed on oxide surface such as silica and alumina to give metallic Ni, Mo and W, respectively, and that such metal dispersions exhibited marked catalytic activities. In addition, similar application of polynuclear metal carbonyls by Mainwaring and Anderson [7], Smith and Basset [8], Smith *et al.* [9] and Ichikawa [10] produced highly dispersed metallic crystallites including bimetallic systems, as illustrated in the general flow chart for catalyst preparation give in Figure 3.

These authors have provided detailed accounts of this method of preparation together with characterizations of the resulting cluster-derived catalysts. Such catalysts offer the following differences/advantages over conventionally prepared supported metal catalysts. (i) Non-aqueous methods of catalyst preparation can be employed by using metal cluster compounds which dissolve homogeneously in inert organic solvents. (ii) It is possible to use halogen-free cluster compounds which may in some instances be an advantage in catalyst preparation. Thus the halogen atoms and ions often present in conventionally prepared metal catalysts (which arise from halide salt contamination and the use of commercial oxide supports) are difficult to remove and may contribute to catalyst poisoning and to the acceleration of metal aggregation and sintering. (iii) The metals in cluster molecules are in formal zero-valent states. As a result, reduced, and hence active, metal aggregates may be produced by mild activation of impregnated metal clusters, avoiding the high-temperature reduction with hydrogen necessary to reduce metal ions and oxides dispersed on a support in conventional catalyst preparation. (iv) The use of mixed metal or heteronuclear clusters, e.g. clusters involving heteroatoms such as S, C, N or P, offers the possibility of preparing

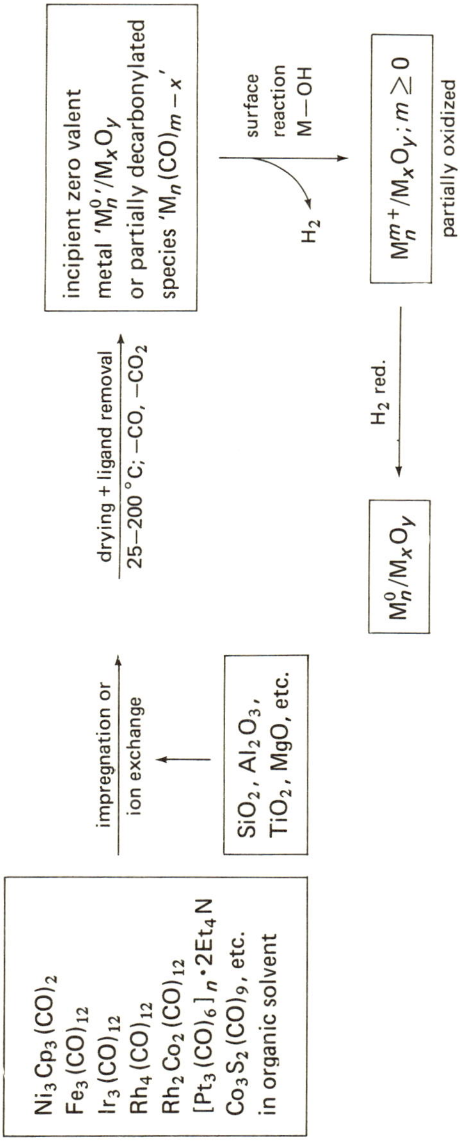

Fig. 3. General procedures for the preparation of cluster-derived metal catalysts.

bimetallic 'alloys' and modified metal catalysts with a defined stoichiometry/composition in a more controlled manner. Supported bimetallic catalysts are of great interest in heterogeneous catalysis, although such materials are often difficult to obtain by conventional methods of preparation. Recently, however, Sinfelt *et al.* [11] have shows that the use of multicomponent metal salt solutions in the conventional impregnation method allows the preparation of highly dispersed alloy particles whose surface compositions were revealed by EXAFS and Mössbauer spectroscopy to be comparatively uniform.

The similarity between very small transition metal particles and molecular metal clusters was first suggested by Mutterties [12], Ugo [13] and Chini [14] on the basis of the chemical behavior of ligands 'coordinated' to a cluster frame of mono- and bi-metallic aggregates and substrate 'chemisorbed' onto the surface of metal particle. As suggested by Ozin [15], this similarity is brought out by the following reaction matrix involved in the formation of cluster molecules from metal aggregates (M, M_2, M_3 . . .) generated by the metal vaporization method.

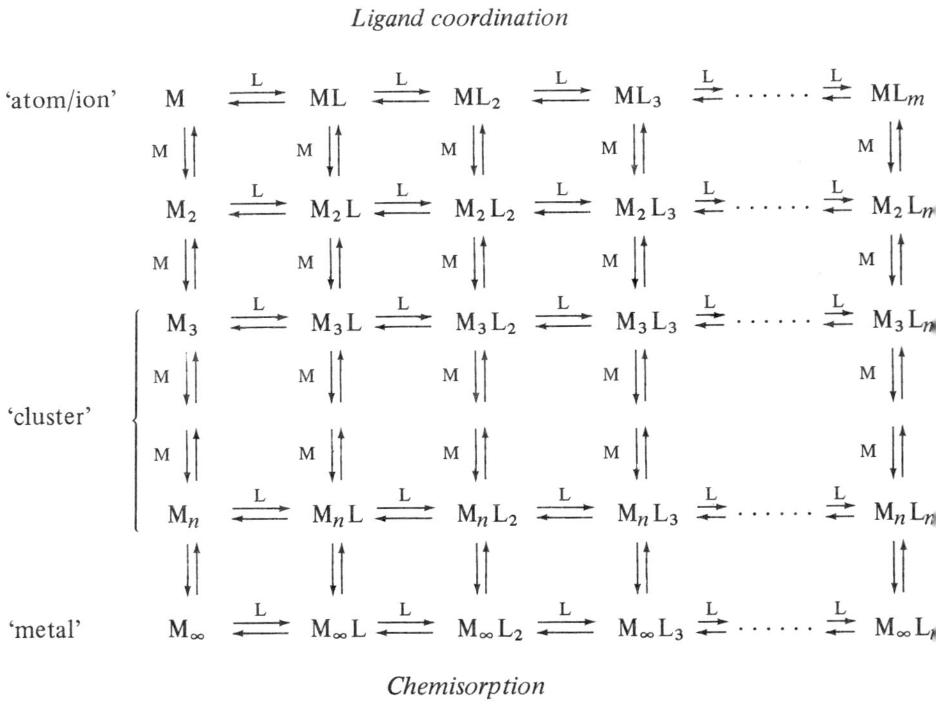

(M = Ni, Co, Rh, Pt, ... ; L = CO, Cp, Olefins, H, . . .)

According to this scheme, surface metal species prepared on oxide supports may be regarded as catalyst intermediates formed via complex reactions involving the removal of ligands (L) and the agglomeration of metal units. Such agglomeration may be accompanied by fragmentation into submetallic species corresponding to those listed in the above matrix. An alternative method of preparation consists of entrapping the cluster molecules (if their size permits) within zeolitic cavities having dimensions of 0.5–1.5 nm, or the precursor for such clusters thereby allowing aggregation to occur within the cavity itself. This approach has been employed successfully by Montovani *et al.* [16] and Naccache *et al.* [17] to produce zeolite-entrapped $Rh_6(CO)_{16}$ and $Ir_4(CO)_{12}$.

Let us consider some of the opportunities offered by clusters as homogeneous and heterogeneous catalysts for CO-based reactions. The close proximity of the metal centers in clusters facilitates multiple bonding of a single reactant molecule such as CO, N_2 or CN^-, offering conversion opprtunities not available with mononuclear metal complexes. Clusters also enable different reaction partners to be bonded in neighboring positions. Since both CO and H (or CH_3) are common ligands on clusters, good prospects should exist for their use as catalysts in synthesis gas conversion processes. In this context, routes for selective conversion into fuels and chemicals as a replacement for petrochemicals are an important topic of current catalysis research. Catalysts offering the greatest potential for improve performance are supported metal clusters with high selectivities capable of producing lower olefins and useful oxygenated carbon compounds such as ethanol, acetic acid and methanol from CO and H_2.

This chapter is devoted to studies focused on opportunities for preparing improved catalysts using well-defined metal clusters through control of the combination of metals employed, and of the particle size of the metal, the metal/support interaction, and the degree of metal segregation. Possibilities exist for the development of selective catalysts with bifunctional oxide supports and shape-selective zeolites thereby influencing the product distribution in the CO/H_2 conversion process. Metal clusters supported on inorganic oxides could also find application in a variety of other processes, for example carbonylation, provided that the clusters could be stabilized sufficiently to withstand the reaction conditions involved in such processes.

2. Preparation and Characterization of Surface-supported Metal Clusters

2.1. RHODIUM AND PLATINUM CARBONYL CLUSTERS IMPREGNATED OF METAL OXIDES

Surface-supported metal clusters and derived metal catalysts have been prepared from precursor cluster compounds through the following principal procedures and treatments.

Initially, the support (silica, alumina or phosphine-substituted polystyrene, etc.) is impregnated with a solution of the polynuclear cluster compound $M_x L_y$, and this is followed either by partial or complete removal of the ligand L by thermal or photochemical decomposition *in vacuo*, or by chemical reaction with hydrogen, oxygen, olefins, etc. Provided that the pretreatment of the support is carried out carefully and that the decomposition of the cluster compound is conducted at a modest temperature, it should be possible to obtain small metal aggregates of similar nuclearity to those in the parent complex attached to the support surface through the following typical reactions:

In vacuo *thermal decarbonylation*

$$[Pt_3(CO)_6]_n \cdot 2Et_4N\ (n = 2\text{–}5) \xrightarrow{\Delta} [Pt_3]_n + 6_n\,CO + Et_3N + H_2C{=}CH_2 + \ldots$$

Treatment with O_2 and H_2 under mild conditions

$$Rh_6(CO)_{16} + O_2 \longrightarrow [Rh_6]_{ox} + CO_2$$

$$[Rh_6]_{ox} \xrightarrow{H_2} [Rh_6]_{red} + H_2O$$

Partial decarbonylation in CO, olefin and/or CO/H_2 atmosphere

$$Ir_4(CO)_{12} \rightleftharpoons Ir_4(CO)_{12-x} + x\,CO$$

Oxide supports with large surface areas have a multifunctional role in that they offer a favorable environment for dispersing the metal clusters as well as providing opportunities for further chemical reaction between the metal aggregate and the support, thus stabilizing the clusters and preventing further agglomeration. Such metal/support interactions allow the formation of intermediate 'surface compounds' where the support behaves as a ligand toward the cluster unit, a process which is sometimes accompanied by more complicated surface reactions, e.g. cluster aggregation, or breakdown to clusters of a higher or lower nuclearity.

The behavior of $Rh_6(CO)_{16}$ and $Rh_4(CO)_{12}$ on SiO_2 and Al_2O_3 has been investigated extensively by Ugo *et al.* [18], who have characterized the resultant surface-supported species by means of IR spectroscopy. Thus, upon impregnation onto SiO_2 gel the supported $Rh_6(CO)_{16}$ was partially decarbonylated probably as a result of reaction with surface OH groups in a similar manner to the situation arising when alumina was impregnated [9]. Interaction at this stage does not lead to a breakdown of the cluster moiety on such oxide supports, since it is possible to regenerate the original $Rh_6(CO)_{16}$ species simply by treatment with CO at 50–200 °C. However, heating the supported samples *in vacuo* above 50–100 °C resulted in almost complete elimination of carbonyl ligands to produce highly dispersed metal crystallites via intermediate carbonyl species which exhibited IR bands corresponding to the carbonyl at 2043 and 1890 cm^{-1} group. The oxidized species generated after treating supported $Rh_6(CO)_{16}$ in air, exhibited intense sharp bands at 2093 and 2038 cm^{-1} in the presence of CO which may be assigined to the symmetric and antisymmetric stretching vibrations of $Rh^I(CO)_2$ species such as $[Rh(CO)_2X]_2$ (X = Cl, Br, OEt, OMe). This oxide species could be reduced by H_2 (1 atm, 50 °C) to regenerate the metal (IR bands again observed at 2043 and 1900 cm^{-1} in CO chemisorption) [19].

The relative intensity and location of such carbonyl absorption bands in the IR spectrum for these oxidized species suggest the presence of local $Rh^I(CO)_2$ surface species similar to $[Rh(CO)_2(OSiPh_3)]_2$ and formed as depicted in Figure 4. If the silica (or Al_2O_3, MgO) support has not been fully dehydrated, or even better if H_2O has been added to CO, the oxidized $Rh^I(CO)_2$ species may be readily converted back to the original Rh carbonyl cluster, i.e. $Rh_6(CO)_{16}$, which may be identified by its characteristic carbonyl bands. The very facile generation of carbonyl clusters with CO/H_2O or H_2 is in agreement with the presence of weakly aggregated, high nuclearity Rh^I units on the surface. In contrast, $RhCl_3$ or $Rh_2(CO)_4Cl_2$ adsorbed on silica is converted by CO/H_2O to $Rh_4(CO)_{12}$ and/or $Rh_6(CO)_{16}$ through surface synthesis but with a lower rate and efficiency.

$$[Rh(CO)_2Cl]_2 + CO + H_2O \xrightarrow{HO^-} Rh_4(CO)_{12} + Rh_6(CO)_{16} + \ldots$$

It is quite probable that direct bonding occurs between Rh and SiO, AlO or MgO units as indicated by the dependence of the ν_{CO} values on the nature of the support (see Figure 4). The nuclearity of the various carbonyl clusters formed is generally dependent upon the surface basicity and acidity of the oxides employed, an increase of OH^- concentration and temperature of the system leading to an increase in the observed nuclearity.

There is evidence to suggest that the original clusters adsorbed onto an oxide surface are attacked by highly nucleophilic OH surface species. The magnitude of the interaction between the surface and carbonyl groups varies depending on whether MgO, Al_2O_3 (basic) or SiO_2 (neutral) is used as the support. When $Rh_4(CO)_{12}$ is supported on silica, it is comparatively stable and even retained in the presence carbon monoxide, but it is readily converted to $Rh_6(CO)_{16}$ if the CO ligands are removed *in vacuo* or at elevated temperatures under an inert atmosphere. The major mechanism of destruction and transformation must be proton attack with hydrogen evolution, at least for surfaces with a high Brönsted acidity.

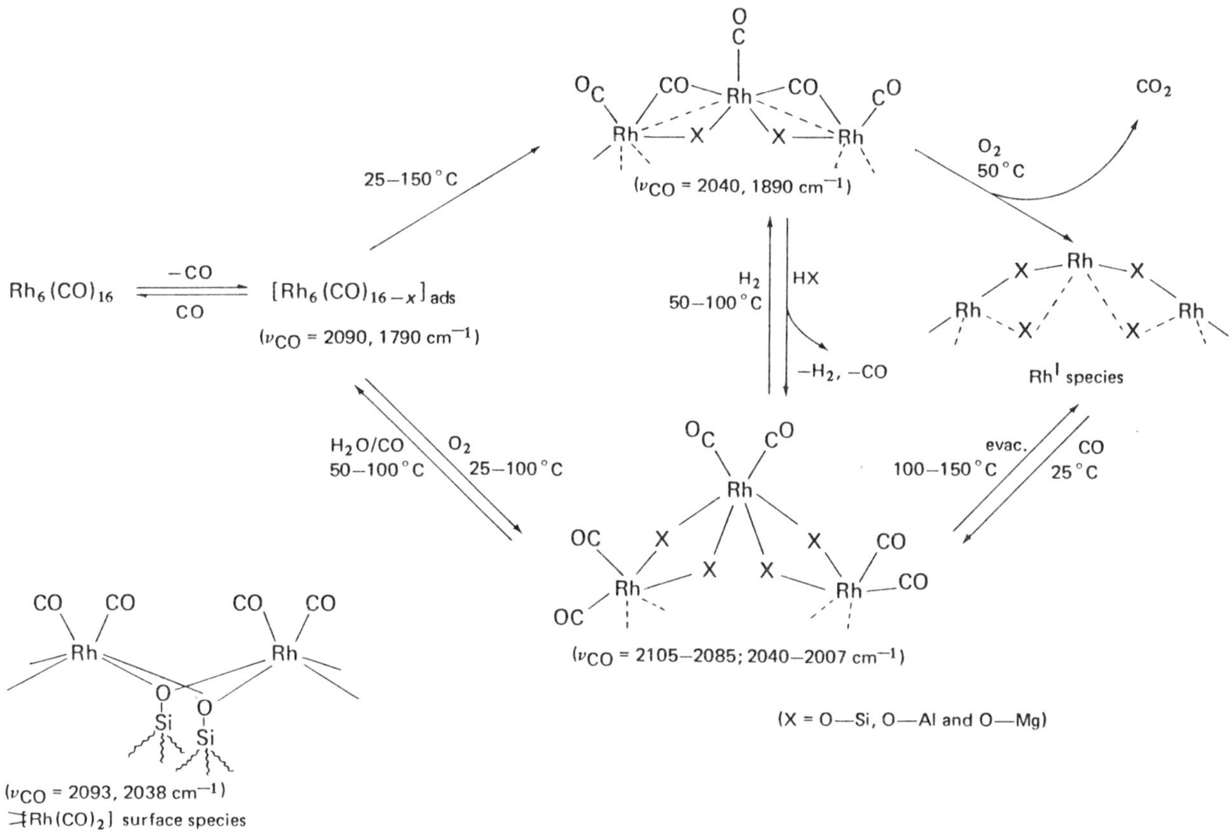

Fig. 4. Surface reactions of $Rh_6(CO)_{16}$ impregnated on metal oxides (SiO_2, Al_2O_3, MgO).

For ideal cases, decomposition of clusters on oxide surfaces to produce metal aggregates and the reversible regeneration of the original carbonyl clusters by adsorption of CO might be visualized as follows:

$$M_x(CO)_y \rightleftharpoons \underset{\text{supported molecular clusters}}{[M_x(CO)_y]_{ads}} \underset{CO}{\overset{-CO}{\rightleftharpoons}} \underset{\text{metal aggregates}}{[M_x(CO)_{y-m}]_{ads}}$$

$$\underset{CO}{\overset{-CO}{\rightleftharpoons}} [M_x]_{ads}$$

where M = Pt, Rh, Ru, Ir, etc., and where the supported metal species could have a structure of the type:

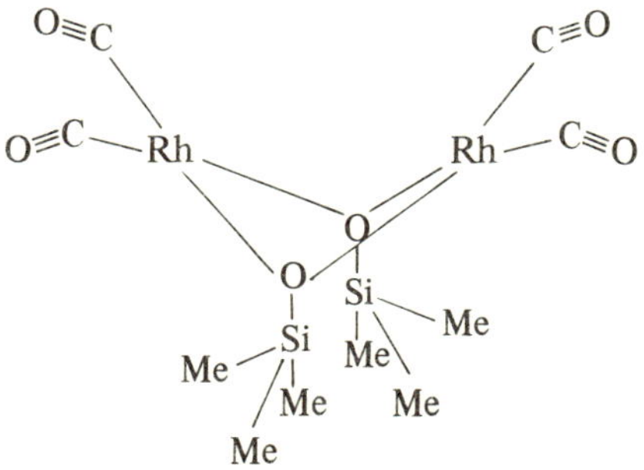

(ν_{CO} 2006, 2075 cm^{-1})

Supported twin carbonyl Rh species of this type have been shown by IR spectroscopy [20] to undergo chemical reactions with the different gaseous reagents. Thus the $\ddagger Rh(CO)_2]_m$ species generated by impregnating Ti- and Zr-oxide containing silica, when heated at 50–100 °C *in vacuo* for several hours, completely loses its twin carbonyl ligands. The resulting materials adsorbed ca. 2 mol CO per mole of Rh when exposed to CO, restoring the original twin carbonyl bands in the IR spectrum. The rates of the reversible dissociation of the different surface Rh carbonyl species generated by impregnating a Ti- oxide containing silica support with solutions of $Rh_4(CO)_{12}$ or $Rh_7(CO)_{16} \cdot 3Et_4N$ on heating at 150 °C and 10^{-4} Torr pressure have been shown to lie in the sequence $\ddagger Rh(CO)_2]_m \ll +Rh(CO)+_m < +Rh-CO+_m$. This decreasing order of stability agrees quite well qualitatively with that observed for conventionally prepared Rh/Al_2O_3 catalysts.

Reactivities towards gases such as O_2 and NO have also bee studied [19] for surface carbonyl species such as $\ddagger Rh(CO)_2]_m$ and $+Rh(CO)-CO+_m$ supported on Ti-oxide containing silica. It was found that oxidized $\ddagger Rh(CO)_2]_m$ was rather inactive up to 100 °C on exposure to O_2 and NO (200 Torr), whereas the linear and bridging carbonyl groups associated with the reduced Rh aggregates $\{Rh(CO)-CO\}_m$ disappeared instantaneously at 25–50 °C, giving a carbonate (CO_3^{2-}) containing species (IR absorption bands at 1580, 1440 and 1470 cm^{-1}) for O_2 treatment, and Rh—NO (1730, 1665 cm^{-1}) and Rh(NO)(CO) species (2100, 1440 and 1470 cm^{-1}) for NO treatment, respectively. These results imply that $\{Rh(CO)_2\}_m$ species are highly hindered through steric limitation and/or coordinative saturation of Rh sites. In contrast, nucleophilic reagents such as OH^- and CH_3O^- react readily with the terminal carbonyl groups of $\{Rh(CO)_2\}_m$ species transforming them into reduced higher subcarbonyl species and ultimately to clusters of Rh^0, possibly through the formation of hydrido-carbonyl complexes with successively greater degrees of coordinative unsaturation, as mentioned above [18, 20].

In contrast to other noble metal clusters such as Pt and Ir, reactive Rh carbonyl clusters do not retain their original nuclearity when the carbonyl ligands are removed in the presence of supports containing strongly basic hydroxyl groups. Oxidative addition of hydroxyl groups causes bond fission and the formation of small metal particles (< 1.5 nm). On more acidic or neutral oxides containing small numbers of surface hydroxyl groups such as ZnO, La_2O_3 and ZrO_2 (which contain Lewis acid sites associated with the higher valency metal cations), Rh carbonyls and other metal carbonyls such as $[Pt_3(CO)_{6n} \cdot 2Et_4N$ (n = 3–5) and $Ir_4(CO)_{12}$ apparently retain their original characteristics upon impregnation and subsequent treatment in CO and CO/H_2 atmospheres at temperatures below 150 °C. When $Rh_4(CO)_{12}$ from a hexane solution was impregnated onto a degassed pellet of silica containing Ti- and Zr-oxide (thin layers of TiO_2 and ZrO_2 having been allowed to develop on silica gel by preparation from the coresponding metal propoxides), the resulting samples gave strong carbonyl bands at 2080, 2035 and 1870 cm^{-1} resembling those of the original Rh_4 cluster, but shifted to higher wavenumbers in comparison, possibly as a result of proton interaction with the oxide surfaces. When $Rh_4(CO)_{12}$ was deposited directly by successive vacuum exaporation (10^{-4} Torr) onto the same Ti oxide containing silica disc, the resulting IR spectra (Fig. 5) exhibited ν_{CO} bands at 2080, 2040 and 1875 cm^{-1}, respectively. During such vacuum evaporation of $Rh_4(CO)_{12}$, the silanol bands (3780–3760 cm^{-1}) in the spectrum arising from the TiO_2/SiO_2 disc decreased considerably in intersity. The remaining IR bands of the impregnated sample were relatively unchanged at 25 °C, but slowly transformed on strong evacuation for 5–15 h to give bands at 2075 and 1810–1780 cm^{-1}, which may be assigned to those of an adsorbed $Rh_6(CO)_{16-x}$ species. In contrast, $[Rh_6(CO)_{15}]^{2-}2Et_4N^+$ (dark green; 1990, 1970 and 1770 cm^{-1} in THF solution) and $[Rh_6(CO)_{14}]^{4-}4Bu_4N^+$ (red–brown; 1995 and 1770 cm^{-1} in acetone solution) when impregnated onto a degassed TiO_2-containing silica changed color rapidly to yellow, the resulting samples exhibiting IR bands at 2030–2020(s), 1820 and 1780 cm^{-1}(s), possibly due to a hydride cluster such as $[Rh_6(CO)_{15}H]^-$ [21]. No fragment resembling an $Rh(CO)_2$ species was detected in this surface reaction.

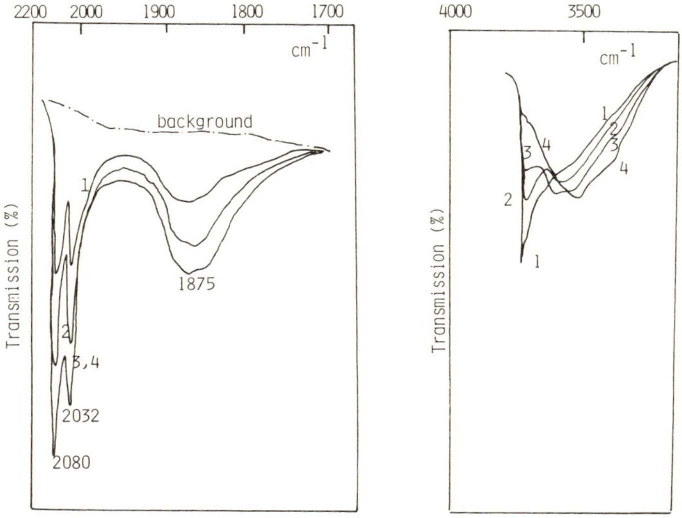

Fig. 5. IR spectra over the ν_{CO} and ν_{OH} regions for $Rh_4(CO)_{12}$ deposited by successive evaporation onto Ti-oxide containing silica pellet (TiO_2 loading 11wt.%, preheated at 400 °C *in vacus*). The amount of $Rh_4(CO)_{12}$ deposited increases in the order sample 1 < sample 2 < sample 3. Spectrum 4 was obtained when sample 3 was exposed to N_2 (100 Torr pressure) for 5 h at 25 °C.

Anionic carbonyl clusters such as $[Rh_7(CO)_{16}]^{3-}$ and $[Rh_{13}(CO)_{23}H_{5-n}]^{n-}$ (n = 2 and/or 3) when impregnated onto TiO_2-containing silica gave IR bands corresponding to the carbonyl group at 2010–2000(s) cm^{-1} and 1810–1840(br) cm^{-1} respectively which resemble those of the corresponding original cluster anions but shifted some 20–30 cm^{-1} to ligher energy levels. These results suggest that the reduced carbonyl cluster anions react with acidic support sites (possibly SiO—H or TiO—H) and as a consequence are converted into the hydride species which is bound onto the oxide surface, in a similar manner to that observed $[HFe_3(CO)_{11}]^-$ [22] and $[HOs_3(CO)_{10}]$ [O—Si] or $[HOs_3(CO)_{10}]$ [O—Al] [23]. In support of this suggestion, we have observed the generation of a considerable number of strong acid sides through the addition of a TiO_2 or ZrO_2 precursor ($Ti(OPr)_4$ or $Zr(OPr)_4$) to silica gel, as suggested by the results depicted in Figure 6. Changes in the distribution of acid sites were observed for all the silica compositions by means of calorimetric measurements of the differential heat of adsorption of ammonia.

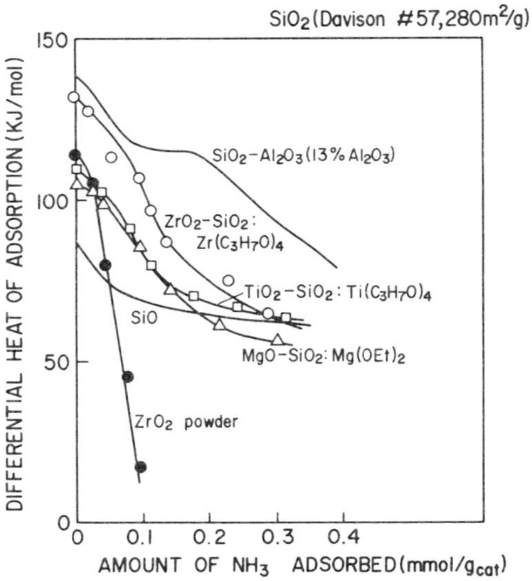

Fig. 6. Surface acidity of silica samples containing ZrO_2, TiO_2 or MgO as determined from the differential heat of adsorption of ammonia.

In all cases, the surface-supported Rh cluster carbonyls were almost completely oxidized by treatment with O_2 at 50–80 °C (demonstrated by the disappearance of the carbonyl absorption bands in the IR spectrum), the resulting oxidized Rh aggregates from each carbonyl cluster yielding an oxidized carbonyl $Rh(CO)_2$ species on exposure to CO at 25–100 °C. As the initial starting material was varied from $[Rh(CO)_2Cl]_2$ to $Rh_4(CO)_{12}$ and then to $Rh_{13}(CO)_{23}$, corresponding twin carbonyl bands in the IR spectrum shifted to lower wavenumbers (2095–2070, 2040–2015 cm^{-1}, as summarized in Table I.

When each oxidized $Rh(CO)_2$ species was heated under an atmosphere of H_2 at 50–100 °C, they were transformed uniformly to reduced forms giving IR carbonyl bands consistent with linear CO groups at 2060–2010 cm^{-1}) and bridging CO groups at 1880–1840 cm^{-1}). The precise ν_{CO} values for the reduced carbonyls $-\!\!\!\{CO-Rh(CO)\}\!\!\!-$ vary with the sizes of the precursor carbonyl clusters.

In the presence of CO (200 Torr) and H_2O (10–20 Torr), the oxidized carbonyl species $\nmid Rh(CO)_2]$ successively regenerate adsorbed Rh_4 and Rh_6 carbonyls, as previously reported by Ugo and Basset for silica or alumina supports. However, the relative proportions of Rh_4 and Rh_6 carbonyls and uncharacterized species of higher nuclearity formed in the surface reaction was found to depend upon the type of cluster employed in the original impregnation.

TABLE I

The ν_{CO} vibrations of the carbonyl species upon impregnation of various supports, and after subsequent treatment with gases such as O_2, CO, H_2 and CO/H_2O

Treatments		Impregnation 25 °C, N_2 atm (200 Torr)	[I] CO admission after O_2 treatment CO (200 Torr), 50 °C	[II] H_2 treatment after [I] H_2 (200 Torr), 100 °C	[III] CO/H_2O treatment after [I] CO (200 Torr), H_2O (20 Torr), 50–100 °C
Precursor ν_{CO} (cm^{-1}) in solution[a]	Support	(cm^{-1})	(cm^{-1})	(cm^{-1})	(cm^{-1})
$Rh_2(CO)_4Cl_2$ (hexane) 2105(m), 2089(s), 2035(s)	TiO_2/SiO_2				2098 2045
$Rh_4(CO)_{12}$ (hexane) 2075(s), 2045(m), 1890(s)	SiO_2 TiO_2/SiO_2	2050(l) 1890(b) 2035(l) 1880(b)	2095, 2040 (tw) 2085, 2035 (tw)	2080 1800(br) 2075 1875(br) 2075 1810(br)	2085 1885(br) 2050 2075 1875(br) 2035
$Rh_6(CO)_{16}(CHCl_3)$ 2080(s), 1800(s)	SiO_2	2050(l) 1880(b)	2090, 2040 (tw)	2080 1800(br)	2075 1800
$[Rh_6(CO)_{15}]\cdot NEt_4$ (THF) 1990(s), 1970(m), 1770(s)	ZrO_2/SiO_2 TiO_2/SiO_2	2035(l) 1883(b) 2032(l) 1880(b)	2090, 2030 (tw) 2085, 2030 (tw)	2075 1810, 1870 (br) 2080 1810, 1780 (br)	2080 1810, 1780 (br) 2030 1790, 1815 (br)
$[Rh_7(CO)_{16}]\cdot 3NEt_4$ (acetone) 2020(w), 1995(s), 1810(w), 1770(s)	TiO_2/SiO_2	2010(l) 1840(b)	2080, 2020 (tw)	2010 1810(br)	2005 1790, 1815 (br)
$[Rh_{13}(CO)_{23}H^{n-}_{5-n}]\cdot nNBu_4$ (THF) (n = 2) 2020, 1840 (n = 3) 1995, 1810	TiO_2/SiO_2	2028(l) 1830(b)	2070, 2015 (tw)	2020, 2080 1810(br)	2020 1810

[a] br = broad peak, s = strong intensity, m = medium, w = weak, l = linear carbonyl, b = bridging carbonyl, tw = twin carbonyl.

This was particularly so when $[Rh_7(CO)_{16}]^{3-}$ or $[Rh_{13}(CO)_{23}H_{5-n}]^{n-}$ (n = 2, 3) was used in the impregnation process. In these cases, the resultant carbonyl species regenerated from the oxidized carbonyl species $\ddagger Rh(CO)_2]_m$ under the influence of CO/H_2O exhibited IR carbonyl bands different from those of adsorbed $Rh_4(CO)_{12}$ and/or $Rh_6(CO)_{16}$, with some similarity to those generated by initial Rh_7 and Rh_{13} clusters. These results suggest that binding Rh ensembles of the type $[Rh_m \ddagger, \ddagger Rh(CO)_2]_m$ or $\ddagger Rh(O{-}M)_2]_m$ (M = Ti, Zr, Si, etc.) to silica supports containing Ti- or Zr-oxide prevents further aggregation and migration. However, they probably still exist in the proximity of the coordination sphere to enable reconstruction of particular cluster species during cyclic procedures consisting of redox reactions such as O_2–CO–CO/H_2O and H_2–O_2–CO–CO/H_2O, etc.

Howe *et al.* [24] have suggested on the basis of IR, TPR and XPS measurements that $Rh_6(CO)_{16}$ impregnated on Al_2O_3 is completely transformed to an oxidized cluster species on prolonged standing in air. The XPS spectra observed could be explained in terms of the retention of the Rh_6 cluster throughout low-temperature decarbonylation and carbonylation cycles:

$$[Rh_6(CO)_{16}]_{ads} \underset{CO}{\overset{O_2}{\rightleftharpoons}} [Rh_6(CO)_{12}(O)_{12}] \underset{CO}{\overset{-CO}{\rightleftharpoons}} [Rh_6(O)_{12}]$$

The oxidized surface species was characterized as being $[(Rh^+)_6(O)^-_{12}]$ from the XPS measurements.

Ichikawa [25] has tried to prepare supported platinum aggregates by pyrolysis *in vacuo* (at 10^{-4} Torr) or by mild oxidation of a series of Pt carbonyl cluster anions $[Pt_3(CO)_3(\mu_2\text{-}CO)_3]_n \cdot 2M$ (n = 2–5, M = Et_4N, Li, Na) impregnated on predehydrated oxide supports such as SiO_2 and Al_2O_3. The resultant impregnated catalysts contained highly dispersed Pt crystallites strongly associated with the supports, the size of these crystallites being confirmed by electron microscopy as being less than 1 nm an average. The trigonal prismatic Pt carbonyl clusters could also be synthesized separately by the reaction of H_2PtCl_6 and CO (1 atm, pressure at 25 °C) in a controlled alkaline methanol solution using the method of Chini *et al.* [26] and isolated as their quaternary alkylammonium salts. The air-sensitive Pt cluster anion obtained was impregnated onto the oxide surface from THF solution, followed by removal of the solvent *in vacuo*. Heating *in vacuo* at 50–180 °C led to almost complete removal of the CO ligands together with pyrolysis of the quaternary alkylammonium cations. From H_2 and CO adsorption studies of the completely decarbonylated samples, chemisorption ratios of 1.95, 1.5 and 1.4 for H/Pt and 0.45, 0.37 and 0.25 for CO/Pt were obtained for Pt_{15}, Pt_9 and Pt_6 carbonyl clusters impregnated on γ-Al_2O_3, respectively. The results suggest that the method leads to a highly dispersed impregnation of Pt, confirmation being provided by the chemisorption ratios greater than unity observed for H_2 absorption and by direct observation via electron micrography. The reason why the CO/Pt ratios are less than would be expected for an atomic dispersion of Pt may be that preferential multisite CO chemisorption occurs on the small Pt crystallites and/or electronic influences arise from the active sites on the alumina surface. The sequence of reactions involved in the method of Chini *et al.* may be depicted as:

$$H_2PtCl_6 \xrightarrow[CO]{NaOH/CH_3OH} \underset{\substack{\text{green} \\ (\nu_{CO}\ 2055,\ 1870\ cm^{-1})}}{[Pt_{15}(CO)_{30}]^{2-}} \xrightarrow{OH^-} \underset{\substack{\text{blue–green} \\ (\nu_{CO}\ 2045,\ 2025\ cm^{-1})}}{[Pt_{12}(CO)_{24}]^{2-}}$$

$$\xrightarrow{OH^-} \underset{\substack{\text{red–violet} \\ (\nu_{CO}\ 2030,\ 1842\ cm^{-1})}}{[Pt_9(CO)_{18}]^{2-}} \xrightarrow{OH^-} \underset{\substack{\text{red–orange} \\ (\nu_{CO}\ 1999,\ 1795\ cm^{-1})}}{[Pt_6(CO)_{12}]^{2-}}$$

the IR absorptions being obtained from studies on THF solutions at 25 °C.

When each Pt carbonyl cluster anion was impregnated onto γ-Al_2O_3, the resulting IR spectra exhibited intense carbonyl absorption bands at 2040, 1850 cm^{-1}, 2025, 1830 cm^{-1}, 2005, 1810 cm^{-1}, and 1970, 1790 cm^{-1} for the Pt_{15}, Pt_{12}, Pt_9 and Pt_6 carbonyl clusters, respectively, (see Figure 7). These have been assigned to the corresponding terminal and bridging carbonyls of the starting clusters. Both carbonyl bands disappeared completely on heating the samples at 50–200 °C to remove the CO ligands. Exposure of the resulting species to CO at 25 °C gave spectra similar to those of the starting carbonyl clusters used for impregnation, although the position and relative intensities of the terminal and bridging carbonyl groups showed some changes; for example, they shifted to lower wavenumbers and showed the preferential appearance of bridging carbonyl bands at ca. 1800–1840 cm^{-1}. This result implies that the Pt impregnated species derived from such clusters of different nuclearity possess a preferred geometric arrangement, in marked contrast to conventional Pt/Al_2O_3 catalysts prepared from Pt salts [27] where the Pt atoms have much lower coordination numbers.

A silica-supported $[Pt_{15}(CO)_{30}]^{2-}$ catalyst (1.5 wt.% Pt) was also readily oxidized with O_2 to remove CO at 25 °C, the carbonyl ligands being converted to CO_2 and CO_3^{2-}. When the oxidized Pt impregnated species was exposed once more to CO, the characteristic terminal and bridging carbonyl bands reappeared in the IR spectrum, being similar to those of the original cluster, as shown in Figure 8.

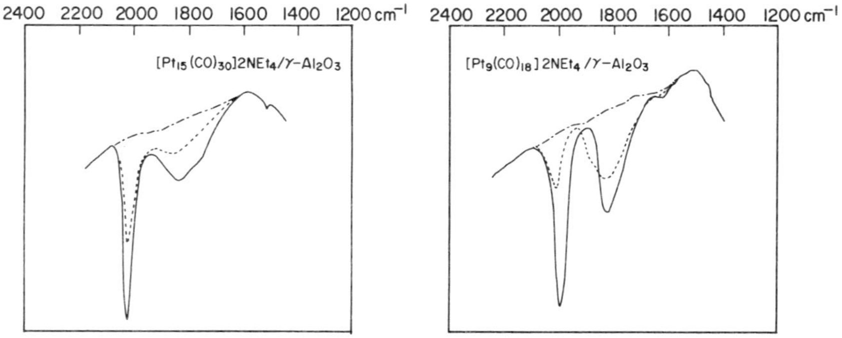

Fig. 7. IR spectra over the ν_{CO} region for $[Pt_{15}(CO)_{30}] \cdot 2Et_4N$ (1.2 wt.% Pt) on γ-Al_2O_3 and $[Pt_9(CO)_{18}] \cdot 2Et_4N$ (1.6 wt.% Pt) on γ-Al_2O_3. The spectra correspond to impregnation of support with Pt carbonyl cluster anion under He (———), followed by evacuation at 250 °C for 0.5 h (–.–.–.–.–) and adsorption of CO at 100 Torr pressure and 25 °C (--------) in each case.

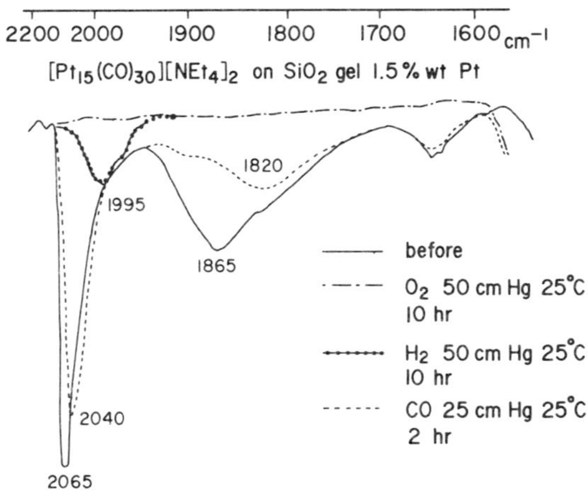

Fig. 8. Changes in the IR spectra of $[Pt_{15}(CO)_{30}] \cdot 2Et_4N$ impregnated onto SiO_2 (1.5 wt.% Pt) (———), followed by mild oxidation with O_2 (400 mmHg, 25 °C (–.–.–) and admission of CO (250 mmHg, 25 °C) (--------) or H_2 (500 mmHg, 25–50 °C) (–o–o–o–o–o–) to the resulting oxidized sample.

When Pt_{15} and Pt_9 carbonyl clusters were deposited onto basic oxides such as MgO, La_2O_3 and TiO_2 (after degassing at 350 °C), the resulting materials (0.5–1.0 wt.% loading) gave similar carbonyl IR bands to the starting cluster complexes but shifted to lower wavenumbers by 40, 20 and 25 cm^{-1} on MgO, La_2O_3 and TiO_2, respectively. No carbonyl fragment corresponding to the oxidized ⫶Rh(CO)$_2$] species was detected during impregnation or subsequent treatment of the Pt carbonyl clusters. The supported Pt cluster anions possibly retain their cluster integrity on these basic oxides, simple electronic or protonic interaction with surface OH^- and O^{2-} groups resulting in the lower carbonyl absorption frequencies observed. When H_2O is present on MgO, the terminal carbonyls of Pt cluster anions interact favorably with the strong basic sites to produce CO_3^{2-} or HCO_2^- $\nu_{C=O}$ 1620–1580 cm^{-1}; ν_{C-O} 1400–1360 cm^{-1}) as formate species at the expense of linear carbonyls.

Ichikawa [28] has studied the dehydrocyclization and skeletal isomerization of *n*-hexane at 250–340 °C in the presence of hydrogen (*n*-hexane/H_2 = 1:10; total pressure, 1 atm) over Al_2O_3 catalysts impregnated with Pt cluster anions. Catalysts derived from smaller Pt carbonyl clusters such as $[Pt_3(CO)_6]_{2,3}\cdot 2Et_4N$ and $Pt_3(CO)_3(PPh_3)_4$ which includes a phosphorus heteroatom in the cluster molecule, act as preferential catalysts for dehydrocyclization with 75% selectivity and reduced hydrogenolysis. Ichikawa also demonstrated that the skeletal isomerization of neopentane proceeded favorably over Pt_{15} and Pt_{12} carbonyl cluster impregnated catalysts with isopentane/isobutane product ratios of 5 : 20 as the hydrogenolyis proceeded, the latter being accompanied by the usual hydrocracking reaction. The results suggest that such preferential dehydrocyclization catalyzed by smaller Pt aggregates prepared from Pt carbonyl clusters might be associated with active sites derived from lower coordinated Pt atoms which favor the formation of adsorbed carbocyclic reaction intermediates as suggested by Anderson, Gault and coworkers [29]. In addition, Maire *et al.* [30] have reported recently that Pt-supported catalysts similarly prepared from $[Pt_3(CO)_6]_5\cdot 2Et_4N$ and $Pt_3(CO)_3(PEt_3)_4$ exhibit particular selectivities for the demethylation of mono- and di-methylcyclopentane. It is interesting that the resultant catalyst derived from $Pt_3(CO)_3(PEt_3)_4$ showed higher selectivities for demethylation relative to hydrogenolysis, possibly due to poisoning effects arising from the phosphine ligands incorporated in the Pt aggregates.

2.2. RUTHENIUM, OSMIUM AND IRIDIUM CARBONYL CLUSTERS IMPREGNATED ON METAL OXIDES

Comparatively stable carbonyl clusters (with higher metal–metal bond energies to retain the cluster frameworks) such as $Ru_3(CO)_{12}$ [31], $Os_3(CO)_{12}$ and

$Ir_4(CO)_{12}$ [32], usually generate essentially the same IR carbonyl bands as those observed in solution when impregnated on the metal oxides such as silica and alumina with or without the presence of adsorbed water. Upon heating the impregnated samples at 50–100 °C *in vacuo*, a minor evolution of CO was observed with a slight decrease in the intensity of the carbonyl IR bands (1990–2100 cm^{-1}). Exposing the resulting samples to a CO atmosphere at 100 °C led to the reversible restoration of the original carbonyl spectra of $Os_3(CO)_{12}$ and $Ru_3(CO)_{12}$, respectively. The Raman spectra of pure $Ru_3(CO)_{12}$ and of the partially decarbonylated species (impregnated on Al_2O_3 at 1 wt.% loading) were virtually identical at this stage over the range of 160 to 205 cm^{-1} as a result of the preservation of the Ru—Ru cluster framework [32].

$$M_3(CO)_{12} \xrightleftharpoons{100\ ^\circ C\ \text{evac.}} x\,CO + M_3(CO)_{12-x}$$

where x is less than 2 and M = Os or Ru impregnated on Al_2O_3 or SiO_2.

On warming to 100–150 °C *in vacuo* or in a helium flow, 2–3 mol of CO (or mixtures of CO and CO_2) are evolved for each mole of $Os_3(CO)_{12}$ and $Ru_3(CO)_{12}$ impregnated on SiO_2 (and also on Al_2O_3) [32, 33], leading to substantial changes in the carbonyl absorption bands which now closely resemble those of hydrido-carbonyl clusters such as $HOs_3(CO)_{10}(O{-}X)$ (X = $SiPh_3$ or Ph) and $HRu_3(CO)_{10}L$ in position (although shifted ca. 20 cm^{-1} to lower wavenumbers), shape and intensity. A similar interaction with the evolution of 3 mol of CO has been reported for the thermal treatment of $Ru_4(CO)_{12}$ on zeolites [34].

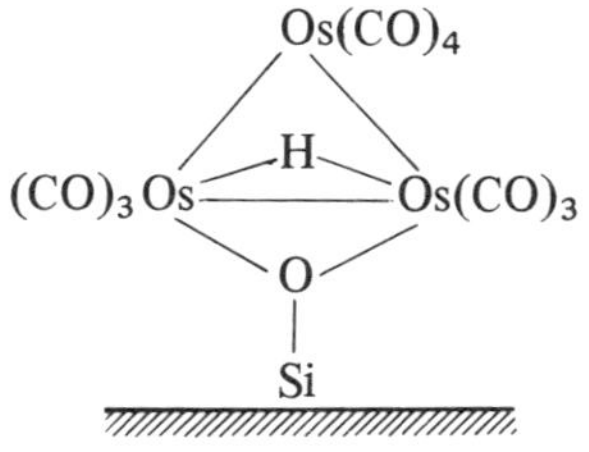

(ν_{CO} = 2112(w), 2088(s), 2073(s), 2037(s), 2010(sh), 1980(sh) cm^{-1})

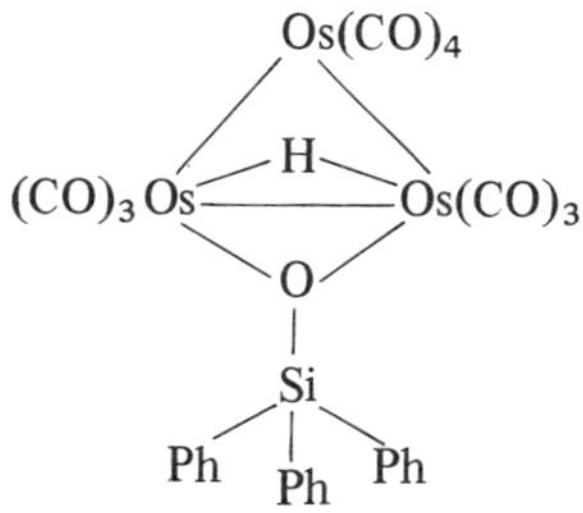

(ν_{CO} = 2107(w), 2069(s), 2055(s), 2017(s), 1997(m), 1980(m) cm^{-1}
^1H—Os NMR, 21.0 ppm (TMS))

EXAFS measurements of the resulting impregnated species revealed that the Os—Os cluster framework was in fact retained, but that average Os—Os nearest-neighbor distance for the adsorbed species was 0.268 nm, shorter than that for the original crystalline $Os_3(CO)_{12}$ as reported by Basset *et al.* [35]. At this stage the impregnated Os- and Ru-carbonyl cluster species could be restored to their original forms by prolonged heating under CO at 100–200 °C. After more severe heating at 150–200 °C *in vacuo* or under or He flow to eliminate carbonyl groups a three-band spectrum (2135(m), 2052(s), 1960(m) cm^{-1}) was obtained for $Os_3(CO)_{12}$ adsorbed on silica or Al_2O_3. This is in agreement with the results reported for $[Os(CO)_3Cl_2]_2$ when dissolved in $CHCl_3$, which also exhinited a three-band IR spectrum (ν_{CO} = 2136(m), 2054(s), 2036(s) and 1965(w) cm^{-1}). This suggests that the origianl supported carbonyl clusters have been destroyed to produce $Os^{II}(CO)_3$ species. Similar carbonyl species have also been obtained from $Os_6(CO)_{18}$ species impregnated onto silica after heating to 200 °C *in vacuo*, such treatment being accompanied by the evolution of CO, CH_4 and higher hydrocarbons [32, 36].

Reaction of $Os_3(CO)_{12}$ and $Ru_3(CO)_{12}$ impregnated on silica or alumina, with H_2 at 200 °C leads to the preparation of metallic catalysts with a high degree of dispersion which act as catalysts for the reduction of CO with the formation of methane and other hydrocarbons [33].

In summary, as for $Ru_3(CO)_{12}$ and $Ir_4(CO)_{12}$, various surface-supported $Os_3(CO)_{12}$ species, have been characterized during different stages of impregnation and subsequent treatment, as depicted in Figure 9. The impregnated carbonyl clusters react with surface OH groups, and this is followed by deformation and cleavage of the cluster frameworks and their connection to oxide surfaces by means of stronger metal/support interactions.

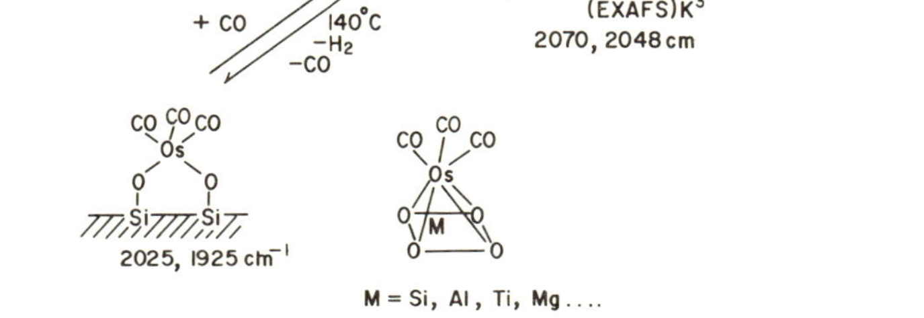

Fig. 9. Various surface-supported osmium species derived from $Os_3(CO)_{12}$ impregnated on SiO_2, followed by thermal activation at different temperatures.

Collier *et al.* [36] have recently shown that $Os_3(CO)_{12}$ and $Os_6(CO)_{18}$ when impregnated on extensively dehydroxylated oxide supports such as SiO_2, Al_2O_3 and TiO_2, and heated *in vacuo* or in a helium flow at 150 °C, are transformed mainly into unique Os_{12} and/or Os_{13} cluster species. The resulting osmium surface species has been characterized by IR and UV – vis reflectance spectroscopies and TPR measurements, which indicated that it contained Os—Os bonding (associated with UV bands above 400 nm) and consisted of 12–13 Os atoms with a particle size less than 0.8 nm (electron microscopy). The Os-impregnated species possibly exists in a formal zero, partially negative and/or partially positive valency state, being affected by the surface properties of the oxides used, e.g. SiO_2, Al_2O_3 or TiO_2.

These workers have suggested that a cluster growth process occurs when the impregnated $Os_3(CO)_{12}$ and/or $Os_6(CO)_{18}$ species is heated, possibly as a result of partial decarbonylation following di- and tri-merization on the oxide surface, as follows:

$$2\,Os_3(CO)_{12} \xrightarrow[-6\,CO]{50-100\,^\circ C} Os_6(CO)_{18}$$

$$\xrightarrow{150\,^\circ C} Os_x(CO)_nC_m + 30(?)\,CO + CH_4 + CO_2 + H_2$$

where x = 12–13, n = 2–3 and m = 0–0.4 in the estimated molecular formula of the resulting complex. The degree of coordinative unsaturation in the final Os species depends upon the type of oxide support employed, with the species supported upon TiO_2 and Al_2O_3 in particular showing a higher coordinative unsaturation for CO chemisorption. These results contrast markedly with those of Basset [33], Gates [32] and Knötzinger [37] who observed considerable fragmentation into $Os(CO)_3$ surface species on the same oxide supports. The discrepancy between these observations may be related to different contents of surface hydroxyl groups and adsorbed water molecules in the supports, which are reactive toward cluster carbonyls as previously mentioned.

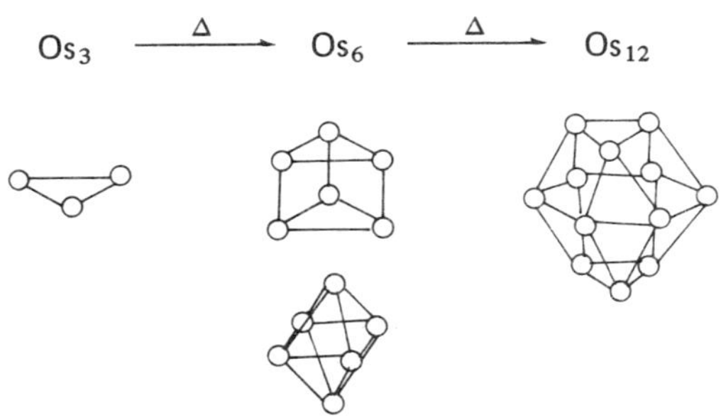

On the other hand, Anderson *et al.* [38] have reported an unusual Ir dispersion for $Ir_4(CO)_{12}$ impregnated on Al_2O_3 after pyrolysis *in vacuo*. From the value of the CO chemisorption ratio CO/Ir = 2.44 for the resulting catalyst it is reasonable to expect that the Ir aggregates produced by decarbonylation of the impregnated $Ir_4(CO)_{12}$ still retain the Ir_4 cluster structure, being converted back to the original $Ir_4(CO)_{12}$ species during CO chemisorption. It is noteworthy that Della Betta and Selef [39] from CO and H_2 chemisorption studies on highly dispersed Ru crystallites in the size range 1.1–2.5 nm (H/Ru = 1) found that the corresponding value for CO/Ru was 2.3–3.8. This suggests that Ru crystallites even bigger than than so-called metal clusters can develop carbonyl clusters such as $Ru_3(CO)_{12}$ and/or $Ru_4(CO)_{12}$ during CO adsorption.

Brenner and Hucul [40] have also claimed unusual chemisorption stoichiometry for very small particles in observing extraordinary values for CO_{ads}/H_{ads} for $Ru_3(CO)_{12}$-derived catalysts on Al_2O_3 (CO_{ads}/H_{ads} = 19–45), those on the traditional Ru catalyst being close to unity in good agreement with the Boudart assumption, i.e. CO/M = 1 and H/M = 1. They suggest that H_2 chemisorption (which being dissociative requires two surface bonds) might itself be a structure-sensitive reaction, whereas CO chemisorption is not. Such a discrepancy in the CO/H adsorption ratio could also be related to the ability of very small metal crystallites to reduce hydrogen activation possibly as a result of their partially oxidized states and metal/oxide support interaction.

In catalyst preparation, the cluster compounds should be dispersed molecularly over large surface area oxide supports before extensive decarbonylation commences. Deposition of a metal crystallite from an impregnated cluster species is undoubtedly a complex process, e.g. ligand degradation is accompanied by chemical interaction with surface functional groups, and surface migration of the resulting subcluster species which may include monometallic complexes or ions formed through metal–metal bond fission accompanied by cluster volatization results in agglomeration to metal crystallites of varying sizes. Preoxidation of the supported cluster species and/or impregnation on supports with reactive surface sites (i.e. acidic or basic oxides) of higher binding energy such as Al_2O_3, MgO, TiO_2 and ZrO_2, may lead to more controlled preparation of highly dispersed metal crystallites by preventing surface migration and volatization of impregnated cluster species.

2.3. IRON, COBALT AND NICKEL CARBONYL CLUSTERS IMPREGNATED ON METAL OXIDES

Reactive metal clusters such as $Fe_3(CO)_{12}$, $Co_4(CO)_{12}$ and $Ni_3Cp_3(CO)_2$ with lower metal–metal bond energies and unstable zero-valent atoms undergo more complicated reactions with metal oxides containing surface OH groups. Thus, the interaction of $Fe_3(CO)_{12}$ (green solid) with the surface of γ-Al_2O_3 containing MgO as a basic oxide results in the formation of an anionic hydrido cluster complex $[HFe_3(CO)_{11}]^-$ [22a] which exhibits characteristic IR carbonyl bands at 2075 and 2005(s) cm^{-1}, respectively. This complex has been separated by ion-exchange extraction of the oxide surface with Et_4NCl as $[HFe_3(CO)_{11}]$ $[Et_4N]$ (IR: 2071, 2002(s), 1970(m), 1945(w) 1715(m) cm^{-1}; UV : 540 nm (deep red)).

The first step in the chemisorption of the carbonyl species onto such a basic surface is nucleophilic attack of $OH^-(O^{2-})$ on a carbonyl ligand leading to the formation of an anionic hydrido cluster linked to cation surface sites such as Al^{3+} and Mg^{2+}, as depicted in the following equation [22]:

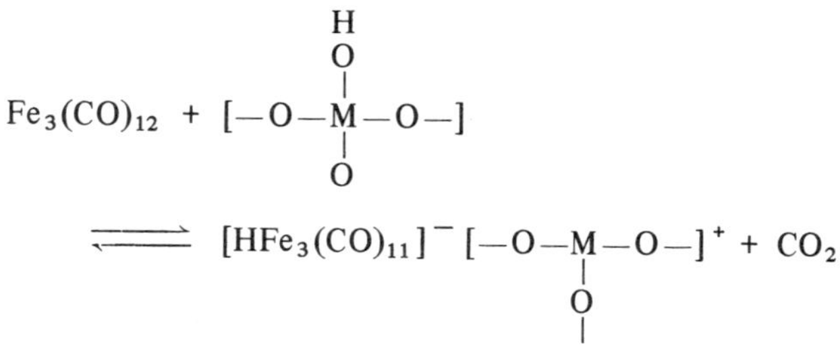

where M = Al^{3+} and/or Mg^{2+}. The CO_2 formed reacts with the surface O^{2-} groups to provide the carbonate CO_3^{2-} (IR: 1470, 1580 cm^{-1}) species. Other subcarbonyl species such as $[HFe(CO)_4]^-$, $[Fe(CO)_4]^-$, $[HFe_2(CO)_8]^-$, $[Fe_3(CO)_{11}]^-$ and $[Fe_4(CO)_{13}]^{2-}$ could not be detected in the IR and UV spectra. In contrast, with acidic oxides such as silica and silica/alumina, simple chemisorption occurred without the formation of hydrido anion cluster species. When the resulting impregnated solid was heated in a CO atmosphere at 50 °C, cleavage of the Fe—Fe bond occurred with the formation of $Fe(CO)_5$. The $[HFe_3(CO)_{11}]^-$ species could also be synthesized [22b] over γ-Al_2O_3 and MgO by adsorbing $Fe(CO)_5$ in the presence of CO at room temperature, when a similar reaction to that occurring from impregnation of $Fe(CO)_5$ or $Fe_2(CO)_9$ in a methanolic KOH solution or an aqueous NaOH solution took place to give the same cluster anion salts.

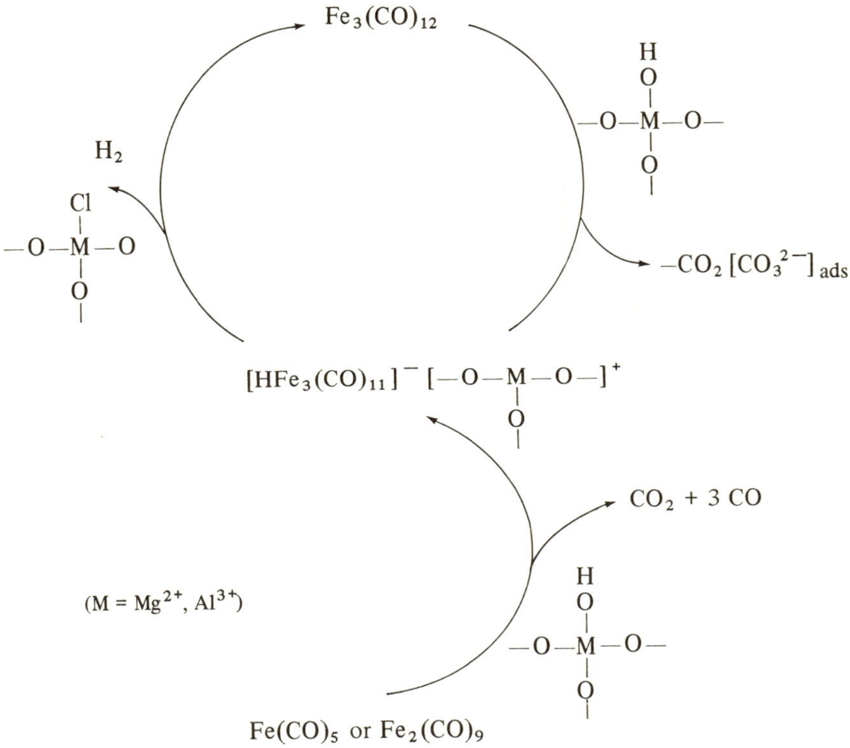

Iwasawa *et al.* [41] have characterized the Fe carbonyl cluster species formed on Al_2O_3 and SiO_2 by means of EXAFS, IR and UV–DRS, and confirmed that the structures of the different adsorbed clusters are as follows:

On Al_2O_3

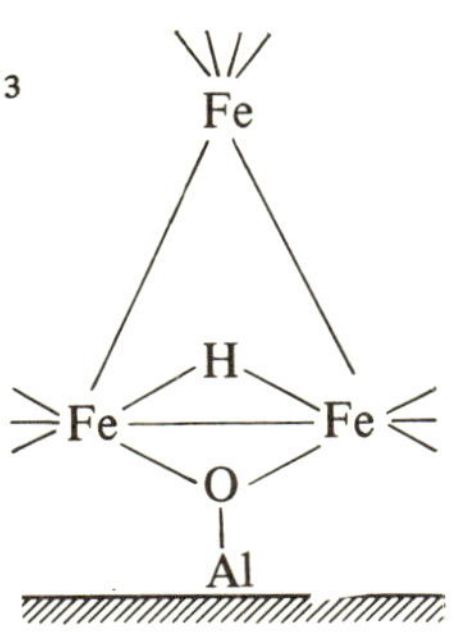

On SiO_2

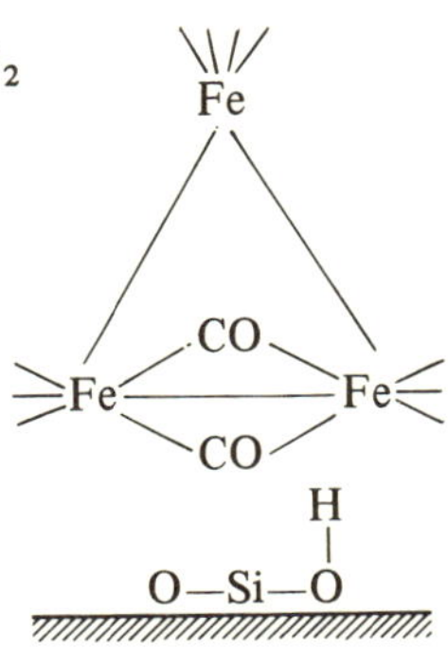

Recently, Iwamoto *et al.* [42] have reported that the $[HFe_3(CO)_{11}]^-$ species is also formed on an NaY zeolite during reaction with $Fe(CO)_5$ or $Fe_2(CO)_9$ as 56 °C in the presence of water vapor. They claim that the cluster species formed (ν_{CO} = 2040, 1988, 1947 and 1600 cm^{-1}) strongly interacts with Al^{3+} ions in the zeolite and is possibly encapsulated in the zeolite cavities.

With cobalt species, Yermakov *et al.* [43] have previously reported that when $Co_2(CO)_8$ is deposited from hexane solution on a pre-evacuated SiO_2 gel, the impregnated species may be readily converted into $Co_4(CO)_{12}$ which exhibits IR carbonyl bands similar to those of the solid crystal. Nakamura *et al.* [44] have also studied the surface reaction of $Co_2(CO)_8$ impregnated both on γ-Al_2O_3 and SiO_2, and found that a variety of surface-supported subcarbonyls such as $[Co_4(CO)_{12}]_{ads}$, $[Co_2(CO)_6B]$ and $[Co(CO)_3B]$ (when B denotes basic sites on the γ-Al_2O_3 surface, such as O^{2-} and OH^-) were formed on controlled heating. Iwasawa *et al.* [45] have conducted further characterization of the Co_4 surface species impregnated on γ-Al_2O_3 and SiO_2, employing data from IR, UV–DRS and EXAFS measurements. They have suggested that the resulting $Co_4(CO)_{12}$ may be adsorbed in C_{3v} symmetry on SiO_2, but transformed to a rectangular structure such as $Co_4(CO)_{10}(PPh_3)_2$ on γ-Al_2O_3 possibly as a result of the stronger support interaction. EXAFS data on Al_2O_3-supported Co_4 species reveal that its Co—Co bonds (average length, 0.254 nm) are elongated by 0.005 nm in comparison to those in crystalline $Co_4(CO)_{12}$. In contrast, the $Co_4(CO)_{12}$ species formed on SiO_2 is retained essentially intact with the Co—Co bond distance the same as that in the original $Co_4(CO)_{12}$. Iwasawa *et al.* have attempted to transform the $[Co_4(CO)_{12}]_{ads}$ species to various other surface cobalt species which could also be characterized by EXAFS, Raman and IR spectroscopies. The surface transformation of such Co carbonyls is described in detail in the article by Iwasawa in this volume.

Burwell and Parkyns [4, 5, 46] have studied the reaction between $Ni(CO)_4$ and a suitably preheated sample of γ-Al_2O_3 as a method of preparing highly dispersed metal catalysts with extraordinary activities for the hydrogenation of ethylene and the hydrogenolysis of ethane. The resulting materials may involve very small Ni aggregates, possibly as clusters and/or as isolated Ni atoms in a reduced state, which are associated with the supporting oxide.

Ichikawa [47] has described the preparation of supported Ni clusters chemically bound to a silica surface by heating impregnated nickel cyclopentadienyl complexes such as Cp_2Ni, $Cp_2Ni_2(CO)_2$ and $Ni_3Cp_3(CO)_2$. When the temperature was raised to 120 °C, impregnated $Ni_3Cp_3(CO)_2$ (loading 0.1–0.5 wt.%) reacted with surface SiOH groups to produce cyclopentadiene and CO, the IR spectrum of $Ni_3Cp_3(CO)_2$ on SiO_2 gel exhibiting peaks at 1750 cm^{-1} (three-centered CO) and 800, 650 cm^{-1} (Cp) which were similar to those for the original cluster. Upon heating the sample *in vacuo*, these adsorption bands gradually decreased together with the intensities of the broad peaks at 3750–3650 cm^{-1} corresponding to surface SiOH groups. When CO (100 Torr pressure) was admitted at 25 °C onto the resulting impregnated sample, approximately 2 mol of CO was chemisorbed per impregnated Ni_3 cluster unit, and hydrogen uptake occurred with the stoichiometry H/Ni = 1.98. These results contrast markedly with those for the conventional Ni-supported catalysts, which generally exhibit average chemisorption H/Ni and CO/Ni stoichiometries less than unity. Furthermore, upon introducing CO (5–100 Torr) onto the pyrolyzed sample produced from impregnated $Ni_3Cp_3(CO)_2$ on silica gel (evacuated at 120 °C and 10^{-4} Torr), a strong ESR signal ($g_\perp$ = 2.026 and $g_\parallel$ = 2.134) was observed over the temperature range from −150 °C to 25 °C. This ESR signal which is illustrated in Figure 10 resembles that of the original cluster compound after its physically adsorption onto silica or in the free solid state. The same ESR signal was also observed when ethyl isonitrile (a σ-electron pair donating molecule) was admitted instead of CO onto the $Ni_3Cp_3(CO)_2$-impregnated silica gel after pyrolysis. These results imply that after pyrolysis of $Ni_3Cp_3(CO)_2$ on silica, the surface Ni atoms may retain the three-centered disposition they exhibit in the original cluster moiety.

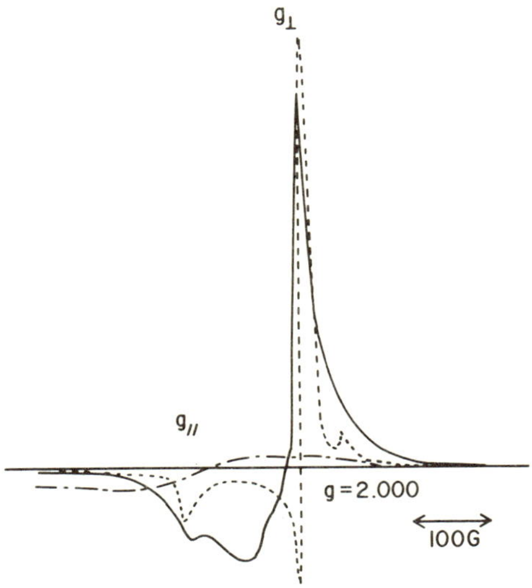

Fig. 10. ESR spectra of [$Ni_3Cp_3(CO)_2$] on silica gel (0.3 wt.% Ni) (——) ($g_\perp$ 2.02_6, $g_\parallel$ 2.12_6); after evacuation at 10^{-4} Torr and 120 °C for 10 min (–.–.–.–.–); and after admitting CO (100 Torr) onto the resulting sample for 5 min at 25 °C (---------) ($g_\perp$ 2.02_6, $g_\parallel$ 2.13_4).

Acetylene readily trimerized to benzene at room temperature (with a 75% yield in 2 h) over a dispersed Ni surface species prepared from $NiCp_2$ impregnated onto silica gel (0.5 wt.% Ni) and pyrolyzed at 120 °C; such trimerization possibly involves as oxidized Ni^{II} state. However, the oligomerization of acetylene did not proceed catalytically over the resulting catalysts obtained from the impregnation of $Ni_3Cp_3(CO)_2$ and/or $Ni_2Cp_2(CO)_2$ followed by pyrolysis. Instead ca. 1 mol of acetylene was absorbed per mole of Ni_3 and/or Ni_2 units present, the impregnated Ni samples which were originally pale gray in color rapidly becoming red or violet implying the formation of acetylene-coordinated cluster complexes similar to $Ni_2Cp_2[CH_3C{\equiv}CCH_3]$ and $Ni_4(CO)_4[CH_3C{\equiv}CCF_3]$. In contrast, H_2–D_2 exchange and the hydrogenations of ethylene and benzene occurred with high efficiency at room temperature over the Ni_2 and Ni_3 cluster-derived catalysts. However, in spite of the presence of Ni cluster surface species, acetylene trimerization (to benzene and/or cyclooctadiene) proceeds catalytically with $Ni_4[CNC(CH_3)_3]_7$ in solution [48].

When a 1:1:1 mixture of H_2, CO and C_2H_4 (1 atm total pressure) was admitted onto the supported Ni catalysts prepared by the pyrolysis of $Ni_2Cp_2(CO)_2$ and $Ni_3Cp_3(CO)_2$ impregnated on silica, propionaldehyde (C_2H_5CHO) was obtained in low yield at 50–100 °C. When the temperature was increased to 120 °C, the activity toward hydroformylation was lost and $Ni(CO)_4$ was evolved. Such hydroformylation reactions did not occur over the catalyst obtained from $NiCp_2$ by pyrolysis *in vacuo* or even in the presence of $Ni(CO)_4$ in the gas phase under similar reaction conditions.

2.4. ZEOLITE-ENTRAPPED CLUSTER CATALYSTS

$Rh_6(CO)_{16}$, $Ir_4(CO)_{12}$ and $Ru_3(CO)_{12}$ have been entrapped in appropriate sized cavities of HY, NaY and 13X zeolites. Because of the size limitation of the zeolite pore windows (ca. 0.8 nm), it is usually difficult to directly introduce bulky carbonyl clusters such as $Rh_6(CO)_{16}$ and $Ir_4(CO)_{12}$ into them. To overcome this difficulty, the metal clusters were either synthesized or rebuilt from exchanged metal ions and/or smaller size subcarbonyls, e.g. $Rh(CO)_2$ and $Ir(CO)_3$, inside the zeolite cavities. Thus, Montovani *et al.* [16] have reported that the ion-exchanged NaY zeolite containing $[Rh(NH_3)_6]^{3+}$ may be converted into a $Rh_6(CO)_{16}$-encapsulated zeolite by treatment with CO and H_2 (1 : 1 volume ratio, 80 atm total pressure at 130 °C). As shown in Figure 11, the entrapped species gives rise to IR carbonyl bands (2095(vs), 2080, 2060 and 1765(s) cm^{-1}) which substantially resemble those of $Rh_6(CO)_{16}$ in solution, but with differences in the exact peak positions, particularly with respect to the bridging carbonyls (which are shifted to lower wavenumbers) and possibly as a result of strong interaction with the acid Al^{3+} sites in the zeolite cavity.

Furthermore, Naccache *et al.* [17] have demonstrated by IR spectroscopy that reaction of CO and CO/H_2O (1 atm, 80–100 °C) with an Ir^{3+}-exchanged HY zeolite leads to the formation of entrapped $Ir_4(CO)_{12}$ exhibiting characteristic IR carbonyl bands at 2085, 2040 and 1815 cm^{-1}, respectively.

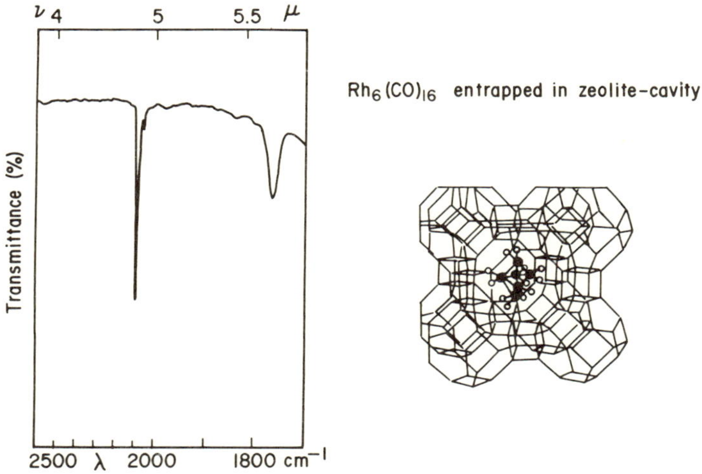

Fig. 11. (a) Infrared spectrum of Rh-exchanged NaY zeolite after heating at 130 °C, under a pressure of 80 atm of CO/H_2 (1 : 1); (b) Structure of $Rh_6(CO)_{16}$ entrapped in the zeolite cavity.

From X-ray diffraction studies, Kim and Seff [49] suggested the existence of an Ag_6 cluster stabilized by coordination to Six Ag^+ ions as the structure $[Ag]_6\,[Ag^+]_6$ entrapped in a zeolite A cavity. These particular cluster species were prepared by vacuum dehydration and partial decomposition of multi-exchange Ag^+ coordinated with ethylene. The entrapped Ag species consists of the octahedral silver cluster Ag_6 complexed with Six Ag^+ ions as the center of the sodalite cavities with an Ag—Ag bond length of 0.285 nm, as shown in Figure 12.

Entrapped $Rh_6(CO)_{16}$ in zeolite has proved [16] to be active as a catalyst in the hydroformylation of hexene-1 with *n*/*iso* selectivities for the aldehyde similar to those obtained with $[Rh(CO)_4]^-$ under homogeneous catalysis conditions. The impregnated catalyst can be easily removed by decantation, and re-used without any noticeable loss in activity. However, zeolite-entrapped $Ir_4(CO)_{12}$ exhibited poor catalytic activity for the carbonylation of methanol compared with homogeneous mononuclear species such as $[Ir(CO)_3I]^+$ [50].

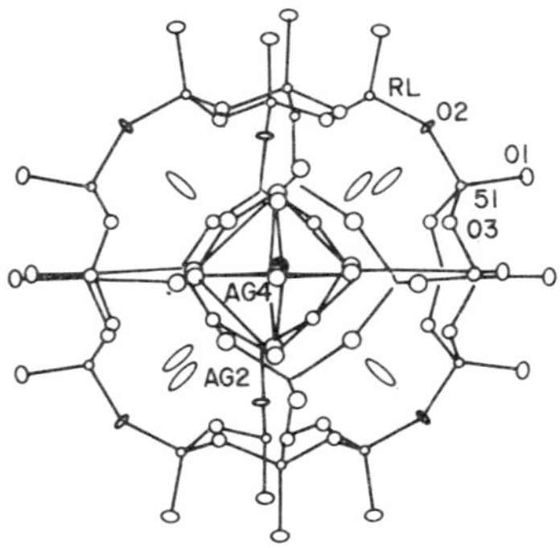

Fig. 12. The disposition of the $[Ag_6][Ag^+]_6$ complex entrapped in a sodalite cavity in zeolite A.

2.5. REACTIVITY AND CATALYSIS OF SURFACE-SUPPORTED METAL CLUSTER SPECIES

Since most organometallic complexes used as metal precursors, including molecular cluster compounds, are coordinatively saturated (with the result that occasionally they are catalytically inactive), it is obvious that the dissociation of ligands such as CO and phosphines is necessary to enable the development of active sites on the metals. To enable the characterization of supported cluster species, particularly metal carbonyl clusters, Brenner *et al.* [51] have measured and analyzed the evolution of CO, CO_2 and other reduced products during activation of the resulting impregnated samples. It is not unreasonable to assume that the development of coordinative unsaturation should be related to the evolution of CO during thermal activation. It has been also shown that during activation the incipient zero-valent metal generated can undergo a redox reaction with the acidic hydroxyl groups of oxide supports such as silica and alumina, causing oxidation of the metal and the evolution of hydrogen. At lower temperatures, the controlled (limited) decarbonylation of the impregnated clusters occurs, sometimes yielding a broad range of particular subcarbonyl metal species:

$$M_x(CO)_y \underset{\text{photochemical activation}}{\overset{\text{thermal or}}{\rightleftharpoons}} M_x(CO)_{y-m} + m\,CO$$

$$M_x(CO)_{y-m} + n\,(O{-}H)$$

$$\rightleftharpoons \quad M_x^{n+}\,(O^{2-})_n + n/2\,[H_2] + (y-m)\,CO + m\,CO_2 + \ldots$$

where O—H are the surface hydroxyl groups of the oxide supports.

Brenner [52] and others [53, 54] have conducted reactive characterization through the use of simple test reactions to measure the catalytic activities of the surface species formed, for example in ethylene hydrogenation, ethane hydrogenolysis and methanation. The combination of catalytic activity and TPDE (temperature-programed decomposition) data lead to the following conclusions:

1. After activation at lower temperatures, there is only slight loss of CO, so the surface metal species are possibly coordinatively saturated or of limited unsaturation, and should therefore have a very low activity.
2. When the impregnated clusters have lost a considerable amount of CO (and other ligands) but only a slight amount of H_2 during the development of coordinative unsaturation, the resulting metal sites will be close to zero-valent.
3. After the complete evolution of CO (or other ligands) accompanied by the evolution of H_2, the surface metal species will be substantially oxidized or exist as ions. As a result they would be expected to have a lower activity.

Reactive characterization of the resultant surface species has revealed that catalysts of metals which are difficult to reduce, e.g. Mo, W, Cr, Mn, V, Fe, Co and Ni, are most active after thermal activation via process 2 above, since this maximizes the coordinative unsaturation but leaves a metal with a lower valency. On the other hand, catalysts of metals which are easily reducable, e.g. Rh, Ir, Pt, Ru and Os, exhibit higher activities after activation via process 2 or 3 above followed by re-reduction with H_2. In all the test reactions, such as ethylene hydrogenation, ethane hydrogenolysis and methanation, the catalysts derived from carbonyl metal precursors such as $Rh_6(CO)_{16}$, $Ru_3(CO)_{12}$ and $Os_3(CO)_{12}$ were found [55] to be at least ten times more active than their traditional counterparts. Moreover, it was found [56] that catalysts of species which were more difficult to reduce, particularly Mo, W and Mn carbonyl cluster-derived catalysts, after activation via process 2 exhibit quite extraordinary activities (up to 10^4 times more than that of the traditionally prepared metal catalysts). Metal ions conventionally impregnated on oxides such as alumina and silica are incapable of complete reduction with hydrogen even above 500 °C. In contrast, noble metals derived from cluster compounds are readily reduced with hydrogen after oxidation of the incipient metal aggregates by surface OH groups or adsorbed water through thermal treatment.

Along with this, Robertson and Webb [57] have demonstrated that SiO_2-supported $Ru_3(CO)_{12}$, which is catalytically inactive, may be partially decarbonylated by heating *in vacuo* at 150 °C when it is transformed into $Ru_3(CO)_5$ chemically bound to the silica. The resulting material was particularly active toward the isomerization and hydrogenation of butene-1. Complete pyrolysis above 200 °C produced a catalyst which was inactive for isomerization, but catalyzed the hydrogenation of butene-1, thus resembling the conventional

Ru/Al_2O_3 catalyst prepared from the salt. Wells *et al.* [58] have extended this work and suggested that the structurally restricted active sites generated from $Ru_6(CO)_{17}(C)$ impregnated on SiO_2 catalyze the isomerization of butene-1, yielding a favorable *cis/trans* ratio as shown in Table II. The reaction proceeded to thermodynamic equibrium (lower values of *cis/trans* ratios) over the impregnated $[Rh_6(C)]$ carbonyl cluster catalyst after the complete removal of CO by heating, similar to the conventional Ru/Al_2O_3 catalyst.

TABLE II
Comparison of product compositions in the isomerization of butene-1 catalyzed by various ruthenium catalysts[a] [58]

Catalyst	Temp./K	Product composition *cis*-Butene-2	*trans*-Butene-2	*cis/trans* Ratio
$Ru_6(CO)_{17}(C)/SiO_2$[b]	253	70	30	2.3
(1.2 wt.% loading)	293	61	39	1.6
After decomposition at 358 K	293	42	58	0.7
Ru/Al_2O_3 (1.0 wt.% loading)	306	37	63	0.6

[a] Reaction conditions: butene-1/H_2 = 7.5; total pressure 15 Torr. Isomerization does not occur in the absence of H_2.
[b] The structure of $Ru_6(CO)_{17}(C)$ may be visualized as

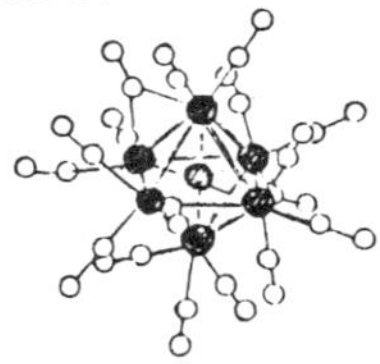

Although the isomerization of butene-1 is a facile reaction, its use together with the hydrogenation of butadiene (as categorized by Boudart [59]) could provide a sensitive chemical probe of the intimate environment of metal sites, thus providing information concerning site congestion and selectivity, and determining whether molecular crowding at active sites has an influence an the catalytic activity.

2.6. BIMETALLIC CLUSTERS IMPREGNATED ON METAL OXIDES

Anderson *et al.* [38] were the first to report that catalysts prepared from $Rh_2Co_2(CO)_{12}$ impregnation on γ-Al_2O_3 rather than on SiO_2 gave much higher dispersions (CO/M = 0.4–1.0) than conventional catalysts prepared from aqueous salt solutions. They claimed that catalysts prepared from the bimetallic cluster carbonyl provided a uniform composition of the metal, possibly as Rh/Co = 1 : 1 as in the ideally mixed alloy. The surface enrichment of Co atoms, χ_s(Co), was evaluated simply from thermodynamic considerations, e.g.

the different surface energies γ_{Co} = 2.37 and γ_{Rh} = 2.74 J m^{-2}, respectively, and mass balance criteria which should logically depend upon the metallic dispersion of the alloy particles. As a consequence and as shown in Table III, at lower dispersion the surface of the metal particles will consist of a larger fraction of cobalt atoms rather than of rhodium.

TABLE III
Characterization of Rh–Co supported catalysts [38]

Catalyst and preparation	Mean metal particle size (nm) H_2 ads.	EM[a]	MS[b]	Metal dispersion	Surface mole fraction of cobalt $x_s(Co)^c$
$Rh_2Co_2(CO)_{12}/SiO_2$ (1 wt.% loading)					
O_2-treated, prior to H_2 red.	2.7	2.7		0.44	0.96–0.99
Pyrolyzed under N_2	2.8	2.8	2.8	0.38	
$Rh_2Co_2(CO)_{12}/Al_2O_3$					
Pyrolyzed under N_2	1.3	1.0–1.5		0.76	0.63–0.64
O_2-treated, prior to H_2 red.	1.2			0.98	0.51
Conventional Rh–Co/SiO_2					
$RhCl_3$–$Co(NO_3)_2$, H_2 red.	5.6		5.0	0.16	0.97–0.99

a Electron microscopy.
b From magnetic susceptibility.
c Calculated values from metal dispersion.

Ichikawa *et al.* [60] have recently conducted *in situ* EXAFS measurements to characterize the surface composition of the catalysts derived from bimetallic clusters such as $RhCo_3(CO)_{12}$ and $Rh_2Co_2(CO)_{12}$ impregnated on γ-Al_2O_3. The catalysts were studied separately upon impregnation of the bimetallic cluster, followed by heating *in vacuo* at 200 °C and reduction in an H_2 flow at 400 °C. The EXAFS spectra were taken at each stage of the proceedings using a synchrotron orbital radiation (SOR) laboratory EXAFS system. The EXAFS data obtained were analyzed by theoretical curve-fitting methods from Fourier-filtered and calculated data. Examples of the analyzed results are presented in Table IV, and show that when dehydrated Al_2O_3 (evacuated at 350 °C for 2 h) was impregnated with $Rh_2Co_2(CO)_{12}$ or $RhCo(CO)_{12}$, the impregnants retained their Rh—Rh and/or Rh—Co bonds in cluster configurations similar to those in the original materials, with the same coordination numbers and bond distances.

It was also demonstrated by IR spectroscopic measurements that $Rh_2Co_2(CO)_{12}$ when impregnated on γ-Al_2O_3 exhibited considerable shifts of the carbonyl bands to lower wavenumbers possibly as a result of the strong interaction between Al^{3+} sites and bridging carbonyls. After heating at 200 °C *in vacuo*

TABLE IV
Curve-fitting results from EXAFS data [60].

	Rh—Rh		Rh—Co	
Sample	N^a	r (nm)	N^a	r (nm)
$Rh_4(CO)_{12}$ crystal	3	0.273	–	–
$RhCo_3(CO)_{12}$ crystal	–	–	3	0.259
$Rh_2Co_2(CO)_{12}$ crystal	1	0.278	2	0.257
Rh_2Co_2 cluster/Al_2O_3 impreg.	0.7	0.279	1.8	0.257
Rh_2Co_2 cluster/Al_2O_3 heated	–	–	1.8	0.260
Rh_2Co_2 cluster/Al_2O_3, H_2 red.	~1	0.271	2.0	0.252
$RhCo_3$ cluster/Al_2O_3, H_2 red.	~0.4	0.273	2.6	0.254
Rh/Co = 1 : 1 salt/Al_2O_3, H_2 red.	~8	0.270	~4	0.255
Rh/Co = 1 : 3 salt/Al_2O_3, H_2 red.	~7	0.269	~5	0.255

[a] N = coordination number.

to remove CO, the resulting samples showed cleavage of the Rh—Rh bonds but retention of Rh—Co bonds, implying that each Rh atom is surrounded by Co atoms to provide a uniform surface composition similar to that in the starting cluster. The bimetallic crystallites prepared by hydrogen reduction of oxidized $Rh_2Co_2(CO)_{12}$ and $RhCo_3(CO)_{12}$ in an oxygen atmosphere at 25 °C showed a substantially higher dispersion of metal with coodination numbers less than 6 with respect to the Rh and Co atoms, and with a homogeneous surface composition for the bimetallics.

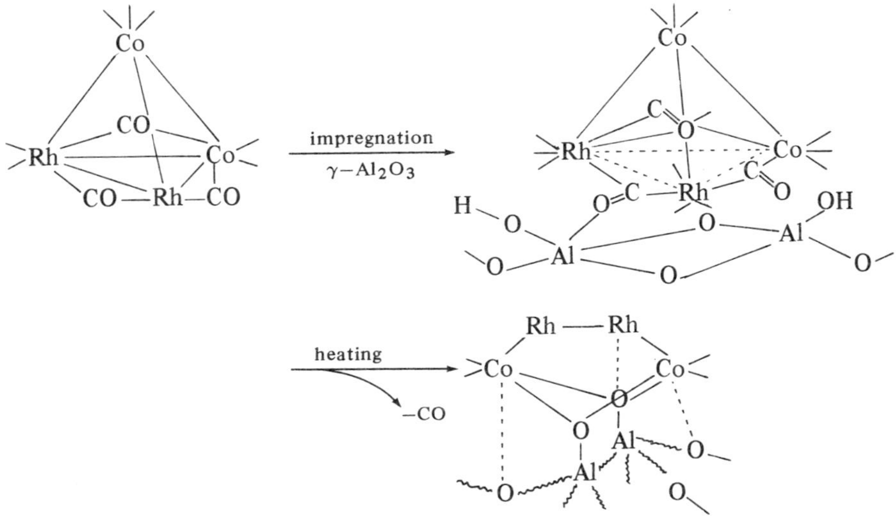

In contrast, conventional impregnation from the Rh and Co salts, e.g., $RhCl_3$ and $CoCl_2$, gave lower metal dispersions (coordination numbers of almost 12 in total, corresponding to those of the bulk metals) implying immisible alloy compositions with considerable enrichment of cobalt atoms at the particle surfaces. In conventionally prepared Rh–Co catalysts, the Rh and Co atoms were segregated from each other.

The resulting catalyst from the pyrolysis of $Rh_2Co_2(CO)_{12}$ on SiO_2 has been applied by Anderson and Mainwaring [7] in the reforming of hydrocarbons such as methylcyclopentane. The bimetallic cluster-derived catalyst showed catalytic activity principally toward skeletal isomerization into 2- and 3-methylpentane, which differs from the behavior of the conventional monometallic Rh and Co catalysts. The latter catalysts enable preferential cracking to lower hydrocarbons such as methane under similar conditions.

Linksey *et al.* [61a] have described the preparation of supported Pt–Sn bimetallic catalysts through the use of $Pt_3Sn_2Cl_6(COD)_3$ or $Pt_3Sn_2Cl_6(NBD)_3$ impregnated on γ-Al_2O_3. After hydrogen reduction, the resulting material led to a lower conversion with the same selectivities in the skeletal isomerization of *n*-hexane in comparison to conventional Pt–Sn/Al_2O_3 catalysts prepared by co-impregnation of the salts.

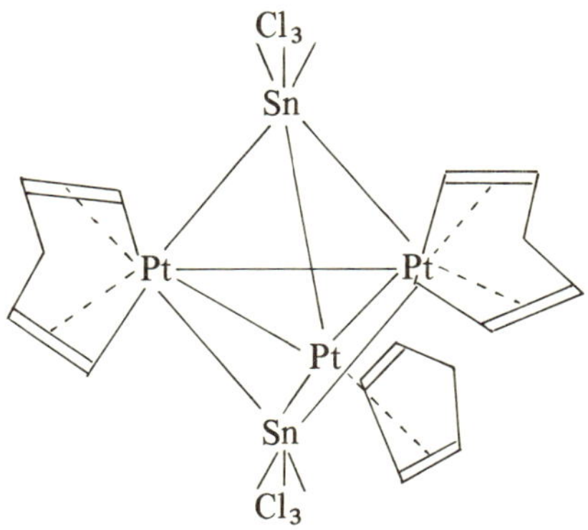

[$Pt_3Sn_2Cl_6(COD)$]

Since the number of bimetallic cluster complexes is limited, the conventional preparation of impregnated bimetallics has been carried out with a mixture of monometallic carbonyl clusters or subcarbonyls as metal precursors. Guzz *et al.* [62] have employed $Fe_3(CO)_{12}$ and $Ru_3(CO)_{12}$ in the preparation of Fe–Ru bimetallic alloy catalysts. Their characterization by TPD, XPS and Mössbauer spectroscopic measurements revealed that some chemical interaction occurs between Fe and Ru in the pyrolyzed materials obtained from Fe and Ru carbonyl clusters even at low metal loading. Nevertheless, the catalysts provided much higher activities for the hydrogenolysis of *n*-butane and in the conversion of a CO/H_2 mixture to methane, although with unchanged selectivities compared with conventional catalysts prepared via the co-impregnation of Fe and Ru salts from aqueous solution.

Barbier and Braunstein [63] have tried to prepare bimetallic catalysts containing Co and Mn by using mixtures of $Mn_2(CO)_{10}$ and $Co_2(CO)_8$ at various metal ratios and measuring the extent of $CO-H_2$ conversion over the pyrolyzed materials. The characterization and the detailed catalytic behavior observed remain open to question at the present moment. In contrast, Sinfelt *et al.* [64] have used EXAFS, Mössbauer and electron microscope measurements extensively to verify the formation of bimetallic clusters during the conventional co-impregnation of metal salts. Thus, supported Ru–Cu and Os–Cu catalysts from co-impregnated salts gave high metal dispersion (1–2 nm particle size) with uniformly mixed compositions, e.g. the coordination numbers of Cu atoms arranged around the Ru or Os atoms were always less than six. Sinfelt *et al.* claimed the formation of bimetallic clusters even during the combination of immiscible metal components with two-dimensional rafts and three-dimensional clusters being formed on different oxide supports such as SiO_2, Al_2O_3 and TiO_2. Additionally, bimetallic Ru–Cu and Os–Cu catalysts at higher metallic dispersions showed a specific inhibition toward hydrogenolysis of ethane through the incorporation of copper, although the activity toward the dehydrogenation of cyclohexane to benzene was only affected marginally. By squeezing into the formed clusters, the inactive copper destroyed the active Ru or Os ensembles and thus caused selective suppression of the hydrogenolysis reaction [65].

3. Selectivity Aspects of Fischer–Tropsch (F–T) Synthesis with Supported Metal Carbonyls

Heat treatment *in vacuo* or in a helium flow of oxide-supported carbonyl clusters such as $Rh_6(CO)_{16}$, $Fe_3(CO)_{12}$ and $Co_4(CO)_{12}$ at temperatures below 200 °C led to a mixture of methane and other higher hydrocarbons with only a slight evolution of CO_2 and H_2 [66]. Other stable carbonyl clusters such as $Ru_3(CO)_{12}$, $Os_3(CO)_{12}$ and $Ir_4(CO)_{12}$ are similarly transformed by heating above 200 °C, particularly when impregnated on hydrated oxide surfaces, to liberate CO or CO_2 and H_2 initially, followed by hydrocarbons consisting mainly of methane with slight amounts of C_2–C_5 hydrocarbons [33, 35].

The efficiencies and selectivities of hydrocarbon formation depend roughly on the temperature, the nature of the oxide support (its water content and surface OH group concentration) and the nature of the starting carbonyl clusters. When a carbonyl species impregnated on a metal oxide is heated, the carbonyl ligand interacts with surface OH groups (or adsorbed H_2O molecules) to produce

H_2 and CO_2 via the water–gas shift reaction. A part of the hydrogen reacts with the carbonyls, surface carbon and/or hydrocarbons CH_x, converting them to methane and other hydrocarbons. Whitmire and Shriver [67] have found that the proton-induced reduction of CO derived from $[Fe_4(CO_{13}]^{2-}$ through reaction with a strong Brönstead acid such as $H_2SO_3CF_3$ at room temperature, produced 0.5 CH_4 molecules per cluster.

$$[Fe_4(CO)_{13}]^{2-} \xrightarrow[3\ d,\ 25\ ^{\circ}C]{H_2SO_3CF_3} \tfrac{1}{2}\ CH_4 + Fe^{2+} + CO + H_2 + \text{Fe clusters}$$

Isotopic experiments demonstrated that the carbon atom in the CH_4 originated from a carbonyl carbon of the starting cluster and that the hydrogen was derived directly from the proton reagent, and not via the intermediate H_2 evolved. In this process the necessary electrons for the reduction of carbonyl ligands may be provided by the metal clusters, with consequent conversion of the clusters to Fe^{2+} ions during the formation of CH_4; as in the following mechanism:

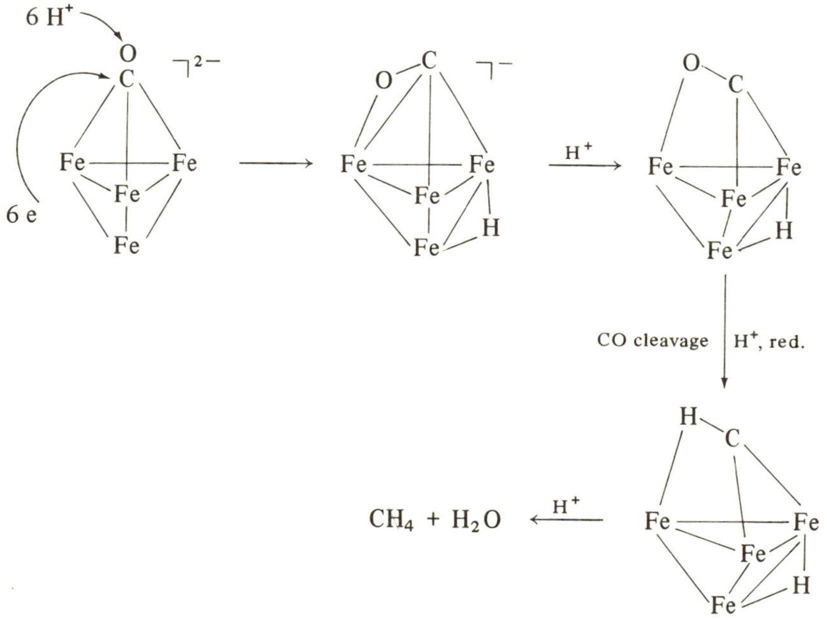

By means of a similar proton-induced CO reduction, carbonyl clusters such as $Ru_3(CO)_{12}$, $Fe_3(CO)_{12}$ and $Co_2(CO)_8$ impregnated onto strong acid oxide supports, e.g. zeolite HY and alumina/silica, have been converted to the metal cations Ru^{3+}, Fe^{2+} and Co^{2+} with accompanying evolution of CH_4 and H_2 through heating, particularly in the presence of water [34, 68].

TABLE V

Gaseous products from the reduction of CO on various carbonyl complexes using excess hydride reagents in THF or toluene

Carbonyl source	Hydride	Conversion (%)	Products (%)					
			CH_4	C_2H_4	C_2H_6	C_3H_6	C_3H_8	$C_4H_8 + C_4H_{10}$
$Ru_3(CO)_{12}(1)$[a]	$AlH_3(6)$[a] in THF	10	28.5	48.6	14.3	5.7	2.8	trace
$Cr(CO)_6(1)$	$AlH_3(16)$ in THF	8	trace	81	16	3	trace	trace
$Cp_2Fe_2(CO)_4(1)$	$LiAlH_4(30)$ in toluene	11	26	41	23	4.7	3.7	trace

[a] Molar parts present in reaction mixture.

It is of interest to note that when $Rh_6(CO)_{16}$ impregnated on γ-Al_2O_3 was subjected to controlled heating under a He flow, selective formation of ethylene and other lower olefins occurred at lower conversions [66]. These results contrast markedly with those from conventionally prepared Rh/Al_2O_3 (1 wt.% loading) obtained from $RhCl_3$, which exhibited no such selective olefin formation during the reaction of CO and CO/H_2O under similar conditions. This implies that the the more selective formation of lower olefins from cluster-derived carbonyl species may be due to the size limitation of the resulting metal aggregates controlling the concentration of surface cabon. In this regard, Masters *et al.* [70] have recently observed the selective formation of lower olefins such as ethylene and propylene in the homogeneous reduction of $Ru_3(CO)_{12}$ with AlH_3 being employed as a homogeneous analog of the conventional Fischer–Tropsch catalysts. As summarized in Table V, a number of carbonyl complexes undergo similar homogeneous reduction with $LiAlH_4$ and $NaBH_4$ to selectively provide C_2 products of ethylene and/or ethane. Demitras and Muetterties [71] have shown that treatment of $Ir_4(CO)_{12}$ with CO/H_2 in the presence of molten $NaCl/AlCl_3$ at 180 °C leads to the formation of hydrocarbons consisting of methane; ethane, propane and butane in the ratio 1:4:trace:trace. Masters *et al.* [70] and Wong and Atwood [72] have suggested that ethylene (and propylene) may be obtained through dimerization of an alkylidene as follows:

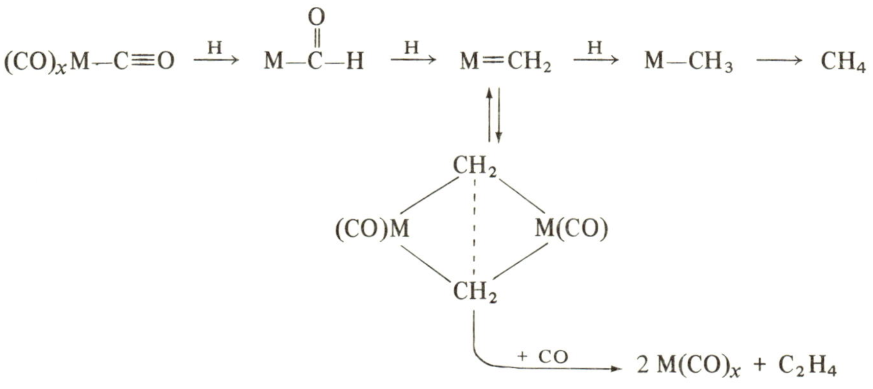

Basset *et al.* [73] have reported that impregnation of $Fe_3(CO)_{12}$ on MgO or Al_2O_3 (possibly leading to the formation of $[HFe(CO)_{11}]^-$ and adsorbed $Fe_3(CO)_{12}$ leads to high selectivity for lower olefins consisting mainly of propylene when the impregnated solids were contacted with a mixutre of CO and H_2 at 200 °C. The hydrocarbon distribution obtained over the resultant catalyst is shown in Figure 13.

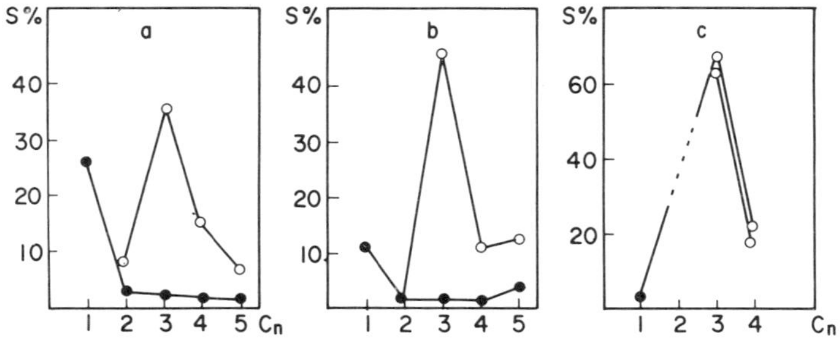

Fig. 13. Selectivities in the reaction of (a) $CO-H_2$ (CO/H_2 = 2; total pressure, 1 atm) at 176 °C, and (b) ethylene at 170 °C with catalysts derived from $Fe_3(CO)_{12}$ on MgO, and (c) of $CO-H_2$ on Al_2O_3 at 270 °C. The catalyst (400 mg) employed in (a) and (b) possessed a loading of 2 wt.% Fe and was decarbonylated irreversibly at 150 °C before each catalytic run. The experimental points refer to olefin (○) and paraffin (●), respectively.

The high selectivities for propylene (> 45%) and the low selectivities for ethylene (< 10%) suggest that ethylene could be the primary product in the Fischer–Tropsch synthesis undergoing a secondary reaction which leads selectively to propylene. This suggestion is supported by the selective formation of propylene when ethylene is added alone to the same catalyst (see Figure 13b). The formation of a source of methlene fragments from ethylene can be accounted for by homolytic cleavage of the C_2H_4 molecule. The results may be explained on the basis of a new mode of C—C bond formation which implied α-olefin coordination to surface methylene fragments or methylene insertion into a methyl–alkyl bond via a metallocyclic intermediate, as depicted in the following reaction scheme:

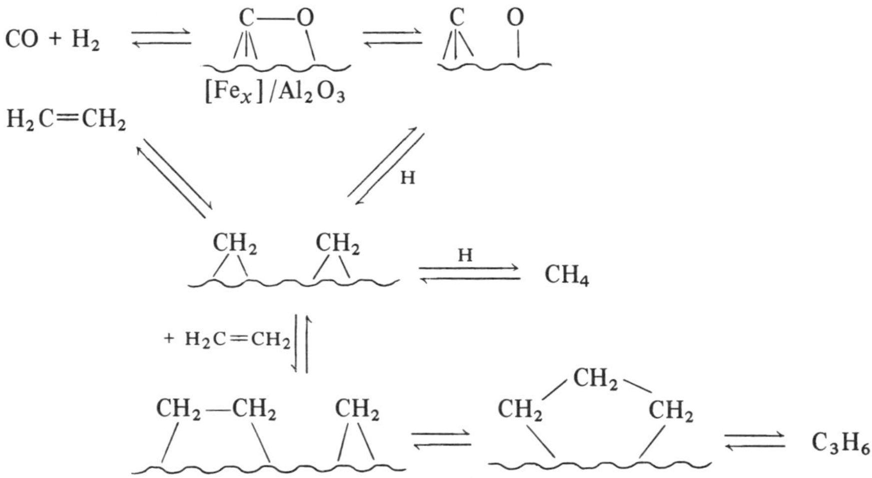

A similar propagation mechanism has been proposed by Pettit *et al.* [74] who obtained propylene selectively from the reaction of ethylene with an octacarbonyl–ν-methylene diiron complex.

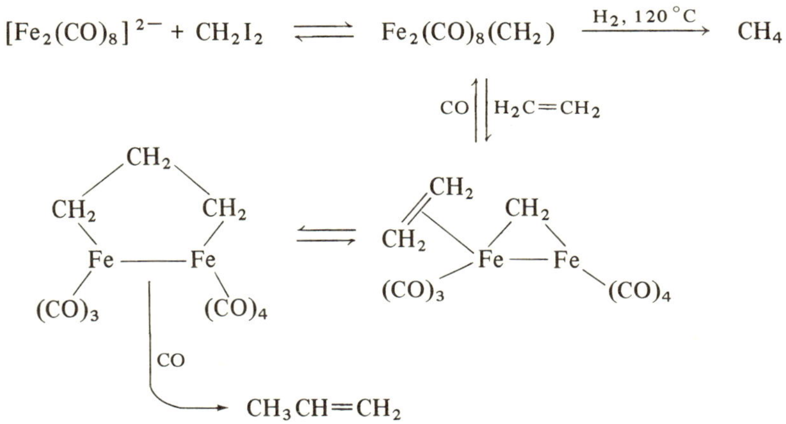

Basset *et al.* [75] have demonstrated by Mössbauer and electron microscopy that after thermal decarbonylation and CO–H_2 reaction at 200 °C, $Fe_3(CO)_{12}$-impregnated catalysts consist of very small Fe particles whose size was estimated to be 1.4–2.0 nm corresponding to ca. 100–150 Fe atoms, which is a much higher nuclearity than that of the starting cluster. It is difficult to speculate why selectivity for propylene should be associated with small iron particles. Possibly the small Fe aggregates strongly adsorb olefins and thus prevent further propagation steps. In addition, it is reasonable to expect that these small metal particles would have lower hydrogenation activities which may prevent the formation of methane from the methylene fragment and of saturated hydrocarbons from the olefins. Nevertheless, the selectivity toward propylene was as high as 45% during the initial stages of the reaction, then declined with time to reach a value close to that observed with conventional catalysts with an acceptable Schulz–Flory distribution. In fact, electron microscopy measurements of the degenerated catalyst have revealed a marked increase in the particle sizes (average 50 nm), possibly due to metal migration during the course of the CO–H_2 reaction.

Recently, Frenkel and Gates [76] have reported that a catalyst with Co crystallites encapsulated within cavities in zeolite A (possibly as clusters) produced propylene catalytically with 80–100% selectivity although at less than 1% conversion from CO–H_2, at a pressure of 1–6 atm and 150–180 °C. The ion-exchanged Co in the zeolite cavities (ca. 0.5 nm pore size) could be reduced directly with Cd vapor. Following CO adsorption, the IR crabonyl bands exhibited by the catalyst, which occur at 2069, 1977, 1968 and 1935 cm^{-1}, resemble those which have been observed from Co carbonyl clusters. The high selectivities mentioned above could be attributed to the retention of the Co (Co–Cd) aggregates of limited size generated within the shape-selective zeolite cavities, allowing mechanistic control of selective propylene formation. When the reaction temperatures were increased above 200 °C, the selectivity toward propylene declined sharply with the formation of methane and higher hydrocarbons, a behavior which is close to that of the conventional Co/Al_2O_3 catalyst. This change is associated with the agglomeration of Co crystallites to produce large crystals outside the zeolite cavities.

4. Carbonylation Reactions catalyzed by Cluster-impregnated Catalysts

Although olefin hydroformylation reactions readily proceed in homogeneous systems catalyzed by Rh and Co monometallic carbonyl complexes, these reactions hardly occur when these same monometallic complexes are simply impregnated on silica, alumina or carbon. Conventionally prepared metal catalysts of Rh and Co also exhibit negligible catalytic activities for olefin hydroformylation in the vapor phase. Recently, Rh and Co ion-exchanged zeolite catalysts have been studied for the vapor-phase hydroformylation of olefins such as ethylene and propylene under atmospheric or elevated pressures [77]. Such catalysts are reported as having relatively low activities compared with those in homogeneous systems and to produce paraffins as the principal products through hydrogenation. These hydroformylation reactions are probably catalyzed by monometallic carbonyl complexes formed on the zeolite surface and/or encapsulated in zeolite cavities.

Ichikawa [78] has reported that surface-supported clusters of Rh and/or Co impregnated on ZnO and other appropriate metal oxides exhibit marked catalysis of the vapor-phase hydroformylation of olefins at atmospheric pressure:

$$H_2C{=}CH_2 + CO + H_2 \rightleftharpoons C_2H_5CHO\ (+\ C_3H_7OH)$$

$$C_3H_6 + CO + H_2 \rightleftharpoons i\text{-}/n\text{-}C_3H_7CHO\ (+\ i\text{-}/n\text{-}C_4H_9OH)$$

A gaseous mixture of C_2H_4 (or C_3H_6), CO and H_2 (1:1:1 molar ratio with a total pressure of 650 mmHg) was admitted onto $Rh_4(CO)_{12}$ (0.12 mmol) or $Rh_6(CO)_{16}$ (0.12 mmol) supported on ZnO powder (Kadox-25/20 g with surface area of ca. 20 $m^2\ g^{-1}$) at ca. 0.5 wt.% loading. Such catalysts were inactive for the hydroformylation reaction at temperatures below 90 °C, but on increaseing the temperature further the corresponding aldehyde was initially produced after an induction period. From volumetric observations it was shown an additional amount of CO was evolved during this induction period, probably due to decomposition of the impregnated carbonyl clusters. The length of this induction period could be reduced by warming the system. The infrared spectrum of $Rh_6(CO)_{16}$ impregnated on ZnO (at 0.75 wt.% Rh loading) was recorded before and after heating the catalyst *in vacuo* or in the presence of a hydroformylation atmosphere consisting of C_2H_4, CO and H_2. As shown in Figure 14, the IR spectrum of the first prepared sample exhibited strong carbonyl bands at 2020 and 2070 (broad, not resolved) and 1795 cm^{-1} which resemble those of terminal and three-centered bridging carbonyls in chloroform solution or in a Nujol mull.

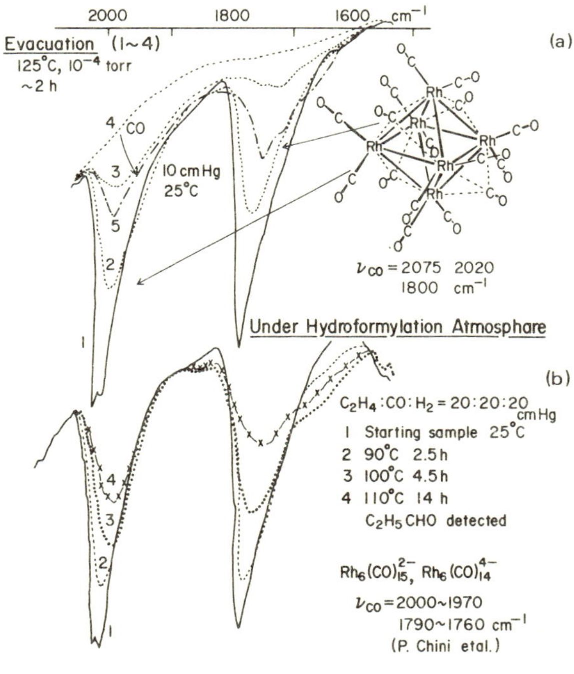

Fig. 14. IR spectra of carbonyl bands associated with $Rh_6(CO)_{16}$ supported on ZnO (0.75 wt.% Rh loading) (a) after initial outgassing at 10^{-4} Torr pressure for *ca*. 2 h at 25 °C and (b) after exposure of the outgassed catalyst to a $C_2H_4/CO/H_2$ mixture (1:1:1 ratio) at a total pressure of 600 mmHg. The individual spectra in Figure 14a correspond to the following: curve 1, initial outgassed sample at 25 °C; curve 2, after heating at 90 °C for 2.5 h; curve 3, after heating at 100 °C for 4.5 h; and curve 4, after heating at 110 °C for 14 h. Curve 5 in (a) was obtained after exposure of the sample outgassed at 110 °C for 14 h to 100 mmHg pressure of CO at 25 °C.

When the impregnated sample was evacuated at 10^{-3} Torr with a slow increase in the temperature up to 150 °C, both carbonyl bands diminished together in parallel and finally disappeared almost completely, probably due to CO elimination. After removal of the carbonyl ligands by such evacuation and heat treatment, exposure of the resulting sample to 50 to 200 mmHg of CO at 25 °C and 50 °C led to the reappearance of relatively broad bands in the 1985 and 1785 cm^{-1} region of the spectrum of about half the intensity of those observed for the initial sample.

In the presence of a $C_2H_4/CO/H_2$ mixture (1:1:1 volume ratio and total pressure of 600 mmHg), the initial carbonyl bands for $Rh_6(CO)_{16}$ adsorbed onto ZnO at 2060 and 1800 cm^{-1} shifted to lower wavenumbers (1985 and 1785 cm^{-1}) when the sample was warmed. Upon heating to 90–110 °C, the intensities of these bands decreased to half their initial values and a weak but distinct new band appeared at 1650 cm^{-1} as shown in Figure 14. At this stage in the procedure propionaldehyde was produced catalytically in the gas phase contained in the IR cell. The resulting IR carbonyl bands resemble those produced by coordinatively unsaturated Rh_6 carbonyl clusters such as $[Rh_6(CO)_{15}]^{2-}$ and $[Rh_6(CO)_{14}]^{4-}$ in solution [121]. The particular band at 1650 cm^{-1} formed in the presence of hydroformylation atmosphere at 110 °C may be due to acyl species analogous to $[Rh_6(CO)_{15}(C_2H_5CO)]^-$ and $[Rh_6(CO)_{15}(C_3H_7CO)]^-$ ($\nu_{(C=O)}$ 1655–1670 cm^{-1}) for example, which were prepared independently by the reaction of $Rh_4(CO)_{12}$ with ethylene or propylene and CO in toluene solution as reported by as Chini *et al.* [79]:

$$Rh_4(CO)_{12} + C_2H_4 + H_2 \xrightarrow[\text{80\% yield}]{} [Rh_6(CO)_{15}(C_2H_5CO)]\ [Et_4N]$$

$$3\ Rh_4(CO)_{12} + 4\ C_3H_6 + 4\ H_2$$

$$\longrightarrow [Rh_6(CO)_{15}(C_3H_7CO)]^-$$

$$\xrightarrow{\text{90–95\% yield}} 1:1\ n\text{-}/i\text{-}C_3H_7CHO + 2\ Rh_6(CO)_{16}$$

$Rh_6(CO)_{16}$ itself was inactive toward effecting the transformation into acyl species. Consequently, these results suggest that Rh carbonyl clusters such as $Rh_6(CO)_{16}$ impregnated onto ZnO are partially decarbonylated but retain their metal frameworks, with the result that they are catalytically active toward olefin hydroformylation in the vapor phase.

4.1. ACTIVITIES AND SELECTIVITIES IN ETHYLENE AND PROPYLENE HYDROFROMYLATION CATALYZED BY CLUSTER-DERIVED CATALYSTS FROM DIFFERENT RHODIUM CARBONYL CLUSTERS ON ZINC OXIDE

It is interesting to find that the activities and selectivities of cluster-derived catalysts toward ethylene and propylene hydroformylation are substantially dependent upon the nuclearities of the precursor Rh carbonyl clusters.

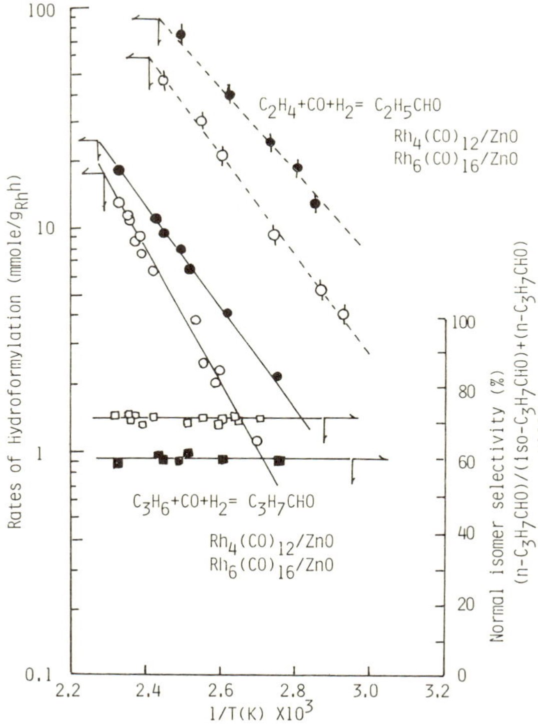

Fig. 15. Temperature dependencies of the rates (mmol (g Rh)$^{-1}$ h^{-1}) and *n*-isomer selectivities in ethylene and propene hydroformylations over $Rh_4(CO)_{12}$- and $Rh_6(CO)_{16}$-derived catalysts. $Rh_4(CO)_{12}$ (0.15 mmol)/ZnO(Kadox 25, 20 g) and $Rh_6(CO)_{16}$ (0.11 mmol)/ZnO(Kadox 25, 20 g), followed by thermal activation *in vacuo* at 160 °C. $C_2H_4/CO/H_2$ = 20:20:20 cmHg; $C_3H_6/CO/H_2$ = 18:18:20 cmHg.

Thus, $Rh_4(CO)_{12}$ and $Rh_6(CO)_{16}$ when impregnated onto ZnO (0.25 wt.% loading) and after removal of their carbonyl ligands by evacustion at 160 °C exhibit distinct activities and selectivities for ethylene and propylene hydroformylation as shown in Figure 15. The specific rates of aldehyde formation over the $Rh_4(CO)_{12}$-impregnated species were 2–3 times larger than those observed over the $Rh_6(CO)_{16}$-impregnated catalyst. Furthermore, the isomer selectivities in propylene hydroformylation (where *n*- and *i*-C_3H_7CHO and the products) were substantially dependent on the precursor clusters, viz. 56–60% *n*-isomer selectivity over the $Rh_4(CO)_{12}$-impregnated ZnO catalyst and 68–70% over the corresponding $Rh_6(CO)_{16}$-derived catalyst over a wide range of applied temperatures. The selectivity was independent of olefin conversion and virtually independent of the partial pressures of propylene, CO and H_2. Moreover, it was found [80] that the *n*-isomer stereoselectivies and relative rates during hydroformylation were consistent over catalysts prepared via different activation processes, e.g. by decarbonylation *in vacuo* at 160 °C, heating in the atmosphere employed for hydroformylation and using different degrees of Rh loading on ZnO. On the other hand, when $Rh_4(CO)_{12}$- or $Rh_6(CO)_{16}$-impregnated ZnO was heated at 200 °C under a hydrogen flow, the formation of the corresponding alcohol during ethylene hydroformylation was catalyzed with an increased yield as was the hydrogenation of ethylene to ethane. This implies that after reduction with H_2, Rh-impregnated catalysts may consist of crystallites which are free from CO and which may control the active sites for olefin hydroformylation.

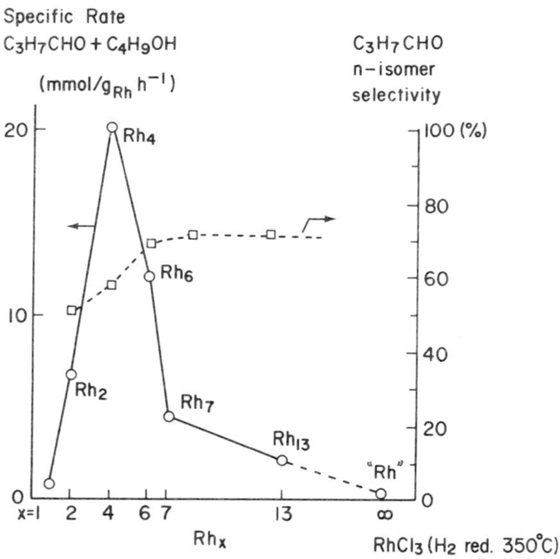

Fig. 16. Influence of size of precursor Rh carbonyl clusters on the specific rates and *n*-isomer selectivities in propylene hydroformylation catalyzed over the resultant cluster-derived catalysts on ZnO (0.5 wt.% Rh loading on Kadox 25 ZnO; $C_3H_6/CO/H_2$ = 18:18:20 cmHg at 158 °C). The following precursors were employed: $RhCp(CO)_2$, $Rh_2Cp_2(CO)_3$, $Rh_4(CO)_{12}$, $Rh_6(CO)_{16}$, $[Rh_7(CO)_{16}]\cdot 3Bu_4N$, and $[Rh_{13}(CO)_{23}H_{2-3}]\cdot 2Bu_4N$. The value at $Rh_x = \infty$ was obtained using $RhCl_3$ as a precursor with H_2 reduction at 350 °C.

As summarized in Figure 16, the following order of specific activities was observed at 158 °C for propylene hydroformylation as the type of Rh carbonyl precursor supported on ZnO was changed: $Rh_4(CO)_{12} > Rh_6(CO)_{16} > [Rh_7(CO)_{16}] \cdot 3Bu_4N \gg [Rh_{13}(CO)_{23}H_{2\sim 3}] \cdot 2Et_4N$. *n*-Isomer selectivities increased over catalysts from larger size Rh carbonyl clusters. However, conventional Rh/ZnO, Rh/SiO_2 and Rh/Al_2O_3 catalysts obtained by impregnation with $RhCl_3$ followed by H_2 reduction at 350 °C showed lower or negligible conversion during ethylene and propylene hydroformylation in the vapor phase under similar reaction conditions, with hydrogenation to paraffins being the dominant reaction. Impregnation of other materials, such as $RhCl_3$, $Rh(CO)_2(\pi\text{-}C_5H_5)$ and $Rh(CO)_2PPh_3$ onto ZnO and silica gel at 0.5–1.5 wt.% loadings gave samples which provided a minute conversion in the initial stages of ethylene hydroformylation, but whose activities diminished rapidly and finally disappeared.

4.2. EFFECT OF METAL OXIDE SUPPORT ON CATALYTIC BEHAVIOR OF RHODIUM CLUSTERS

In addition to ZnO, other basic oxides such as MgO, La_2O_3, TiO_2 and ZrO_2 have also been shown to act as favorable supports in the preparation of Rh cluster-derived catalysts exhibiting activities for olefin hydroformylation, as summarized in Table VI.

It was found that further hydrogenation to alcohol proceeded simultaneously in reactions conducted over these catalysts and, rather surprisingly, that the isomer selectivities in propylene hydroformylation were affected by the properties of the oxide surface, i.e. whether the surface was acidic or basic. In particular, the iso-aldehyde was obtained preferentially when $Rh_4(CO)_{12}$ or $Rh_6(CO)_{16}$ was impregnated on MgO and TiO_2. In contrast to these basic oxides, acidic oxides such as SiO_2, $\gamma\text{-}Al_2O_3$, SiO_2/Al_2O_3, zeolite 13X, V_2O_5 and P_2O_5, provided inactive or marginally active catalysts when impregnated with Rh carbonyl clusters. On heat treatment at 90–150 °C in the presence of a mixture of C_2H_4, CO and H_2, the $Rh_6(CO)_{16}$ complex impregnated on SiO_2 or Al_2O_3 was transformed into the metal. This was implied by the irreversible change in the IR carbonyl bands from those characteristic of the cluster at 2060 and 1800 cm^{-1} to higher energy bands at 2050 and 1860 cm^{-1}, and a weak band at 2090 cm^{-1}. The latter IR bands were observed during CO adsorption on conventional Rh/Al_2O_3 catalysts. When the Rh carbonyl clusters were impregnated on more acidic oxides their unique cluster configuration was barely maintained, possibly as a result of the proton-induced reduction of the carbonyl ligands resulting in fragmentation and/or oxidation of Rh and allowing its conversion to metal crystallites on reduction with hydrogen.

TABLE VI
Propylene hydroformylation over various metal carbonyl clusters impregnated on metal oxides compared with that over conventional rhodium supported catalyst[a]

Catalyst (0.5 wt.% loading)		Hydroformylation characteristics	
Metal carbonyl	Metal oxide	Rate, V[b]	n-Isomer selectivity (%)[c]
$Rh_4(CO)_{12}$	ZnO	21	59
$Rh_6(CO)_{16}$	ZnO	11	71
$Rh_6(CO)_{16}$	MgO	5	38
$Rh_6(CO)_{16}$	TiO_2	2	62
$Rh_6(CO)_{16}$	ZrO_2	3.8	72
$Rh_6(CO)_{16}$	La_2O_3	3.5	75
$Rh_6(CO)_{16}$	SiO_2	0.4	63
$Rh_6(CO)_{16}$	Al_2O_3	0.01	50
$Co_4(CO)_{12}$	ZnO	1.2	91
$[Pt_{15}(CO)_{30}]\cdot 2Et_4N$	ZnO	–	–
$Ir_4(CO)_{12}$	ZnO	trace	–
Rh[d]	ZnO	0.01	–

[a] Reaction conditions: $C_3H_6/CO/H_2$ = 18:18:20 cmHg at 158 °C.
[b] V expressed in mmol (g Rh)$^{-1}$ h^{-1}.
[c] n-Isomer selectivity = n-$C_3H_7CHO/(n$-$C_3H_7CHO + i$-$C_3H_7CHO) \times 100$.
[d] $RhCl_3/ZnO$ reduced in hydrogen (1 atm) at 350 °C.

4.3. BIMETALLIC RHODIUM–COBALT AND COBALT CARBONYL CLUSTERS SUPPORTED ON ZINC OXIDE AND CARBON

Ethylene and propylene hydroformylation reactions also proceed over bimetallic Rh–Co and Co carbonyl clusters, e.g. $Rh_2Co_2(CO)_{12}$, $RhCo_3(CO)_{12}$ and $Co_4(CO)_{12}$, which may be considered as homologs of $Rh_4(CO)_{12}$ when impregnated on ZnO [81]. The typical specific activities and n-isomer selectivities presented in Figure 17 show the following dependency on the metal composition: $Rh_4(CO)_{12}(100) > Rh_2Co_2(CO)_{12}(60) > RhCo_3(CO)_{12}(42) \gg Co_4(CO)_{12}(5)$, where the figures in parentheses are the relative rates of aldehyde formation per unit weight of metal.

For the bimetallic Rh–Co cluster-derived catalysts, the specific hydroformylation activities per Rh atom were virtually the same as those for the $Rh_4(CO)_{12}$-derived catalysts. This suggests that each rhodium atom in Rh and Rh–Co clusters impregnated on ZnO has an equal facility for promoting hydroformylation. On the other hand, catalysts derived from clusters having larger Co contents generated higher proportions of linear (n-) aldehydes. The n-isomer selectivities

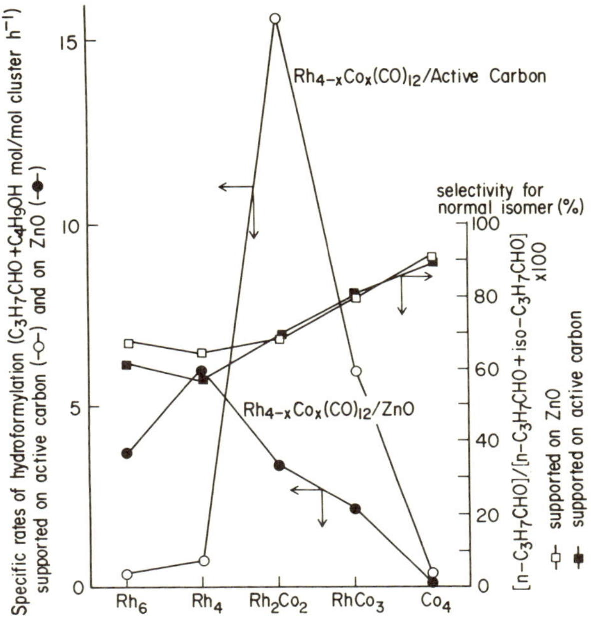

Fig. 17. Influence of metal compositions in bimetallic clusters, $Rh_{4-x}Co_x(CO)_{12}$, and $Rh_6(CO)_{12}$ impregnated on ZnO and active carbon on the specific activation rates and n-isomer selectivities for propylene hydroformylation (0.15 mmol cluster impregnated on 20 g ZnO or 7 g active carbon; $C_3H_6/CO/H_2$ = 1:1:1; total presure, 0.8 atm; 152 °C). The following precursors were employed: $Rh_6(CO)_{16}$, $Rh_4(CO)_{12}$, $Rh_2Co_2(CO)_{12}$, $RhCo_3(CO)_{12}$ and $Co_4(CO)_{12}$. For specific rates, ○ relate to active carbon-supported and ● relate to ZnO-supported catalysts. For n-isomer selectivities, □ relate to ZnO-supported and ■ relate to active carbon-supported catalysts.

observed were obviously associated with the Rh–Co compositions, being independent of kinetic variables such as temperature, pressure and the method of catalyst preparation. Although the $Co_4(CO)_{12}$-impregnated catalyst gave n-isomer selectivities greater than 90% in propylene hydroformylation, the specific rates were lower (by ca. 1/50 per metal atom) than those over the $Rh_4(CO)_{12}$-impregnated catalyst. For this reason, the higher n-isomer selectivity of the mixed Rh–Co cluster impregnated species cannot be explained simply in terms of additional aldehyde formation catalyzed by individual Rh and Co sites. In fact, when ZnO was impregnated with $Rh_4(CO)_{12}$ or $Co_4(CO)_{12}$ (1:1 on 1:3 molar ratio), the resultant catalyst produced comparatively lower n-isomer selectivities than catalysts derived from bimetallic Rh–Co carbonyl clusters.

As has been suggested by EXAFS measurements for bimetallic Rh–Co carbonyl clusters impregnated on Al_2O_3 [60], supported clusters on ZnO also behave as uniformly dispersed alloys having metal compositions similar those of the corresponding precursor Rh–Co carbonyl clusters. Furthermore, it is possible that the cobalt atoms in impregnated Rh–Co aggregates act as electronic donor ligands akin to phosphine ligands modifying the *n*-isomer production during catalysis of olefin hydroformylation by Rh active sites associated with the oxide supports.

When carbon (active carbon; apparent surface area, 500 m^2 g^{-1}) was employed as a support for Rh, bimetallic Rh–Co and Co carbonyl clusters during catalyst preparation, the resultant catalysts exhibited marked activities and selectivities for olefin hydroformylation, as depicted in Figure 17. In particular, catalysts from mixed metal clusters such as $Rh_2Co_2(CO)_{12}$ provided extraordinarily high activities in comparison to simple Rh and Co cluster-derived catalysts, while the *n*-isomer aldehydes selectivities followed trends similar to those observed for the corresponding cluster-derived catalysts on ZnO. This suggests that metal–support interaction is also important in controlling both the geometry and the surface states of the impregnated cluster species, thus leading to improved catalysis of the reaction.

5. Synthesis of Oxygenated Carbon Compounds from Synthesis Gas

According to the Vannice classification of transition metals as catalysts in CO hydrogenation, since Ru, Fe and Co are typical F–T metals they favor the catalytic dissociation and methanation of CO. In contrast, Pt, Pd and Ir are virtually inactive toward CO dissociation and methanation at temperatures below 300 °C. In this context, the behavior of Rh metal itself is intermediate between those of these two extreme groups, as illustrated by the following sequence of relative activities for CO methanation and dissociation during chemisorption:

(*a*) *CO methanation* [82]

$$CO + 3H_2 \rightleftharpoons CH_4 + H_2O \; (CO/H_2 = 0.5;\ 1\ \text{atm};\ 275\ °C)$$

Ru ≫ Fe > Ni > Co > Rh > Pd > Pt > Ir when employed as traditional catalysts supported on Al_2O_3.

(*b*) *CO dissociation* [83]

$$CO \rightleftharpoons [C] + [O] \; (2\,CO \rightleftharpoons [C] + CO_2)$$

Ti, Mo, W, Fe ≫ Ni, Co > Ru > Rh ≫ Pt, Pd, Ir.

< 25 °C   120 °C   130–200 °C   > 300 °C

The trend in activities for CO dissociation is reasonably consistent with that for the methanation reaction catalyzed by the corresponding transition metals. Noble metals such as Pt, Pd and Ir, which dissociate CO with difficult but have high activities toward hydrogenation, catalyze the selective conversion of CO to methanol as high pressures [84].

It has been shown recently that Rh metal can catalyze the conversion of $CO-H_2$ into useful organic compounds, particularly C_2 oxygenates such as C_2H_5OH, CH_3CHO and CH_3COOH, with high efficiencies when a high-pressure synthesis gas atmosphere is employed [85]. In addition, monometallic Rh carbonyls exhibit strong tendencies toward homologation catalysis, yielding various C_2 oxygenates and C_1 homologs such as CH_3OH and HCHO through CO insertion, although they are inactive for CO dissociation and methanation reactions in solution, as shown by the data listed in Table VII.

For this reason, we would expect that the metal nuclearity of the surface species, e.g. ion/atom, cluster or metal, would influence their behavior as catalysts for CO dissociation in chemisorption, CO insertion in, for example, carbonylation and for the propagation of carbon–carbon bonds possibly catalyzed by Rh and other metals. The metal–support interaction might also affect the surface states and compositions of Rh crystallites associated with supports, resulting in subtle changes in the product distribution in, for example, $CO-H_2$ conversion, as shown for Rh-supported catalysts (Figure 18).

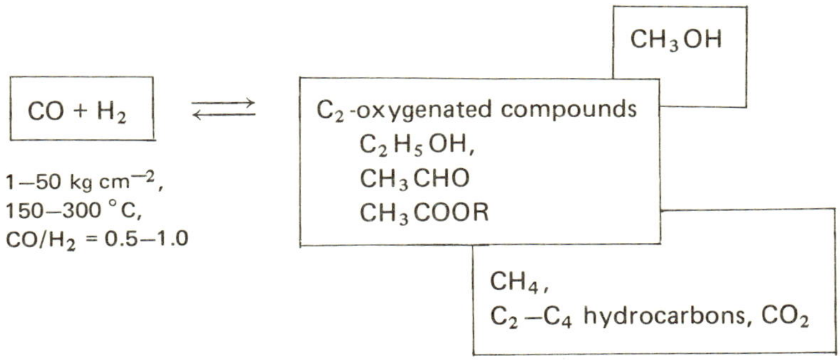

(i) Particle size effect (Rh dispersion, Rh_{4-13} carbonyl clusters)
(ii) Promoting effect of metal oxides (ZrO_2, TiO_2, MnO_2, La_2O_3 . . .)
(iii) Kinetic parameters (temperature, pressure, surface area, CO/H_2 ratio . . .)

Fig. 18. Effective controlling factors for rates and selectivities in $CO-H_2$ conversion catalyzed by rhodium-supported catalysts.

TABLE VII

Typical examples of metal catalysts, metal complexes, metal clusters and metals effective for CO-based reactions

CO-based reaction	Monometallic complex $[M]L_m$	Metal clusters $[M]_nL_m$	Metal [M]
Methanol synthesis			
$CO + 2H_2 \rightleftharpoons CH_3OH$	$HRh(CO)_4$ [89]; $HCo(CO)_4$ [90]; $HRu(CO)_4$; $HMn(CO)_5$. All at 1–300 atm, 200 °C	$(CH_3CN)Cu_2Ru_6(CO)_{16}C$ [91] at 200 atm, 250 °C; $Rh_4(CO)_{12}$/MgO, CaO [86]; $[Pt_3(CO)_6]_{2-5} \cdot 2Et_4N$/MgO, La_2O_3 at 1–20 atm, 200–250 °C [88]	Cu–ZnO; ZnO–Cr_2O_3 at 50–300 atm, 300–350 °C; Pd/SiO_2 [84]; Pd/La_2O_3, Nd_2O_3 [114, 119] at 1–50 atm, 200–300 °C
Methanation, F–T synthesis			
$CO + H_2 \rightleftharpoons CH_4 + C_2\text{–}C_m$ hydrocarbons		$Ru_3(CO)_{12}$, $Ir_4(CO)_{12}$/NaCl–$AlCl_3$ [71]; $Os_3(CO)_{12}$ at 1–2 atm, 150–180 °C; $[HFe_3(CO)_{11}]$/Al_2O_3, MgO, TiO_2 [73, 75]; $Ru_3(CO)_{12}$, $Os_3(CO)_{12}/M_xO_y$ [66]	Ru, Co, Fe, Ni/Al_2O_3, SiO_2 [82] at 1–20 atm, 200–350 °C
Carbonylation			
(i) Hydroformylation	$HRh(CO)_2(PR_3)_2$; $HCo(CO)_4$–HI [92]	$Co_3(CO)_6(PR_3)_3$; $Co_3(CO)_9CPh$ [93]; $Co_4(CO)_{10}(PPh)_2$; $H_4Ru_4(CO)_{12}$ [94]	
$H_2C{=}CH_2 + CO + H_2 \rightleftharpoons C_2H_5CHO + C_3H_7OH$	$Ni(CO)_4$; $Fe(CO)_5$ [92]; Pt–Sn complex	$Rh_6(CO)_{16}$/zeolite [16]; $Rh_4(CO)_{12}$; $Rh_6(CO)_{16}$/ZnO, MgO, La_2O_3 [78]; $Rh_{4-x}Co_x(CO)_{12}$/ZnO, carbon [81]	
(ii) Methanol–acetic acid synthesis			
$CH_3OH + CO \rightleftharpoons CH_3COOH$	$HCo(CO)_4$–PR_3–I [96]	$M[Co_3Ru(CO)_{11}]M^+$–I_2–PR_3 (M = Na; K, Cs . . NR_4) [95]; $M[Co_3Fe(CO)_{11}]$–I_2 at 200 atm, 200–250 °C	

(iii) Homologation reactions			
$CH_3OH + CO + H_2 \rightleftharpoons C_2H_5OH\ (CH_3CHO) + H_2O$	$HRu(CO)_4-PR_4$; $HRh(CO)_4$ [97]		
$HCHO + CO + H_2 \rightleftharpoons HOCH_2CH_2OH\ (HOCH_2CHO)$			
Direct synthesis of C_2 oxygenates			
$2\,CO + 2, 3\,H_2 \rightleftharpoons C_2H_5OH + CH_3OH + + CH_3COOH$	$HRu(CO)_4-NaI$ 300–1000 atm, 300 °C [98]	$Rh_{4-13}(CO)_m/M_xO_y$ (M = La, Zr, Ti, Th, Nd) at 1–20 atm, 200–250 °C [86]	Rh/SiO_2; Rh–Fe, Rh–Mn, $Rh-Mg/SiO_2$ at 50–300 atm, 200–350 °C [85]
Ethylene glycol synthesis			
$2\,CO + 2\,H_2 \rightleftharpoons HOCH_2CH_2OH$	$HRu(CO)_4-PR_3$ [99]; $HCo(CO)_4-PR_3-MX_m$ (M = Na, Cs; X = I, Br, Cl) at 1000–3000 atm, 200–300 °C	$[Rh_{12}(CO)_{30}]M_2$ (M = Cs, K, Mg, NR_4 [100], $N(PR_3)_2$); $[Rh_6(CO)_{15}C]^{2-}$; $[Rh_{13}(CO)_{23}H_{2-3}]^{2-}$; $[Rh_9(CO)_{21}P]^{2-}$; $[Rh_{17}(CO)_{32}S_2]^{3-}$ [101]; $[HRu_3(CO)_{11}]$ (?) all at 500–3500 atm, 250 °C	
Water-gas shift reaction			
$CO + H_2O \rightleftharpoons CO_2 + H_2$	$Fe(CO)_5$; $Cr(CO)_6-KOH$ [120]; $RhH(CO)_2(PR_3)_2-KOH$; $Pt(PR_3)_3$ at 1–20 atm, 100–200 °C	$Ru_3(CO)_{12}-KOH$ [103]; $H_2FeRu_3(CO)_{13}-KOH$; $Rh_6(CO)_{16}-KOH$, at 1–50 atm, 100–150 °C; $[HFe_3(CO)_{11}]$/NaY zeolite at 1 atm, 170 °C [89]	$Co_2O_3-MoO_3$; $Co_2O_3-Fe_2O_3$ at 50–200 atm, 300–400 °C

5.1. RHODIUM CARBONYL CLUSTER-DERIVED CATALYSTS IN $CO-H_2$ REACTIONS

Ichikawa [86] has recently demonstrated that marked selectivity dependencies exist in conversion of $CO-H_2$ at atmosphereic pressure into methanol, ethanol and methane over Rh carbonyl cluster-derived catalysts on different oxide supports. Thus, rhodium carbonyl cluster compounds such as $Rh_4(CO)_{12}$, $Rh_6(CO)_{16}$, $[Rh_7(CO)_{16}]\cdot 3Et_4N$ and $[Rh_{13}(CO)_{23}H_{2-3}]\cdot 2Bu_4N$ were empolyed [122] as metal precursors of different nuclearity and were similarly deposited onto various metal oxides from organic solutions. The resulting impregnated materials (0.25–0.75 wt.% Rh loading) were subsequently heated *in vacuo* at 50–200 °C to remove CO and organic residues, e.g. hydrocarbons and amines from tertiary alkylamine salts. A mixutre of CO and H_2 (1:1–2 volume ratio; total pressure 1 atm) was then admitted onto each impregnated catalyst, and the rates and product distributions were measured in a closed circulating reactor.

It was found that methanol was produced catalytically with higher selectivities at 150–220 °C over the $Rh_4(CO)_{12}$ or $Rh_6(CO)_{16}$-derived catalyst when typical basic oxides such as ZnO, MgO and CaO were used as the supporting carriers. Methane and ethanol were also formed as a byproduct at a lower yield. Ethanol selectivities increased relatively at higher temperatures over the $Rh_4(CO)_{12}$-impregnated MgO catalyst. On the other hand, the use of a number of amphoteric oxides such as La_2O_3, CeO_2, ZrO_2 and TiO_2 as carriers led to marked promotion and an increase in the selectivities of C_2 oxygenated products which mainly consisted of ethanol and to the suppression of methanol formation. Over such catalysts, the yield of hydrocarbons (e.g., methane, C_2–C_4 olefins and paraffins) became appreciable when the $CO-H_2$ reaction was conducted above 220 °C. The data in Table VIII illustrate the marked variation in the selectivity during $CO-H_2$ conversion over Rh cluster-impregnated catalysts when the nature of the metal oxide support is changed.

In summary, methanol was produced preferentially over catalysts in combination with Group IIa and IIb metal oxides such as ZnO, BeO, MgO and CaO but at lower CO conversions [86]. Higher ethanol selectivities (35–56% an a carbon basis) were obtained using of selected oxides of Group IIIa, IVa and Va metals such as La_2O_3, Nd_2O_3, Y_2O_3, CeO_2 (rare earths), TiO_2, ZrO_2, ThO_2, NbO_2, TaO_2 and MnO_2 (which gave a high selectivity for acetaldehyde instead of ethanol). These oxide supports are all refractory at high temperature while those of the higher valency metal oxides exhibit Lewis acid-base properties [102]. In contrast, Rh clusters impregnated on silica, γ-Al_2O_3, zeolites HY and 13X, P_2O_5 or WO_3 as acidic oxides produced mainly methane and higher hydrocarbons, but exhibited poor selectivities (less than 10%) for oxygenated carbon products,

TABLE VIII

Product distribution in CO–H_2 conversion at 1 atm pressure over $Rh_4(CO)_{12}$-derived catalysts impregnated on vaious metal oxides[a]

Catalysts[b]	Temp. (°C)	CO conversion (% h^{-1})	Carbon basis selectivity ($iC_i/\Sigma iC_i \times 100$) (%)					
			CH_3OH	C_2H_5OH	CH_3CHO + CH_3COOR	CH_4	C_2–C_4 hydrocarbons	CO_2 + others
$Rh_4(CO)_{12}$–ZnO	220	1.6	94	+[d]	–	4	–	2
$Rh_4(CO)_{12}$–MgO	220	2.6	88	2	+	7	+	3
$Rh_4(CO)_{12}$–CaO	230	0.8	92	1	–	2	+	5
$Rh_4(CO)_{12}$–La_2O_3	205	3.0	38	42	1	8	2	9
$Rh_4(CO)_{12}$–Nd_2O_3	210	3.8	24	47	1	17	4	7
$Rh_4(CO)_{12}$–ZrO_2	215	4.4	13	45	2	31	5	4
$Rh_4(CO)_{12}$–TiO_2	210	6.0	6	32	8	30	21	3
$Rh_4(CO)_{12}$–Nb_2O_5	195	5.8	7	30	9	32	19	3
$Rh_4(CO)_{12}$–Ta_2O_5	190	4.2	5	22	8	43	20	2
$Rh_4(CO)_{12}$–MnO_2[c]	205	1.2	4	10	15	35	28	8
$Rh_4(CO)_{12}$–SiO_2	235	1.7	1	3	3	66	26	1
$Rh_4(CO)_{12}$–γ-Al_2O_3	250	8.6	+	+	+	67	32	1

[a] Reaction conditions: CO/H_2 = 20:45 cmHg in closed circulating reactor of 420 cm^3 capacity.
[b] $Rh_4(CO)_{12}$ was deposited from hexane solution onto each oxide powder (20 g) at 0.5 wt.% Rh loading, followed by heat treatment at 120–200 °C *in vacuo* or an H_2 atmosphere.
[c] MnO_2 was partially reduced to MnO in the CO–H_2 reaction.
[d] + indicates formation of trace.

at least when the CO–H_2 reaction was carried out at atmospheric pressure. In this regard, the previous results of Smith *et al.* [66] indicating the substantial formation of hydrocarbons during CO–H_2 conversion over pyrolyzed $Rh_6(CO)_{16}$ impregnated on γ-Al_2O_3 and SiO_2 are relevant. Demitras and Muetterties have also demonstrated that $Rh_6(CO)_{16}$, $Ru_3(CO)_{12}$ and $Ir_4(CO)_{12}$, when dissolved in molten $AlCl_3$–NaCl as a Lewis acid medium, provided methane and ethane with no oxygenated carbon products in the CO–H_2 reaction [71].

On the other hand, differences in the nuclearity of the starting clusters in Rh_4–Rh_{13} carbonyls led to moderate differences in product selectivities, (Table IX). As shown in the table, the resultant catalysts prepared from smaller Rh carbonyl clusters such as Rh_4 and Rh_6 exhibited higher selectivities toward oxygenated compounds (mainly ethanol and methanol) when impregnation was conducted on suitable oxide supports such as La_2O_3 and ZrO_2 [87]. Catalyst derived from larger Rh clusters, i.e. Rh_7–Rh_{13}, exhibited relatively lower selectivities for oxygenated compounds but higher proportions of C_2–C_4 hydrocarbons were formed. Methanol and ethanol were also obtained over conventional Rh metal catalysts prepared from salts such as $RhCl_3$ and $Rh(NO_3)_3$ at similar metal loading when La_2O_3, ZrO_2 and TiO_2 were used as the oxide supports. However, they exhibited poorer selectivities toward oxygenates compounds (15–30% in conversion efficiency) then the corresponding cluster-derived catalysts.

Methanol is also produced in 95–99% selectivity over catalysts obtained by impregnating $[Pt_3(CO)_6]_n \cdot 2Et_3N$ (n = 2–5) onto MgO, La_2O_3 and Nd_2O_3 in the CO–H_2 reaction at atmospheric pressure and a temperature of 150–220 °C. Catalysts obtained by $Ir_4(CO)_{12}$ impregnation onto ZnO, MgO or La_2O_3 yielded methanol but less selectively and at lower conversion. In contrast to the Rh cluster catalysts, those derived from Pt and Ir catalyzed methanol formation preferentially (with traces of ethanol and higher oxygenates) regardless of the type of oxide support used, as indicated in Table IX. This behavior may be a reflection of their inherently poor abilities to produce C_2 oxygenated products in CO–H_2 conversion.

Further detailed kinetic studies have suggested that Rh cluster-derived catalysts impregnated on basic oxides such as MgO, La_2O_3 and ZrO_2 exhibit higher CO conversions and selectivities for methanol and C_2 oxygenates (mainly ethanol) than catalysts prepared from Rh salt and/or acidic supports such as silica and alumina. The carbon distribution of oxygenated compounds in a CO–H_2 reaction catalyzed by Rh cluster-impregnated catalysts indicated that the proportion of C_2 oxygenates (e.g. ethanol and acetaldehyde) in the total yield of oxygenated products was extraordinarily high (75–80% on a carbon basis) but with only negligible proportions of C_3 and C_4 alcohols, although the corresponding C_2–C_4 hydrocarbons were produced in proportions almost con-

TABLE IX

Product distribution in $CO-H_2$ conversion at 1 atm pressure over catalyst prepared from various Rh, Pt, Ir and Pd sources[a]

Catalyst (20 g)	Metal wt.%	Temp. (°C)	CO conversion (% h^{-1})	Carbon basis selectivity ($iC_i/\Sigma iC_i \times 100$) (%)					
				CH_3OH	C_2H_5OH	CH_3CHO + CH_3COOR	CH_4	C_2-C_4 hydrocarbons	CO_2 + others
$Rh_4(CO)_{12}-La_2O_3$	0.5	220	4.6	19	49	2	14	6	8
$Rh_6(CO)_{16}-La_2O_3$	0.5	220	3.2	15	44	3	18	8	10
$RhCl_3-La_2O_3$[b]	1.0	220	1	10	17	+[c]	56	13	4
$[Pt_{15}(CO)_{30}]\cdot 2Et_4N-La_2O_3$	0.6	210	3	90	2	–	5	1	2
$[Pt_9(CO)_{18}]\cdot 2Et_4N-La_2O_3$	0.5	210	1.8	78	4	+	14	1	3
$Ir_4(CO)_{12}-ZrO_2$	0.5	230	0.8	47	2	–	46	3	2
$PdCl_2-La_2O_3$[b]	3.0	200	6.2	97	+	–	2	+	1

[a] Reaction conditions: CO/H_2 = 20:45 cmHg.
[b] Reduced by H_2 treatment for 15 h at 350 °C.
[c] + indicates formation of trace.

sistent with the Schulz–Flory distribution. Ichikawa and Shikakura [88] have found that, in contrast to an activation energy of ca. 58.5 ± 4 kJ mol^{-1} necessary for methanol formation, higher values of 109 ± 8 and 113 ± 8 kJ mol^{-1}, respectively are required for ethanol and methane formation. However, the pressure dependencies of the rates for each product exhibited different trends for variations in the CO partial pressure; a small positive order was observed with respect to CO for ethanol production, but a negative order for methane production. The kinetic parameters for CH_4 formation over the Rh cluster-derived catalysts employed were essentially consistent with those reported by Vannice [103] over conventional Rh/Al_2O_3 catalysts.

For ethanol formation in a CO–H_2 reaction, Ichikawa and Shikakura [88] have demonstrated that the yield of ethanol and other C_2 oxygenated compounds is negligible if a CO_2 and H_2 mixture (1:2 volume ratio; 1 atm total pressure) is employed instead of CO and H_2 over the same $Rh_4(CO)_{12}/ZrO_2$ catalyst, although methane is produced with high selectivity. As indicated in Table X, ethanol yields were retained almost unchanged even on reducing the ratio of CO to CO_2 in the feed gas from 20:0 to 5:15 (at a total pressure of 20 cmHg). These results suggest that methanol and methane are produced during the hydrogenation of CO_2 rather than CO, whereas the presence of CO is necessary in ethanol formation in order to generate a carbon–carbon bond.

TABLE X

Product distribution in CO/CO_2–H_2 reaction at 1 atm pressure and 220 °C over $Rh_4(CO)_{12}$ (0.40 g)–ZrO_2 (30 g) catalyst

Feed (ml min^{-1})			Product yield (mmol h^{-1})				
CO	CO_2	$H_2 + He$[a]	CH_3OH	C_2H_5OH	CH_3CHO + CH_3COOR	CH_4	C_2–C_4 hydrocarbons[b]
20	0	70	0.07	0.46	0.05	1.25	0.18
10	10	70	0.07	0.43	0.03	1.50	0.15
5	15	70	0.03	0.31	0.02	1.46	0.11
0	20	70	0.03	+[d]	+	4.3[c]	+

[a] H_2/He = 40:30.
[b] $C_2/C_3/C_4$ = 1:1.5–2.0:0.1.
[c] a small amount of CO was also detected simultaneously in the effluent gas.
[d] + indicates formation of trace.

In contrast, the effects of a controlled pulse feed of methanol (or ethanol) on the product selectivities in the steady-state CO–H_2 reaction over $Rh_4(CO)_{12}$-impregnated ZrO_2 have been investigated. As shown in Figure 19, with a methanol feed CH_3COOCH_3 was abruptly produced ($CH_3COOC_2H_5$ was obtained with an ethanol feed), but the subsequent yield decreased as the methanol feed continued. Upon addition of methanol to the CO–H_2 reaction, the ethanol yields initially increased but returned to their original values when the reaction was continued for some time, while those of methane and of the C_2–C_4 hydrocarbons changed only marginally from their stationary values, as shown in Figure 19. As a result, it has been suggested that the catalyst surface may accumulate an acetyl species which acts as the precursor of the C_2 oxygenate formed in the CO–H_2 reaction.

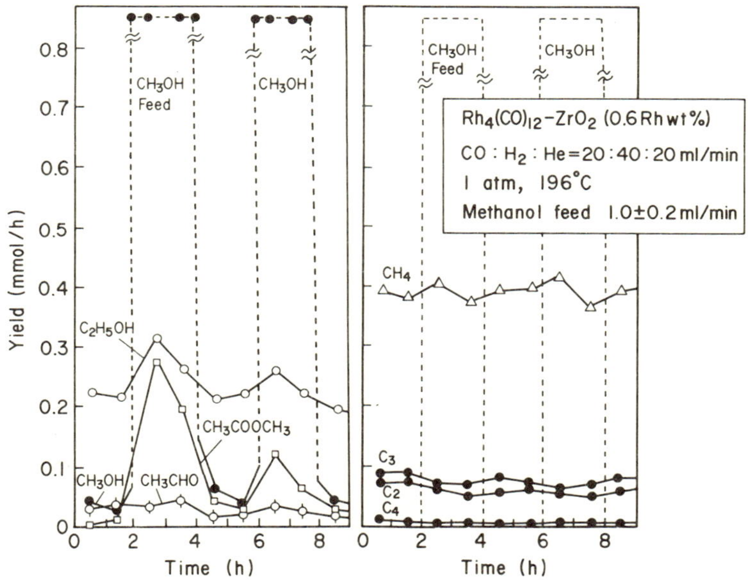

Fig. 19. Influence of controlled methanol feed on product distribution in the CO–H_2 reaction carried out in an open flow reactor over an $Rh_4(CO)_{12}$–ZrO_2 catalyst. (Reaction conditions: catalyst loading, 0.6 wt.% Rh; CO/H_2/He = 20:40:20 at a total flow rate of 80 ml min^{-1}; total pressure, 1 atm; temp. = 196 °C; methanol feed rate, 1.0 ± 0.2 ml min^{-1}.)

In order to investigate the role of selected metal oxides such as ZrO_2, TiO_2 and La_2O_3 in enhancing the production of C_2 oxygenates such as ethanol, composite silica (or alumina) supports containing different metal oxides have been prepared by reaction of the corresponding organometallic compounds as oxide precursors with surface silanol (or Al—OH) groups [104].

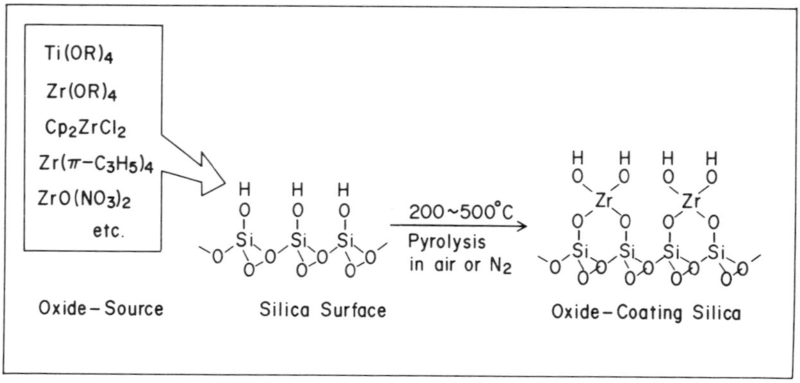

Zirconium and titanium oxides were developed uniformly on the silica surface (at 5–20 wt.% loading) by pyrolysis in air (at 200 °C for 2 h or at 500 °C for 15 h) after impregnation of $Zr(n\text{-}C_3H_7O)_4$ [or $Zr(\pi\text{-}C_3H_5)_4$] or $Ti(i\text{-}C_3H_7O)_4$ from hexane solution onto silica gel (Aerosil 300; apparent surface area, 280–300 $m^2\ g^{-1}$) or porous silica pellets (Davison #57; apparent surface ares, 280 $m^2\ g^{-1}$) which had been preheated at 400 °C. No large crystallites of ZrO_2 or TiO_2 were observed on the composite samples when the latter were examined by high resolution (×600 000) electron microscopy [105].

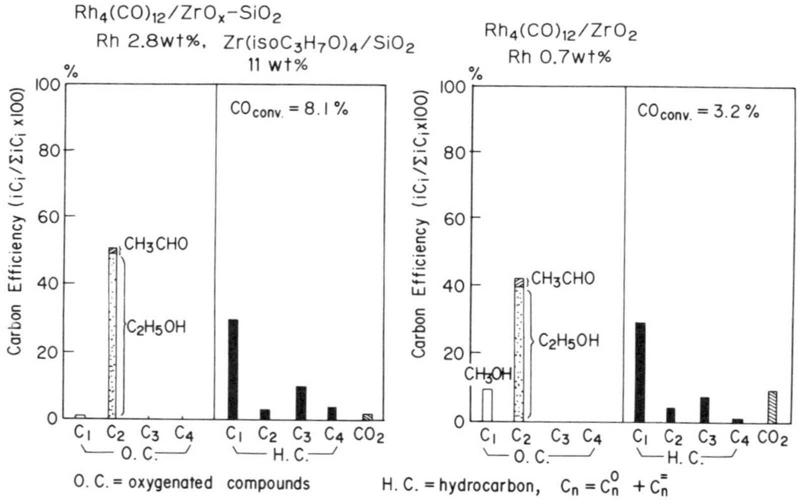

Fig. 20. Product distribution during $CO-H_2$ reaction at 1 atm pressure over catalysts obtained by $Rh_4(CO)_{12}$ impregnation of ZrO_x-SiO_2 or ZrO_2. Reaction conditions: $CO/H_2/He$ = 1:2:1; temp., 200 °C; SV = 200 h^{-1}. (a) $Rh_4(CO)_{12}/ZrO_x-SiO_2$ catalyst with 2.8 wt.% Rh loading on 11 wt.% Zr content support; (b) $Rh_4(CO)_{12}/ZrO_2$ catalyst with 0.7 wt.% Rh loading. O.C. = oxygenated compounds; H.C. = hydrocarbons, C_n = $C_n^0 + C_n^=$.

Impregnated $Rh_4(CO)_{12}$ catalysts have also been prepared using the Ti- and Zr-oxide containing silica supports, followed by heating *in vacuo* at 150 °C and/or H_2 reduction at 400 °C. The resulting impregnated catalysts of higher Rh dispersion (CO/Rh = 0.65–0.84; average particle size, 1–2 nm by electron microscopy) exhibited relatively higher CO conversions and selectivities for the production of ethanol compared with catalysts impregnated on TiO_2 and ZrO_2 alone. As shown in Figure 20, the yields of methanol and CO_2 during a CO–H_2 reaction were substantially suppressed over the Ti- or Zr-oxide containing silica-supported catalysts. This implies that the basic sites on ZrO_2 and TiO_2 surfaces, which as previously mentioned may be associated with methanol synthesis in reactions catalyzed by Rh, Pt and Ir [86], are considerably suppressed when in combination with silica. In fact, this follows from the type of acidity titration exhibited by such coated supports with ammonia, as illustiated in Figure 6. In comparison with ZrO_2, much stronger acid sites are generated on Zr- and Ti-oxide containing silicas, whereas the strength of basic sites decreases. On the other hand, XPS measurements for freshly reduced Rh-impregnated catalysts have revealed that Ti and Zr additives exist as Zr^{4+} and Ti^{4+} oxidation states, while the limited amount of Ti localized in the vicinity of an Rh particle is partially reduced to Ti^{3+}, as indicated by the ESR spectra for the sample given in Figure 21 [106].

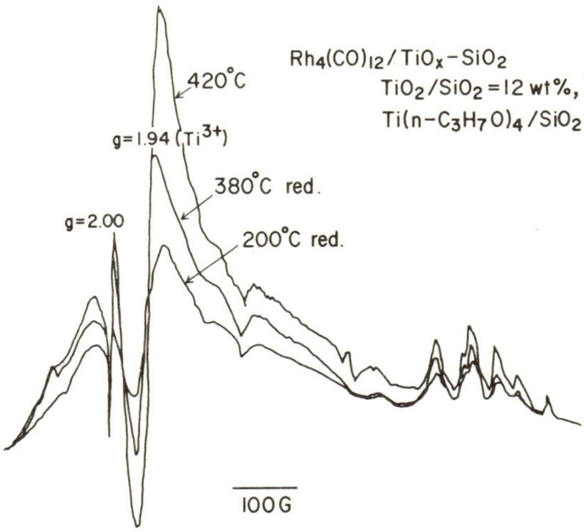

Fig. 21. The ESR signals for the catalyst obtained by impregnating $Rh_4(CO)_{12}$ onto Ti-oxide containing silica followed by hydrogen reduction at 200–420 °C. The particular sample illustiated in the figure possessed a 1.4 wt.% Rh loading on a 12 wt.% Ti content support, the latter having been prepared by pyrolysis of a Ti $(n\text{-}C_3H_7O)_4$-impregnated silica gel at 500 °C. The individual traces were obtained following H_2 reduction at the temperatures indicated.

Additionally, it is of interest to note that when a ZrO_2 precursor such as $Zr(n\text{-}C_3H_7O)_4$ was added successively onto the $Rh_4(CO)_{12}$-derived silica catalyst (original Rh particle size, ca. 2 nm), both CO conversion and ethanol yields were markedly enhanced up to values ca. 50 times greater than for a ZrO_2-free Rh/SiO_2 catalyst, as shown in Figure 22. During successive impregnation of Zr alkoxide, and even following CO–H_2 reaction, the initial Rh dispersion remained virtually unchanged. The addition of the ZrO_2 precursor onto the Rh/SiO_2 catalyst also affected the selectivity toward C_2 oxygenated products by increasing the ethanol selectivity at the expense of acetaldehyde. The activation energies for the formation of C_2 oxygenates such as ethanol in the CO–H_2 reaction decreased considerably, from 109 kJ mol^{-1} on Rh/SiO_2 alone to a value of 67 kJ mol^{-1} on addition of a Zr oxide precursor such as $Zr(n\text{-}C_3H_7O)_4$.

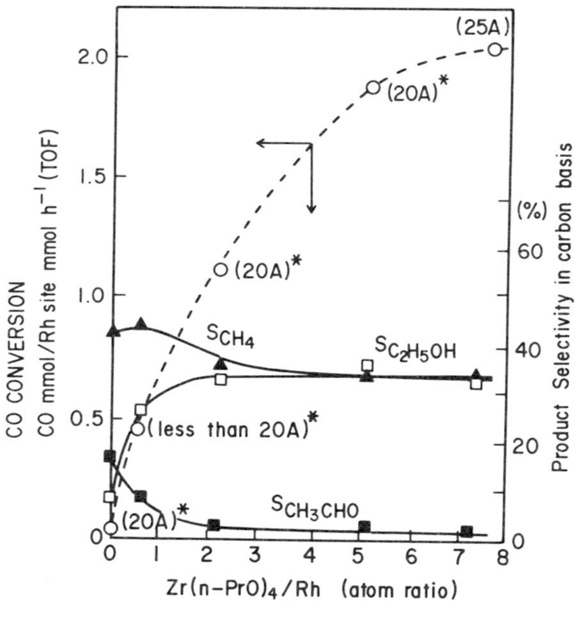

Fig. 22. Enhancement of CO conversion and ethanol selectivity by successive deposition of ZrO_2 precursor [$Zr(n\text{-}PrO)_4$] from hexane solution onto Rh/SiO_2 catalyst, followed by H_2 reduction of 400 °C. Reaction conditions: $CO/H_2/He$ = 20:40:20; total pressure, 1 atm; temp., 205 °C; SV = 200 h^{-1} using a fixed bed flow mode reactor (ϕ, 18 × 500 nm). The Rh site numbers were estimated from the stoichiometry of CO adsorption as 25 °C (P_{CO} = 100 mmHg). Data indicated thus ( )* depict average Rh particle sizes estimated from electron microscopy (×600 000).

Furthermore, Ichikawa and Kawai [105] have observed that the activities and selectivities of Rh cluster-impregnated catalysts in ethylene hydroformylation reactions (which yield C_2H_5CHO and/or C_3H_7OH) are substantially modified by the addition of a Ti- or Zr-oxide precursor, as shown by the data listed in Table XI. The proportion of consecutive hydrogenated product in the reaction (C_3H_7OH) was considerably increased on the catalysts impregnated on Ti- and Zr-oxide containing silica. On the other hand, the addition of Ti- and Zr-oxide considerably enhanced the extent of CO dissociation promoted by the Rh-impregnated catalyst, together with hydrogenation of the resulting surface carbon which was converted to CH_4. In contrast, the addition of ZnO dramatically suppressed CO dissociation but only marginally affected olefin hydroformylation, the latter being a test reaction for evaluating CO-insertion activity.

5.2. THE ROLE OF BIFUNCTIONAL METAL OXIDES IN MODIFYING THE FORMATION OF OXYGENATED SPECIES BY RHODIUM CLUSTER-IMPREGNATED CATALYSTS

In order to study changes in the valency states of Rh impregnated on different oxide supports, *in situ* XPS measurements of the $3d_{3/2,5/2}$ lines of Rh have been made [123] both before and after the use of the oxide supports in the $CO-H_2$ reaction. The observed binding energies were normalized to the 4f line of Au on ZnO and to the Si 2p line for Mn-, Zr- and Ti-containing silica-supported catalysts. As shown in Table XII, the Rh 3d lines are shifted to relatively higher binding energy volunes of 307.0, 307.3, 307.7, 307.8 and 308.4 eV on SiO_2, TiO_2-SiO_2, ZrO_2-SiO_2, ZrO_2 and ZnO, respectively. On comparison with the average binding energies of Rh in the metallic, Rh^I and Rh^{III} oxidation states, the observed results suggest that the impregnated Rh metal exists in a partially oxidized state on Ti- and Zr-oxide containing supports, this being particularly so on impregnation onto ZnO where a higher valency state such as Rh^I in suggested. As summarized in Table XII, Rh metal particles derived from carbonyl clusters strongly interact with supporting oxides, thus modifying their oxidation states which is reflected in the product selectivities observed in the $CO-H_2$ reaction. When Rh is impregnated onto ZnO it is dispersed as the higher oxidation state Rh^I, loading to inhibition of CO dissociation whilst enabling an increased activity toward olefin hydroformylation (e.g. CO insertion) and methanol synthesis via hydrogenation of associated CO molecules. When impregnated onto silica, Rh exists in the metallic state and exhibits preferential catalysis of methanation in the $CO-H_2$ reaction but olefin hydroformylation with poor efficiency. The addition of ZrO_2, TiO_2 and MnO_2 modifies the oxidation state of Rh to some extent, resulting in improved selectivities toward certain oxygenated carbon products during $CO-H_2$ conversion, together with an increase in CO-insertion activity in hydroformylation.

TABLE XI

Promoting effect of Zr- and Ti-oxides for $CO-H_2$ conversion and olefin hydroformylation catalyzed by Rh-supported catalysts

Catalyst	$Rh_4(CO)_{12}/ZrO_2$ 0.7 wt.% 20 g	$Rh_4(CO)_{12}/ZrO_2-SiO_2$[a] 2.8 wt.% 10 g	$Rh_4(CO)_{12}/TiO_2-SiO_2$[a] 2.8 wt.% 10 g	$Rh_4(CO)_{12}/SiO_2$[a] 2.5 wt.% 10 g
Rh dispersion (CO_{ads}/Rh)	0.65	0.54	0.62	0.35
$CO-H_2$ conversion ($CO/H_2/He$ = 20:40:20; total pressure, 1 atm; flow mode reactor (ϕ, 18 × 100 mm)				
CO conversion (%)	2.1 (200 °C)	5.1 (200 °C)	7.6 (200 °C)	1.3 (245 °C)
Carbon basis selectivity ($iC_i/\Sigma iC_i \times 100$) (%)				
CH_3OH	12	1	1	1
C_2H_5OH	42	33	30	5
$CH_3CHO + CH_3COOR$	2	6	3	8
CH_4	32	42	43	69
C_2-C_4 hydrocarbons	6	19	22	12
CO_2 + others	6	trace	2	trace
Olefin hydroformylation ($C_2H_4/CO/H_2$ = 20:20:20; total pressure, 60 cmHg; temp., 110 °C; closed circulation reactor (420 ml)				
Reaction rate[b] (mmol (mmol Rh site)$^{-1}$ h^{-1})	3.6	0.43	0.74	0.12
Alcohol selectivity[c] (%)	56	65	35	1

[a] $Rh_4(CO)_{12}$ impregnated from hexane solution onto silica supports, followed by H_2 reduction at 400 °C. Zr(*n*-PrO)$_4$ and Ti(*i*-PrO)$_4$ supported on silica gel (Davison #57; apparent surface area, 280 m^2 g^{-1}) and pyrolyzed in air at 500 °C. ZrO_2/SiO_2 = 11 wt.% ZrO_2; TiO_2/SiO_2 = 14 wt.% TiO_2.

[b] Number of Rh sites estinated from CO adsorption at 25 °C.

[c] $[C_3H_7OH]/([C_2H_5OH] + [C_3H_7OH]) \times 100$.

TABLE XII

Changes in Rh $3d_{5/2}$ binding energies of $Rh_4(CO)_{12}$-impregnated oxide supports associated with catalysis of $CO-H_2$ conversion, olefin hydroformylation and CO dissociation together with charateristic binding energies for other Rh-supported catalysts. [123]

$Rh_4(CO)_{12}/M_xO_y$	Rh $3d_{5/2}$	Catalytic behaviour and/or products		
Support	binding energy (eV)	$CO-H_2$ conversion	Hydroformylation	CO dissociation
SiO_2	307.01	hydrocarbons	low	moderate
MnO_x-SiO_2[a]	307.52	CH_3CHO	↓	high
ZrO_2-SiO_2[b]	307.67	C_2H_5OH	↓	↑
TiO_2-SiO_2[b]	307.32	CH_4	↓	↑
ZnO	308.4	CH_3OH	high	low
Rh metal	307.00			
Rh evapovated onto SiO_2	307.00			
$RhCl_3$ on ZrO_2-SiO_2	309.8			
Rh (I)[c]	308.8			

[a] $Mn(OBu)_4$ impregnated onto SiO_2, followed by pyrolysis at 500 °C.
[b] ZrO_2-SiO_2 support prepared by impregnation of $Zr(n\text{-PrO})_4$ onto SiO_2; TiO_2-SiO_2 support by impregnation of $Ti(i\text{-PrO})_4$ onto SiO_2.
[c] Data from Nefedov *et al.; Russ. J. Inorg. Chem.* **18**, 444 (1973).

Accordingly, it is suggested that ZrO_2 and TiO_2 as support additives act essentially as promoters enhancing CO dissociation (the increasing the surface concentration of methylene species) and enhancing CO-insertion reactions catalyzed by Rh aggregates to yield C_2-oxygenate precursors, which are converted by hydrogen to ethanol during the $CO-H_2$ reaction.

Shriver [107] has recently proposed bifunctional promotion roles for acids (Lewis and Brönsted) leading to an enhancement (a) in CO bond cleavage, e.g. in CO dissociation, and (b) in alkyl migration with coordinated CO molecules, e.g. in CO insertion to form C_2-oxygenated intermediate species such as acetyl in homogeneous solution. Metal carbonyls, e.g. $Fe_4Cp_4(CO)_4$, $Ru_4(CO)_{12}$ and $Fe_2Cp_2(CO)_4$, form stoichiometric adducts with Lewis acids such as $AlBr_3$, $Al(C_2H_5)_3$ and BF_3 as shown by the data in Table XIII. The marked shift of the CO bridging absorption bands to lower wavenumbers arises from attack on the oxygen of the CO ligand by nucleophiles of higher valency and/or by small-size metal cations such as Al^{3+}, Mo^{6+}, Ti^{4+}, Zr^{4+} and Li^+. As a result a bifunctional acceleration of C—O bond cleavage in CO activation catalyzed by metals and by the acid sites of metal oxides is to be expected, due to the large reduction of CO bond order during chemisorption on metal sites.

TABLE XIII
Metal carbonyl cluster adducts with Lewis acids and changes in IR carbonyl band absorption frequences of clusters during adduct formation

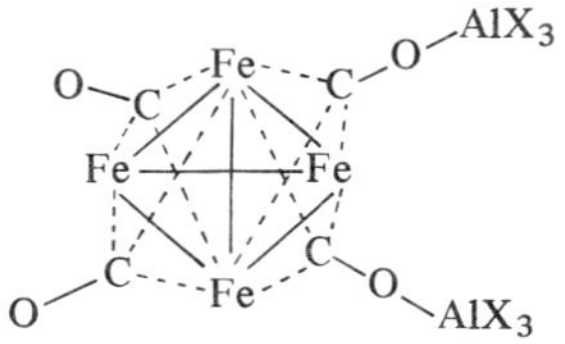

Carbonyl cluster	Lewis acid	Absorption frequency ν_{CO}(cm^{-1}) Terminal	Bridging	Metal-linked
$Cp_2Fe_2(CO)_4$	–	1930, 1970	1765	
	$Al(C_2H_5)_3$	1985	1828	1682
	BBr_3	1984	1848	1438
$Cp_4Fe_4(CO)_4$	–		1640	
	1:1 $AlBr_3$		{1637, 1673}	1392
	1:2 $AlBr_3$		{1675, 1715}	1365, 1395
	1:3 $AlBr_3$		1755	1439, 1470
$Ru_3(CO)_{12}$	–	1998, 1987		
	$AlBr_3$	{1987, 2000}		1535

On the other hand, it has been suggested by Shriver *et al.* [108] that small cations such as Li^+ and Na^+, or Lewis acid reagents such as AlX_3, greatly accelerate CO insertion reactions such as CO uptake by $Mn(CO)_5(CH_3)$, possibly as a result of electrophilic interaction between the Lewis acid or metal cations and coordinated CO weakening the metal–methyl bond and facilitating methyl migration:

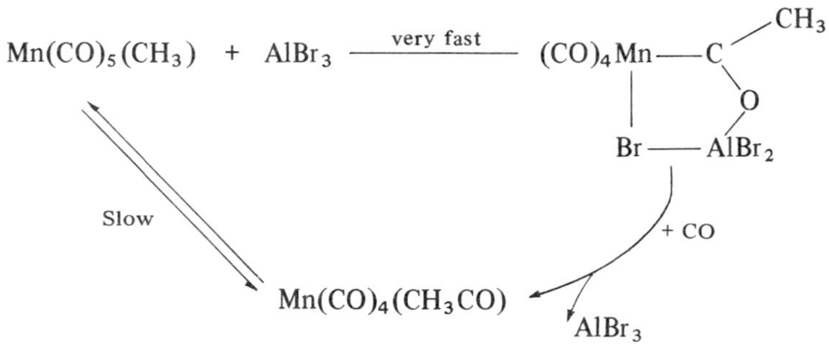

Burwell *et al.* [109] have demonstrated that metal oxides such as γ-Al_2O_3 having Lewis acid sites also promote CO insertion. When $Mn(CO)_5(CH_3)$ and $Fe(CO)_2Cp(CH_3)$ were impregnated onto a γ-Al_2O_3 surface, the resulting IR spectrum indicated a similar surface acetyl species with a cyclic structure was formed as follows:

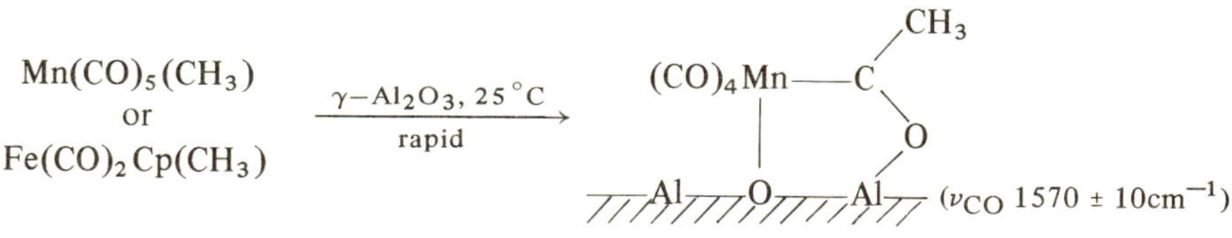

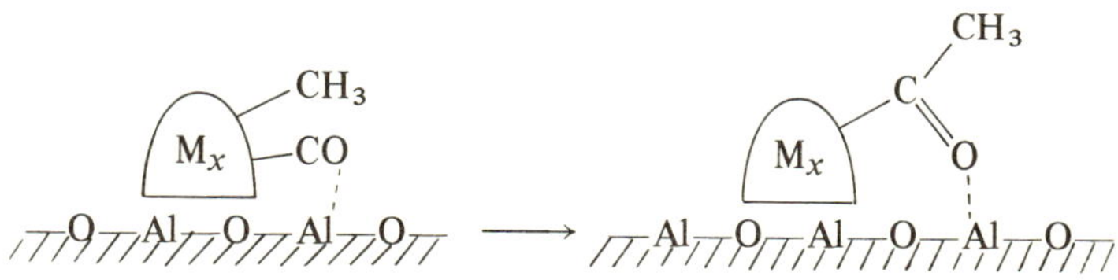

The Al_2O_3 surface withdraws electron density from the metal particle thereby changing its reactivity. On this basis it is possible to suggest that, in addition to Al_2O_3, the support effects of TiO_2, ZrO_2 and La_2O_3 also lead to an enhancement of CO-conversion and -insertion reactions catalyzed by Rh metal particles associated with these particular metal oxides. Such promoting effects may be linked to the high affinities for oxygen exhibited by such of 'oxygen-loving metals', which parallels the larger enthalpies for metal oxide formation listed in Table XIV.

TABLE XIV
Representative heats of formation of some binary d- and f-element oxides [111][a]

TiO_2	VO_2	CrO_2	MnO_2	Fe_2O_3	CoO
(−469)	(−393)	(−292)	(−259)	(−271)	(−238)
ZrO_2	NbO_2	MoO_2		RuO_2	RhO
(−544)	(−395)	(−278)		(−114)	(−82)
ThO_2	UO_2			OsO_2	IrO_2
(−598)	(−523)			(−147)	(−76)

[a] Values expressed in kJ (mol O atoms)$^{-1}$.

In order to provide information about the surface intermediates present in CO–H_2 reactions, particularly the C_2-oxygenate precursor on Rh aggregates (or clusters) impregnated on different oxide supports, Ichikawa and Fukushima [112] have recently employed high-pressure IR spectroscopy. As illustrated in Figure 23, catalyst wafers prepared from $Rh_4(CO)_{12}$ impregnated onto Mn-, Ti- or Zr-oxide containing silicas, which were obtained by the pyrolysis of $Mn(OBu)_4$, $Zr(n\text{-}C_3H_7O)_4$ or $Ti(i\text{-}C_3H_7O)_4$ respectively on silica gel, developed two distinct IR carboxyl bands at 1550–1560 cm^{-1} and 1430–1445 cm^{-1}

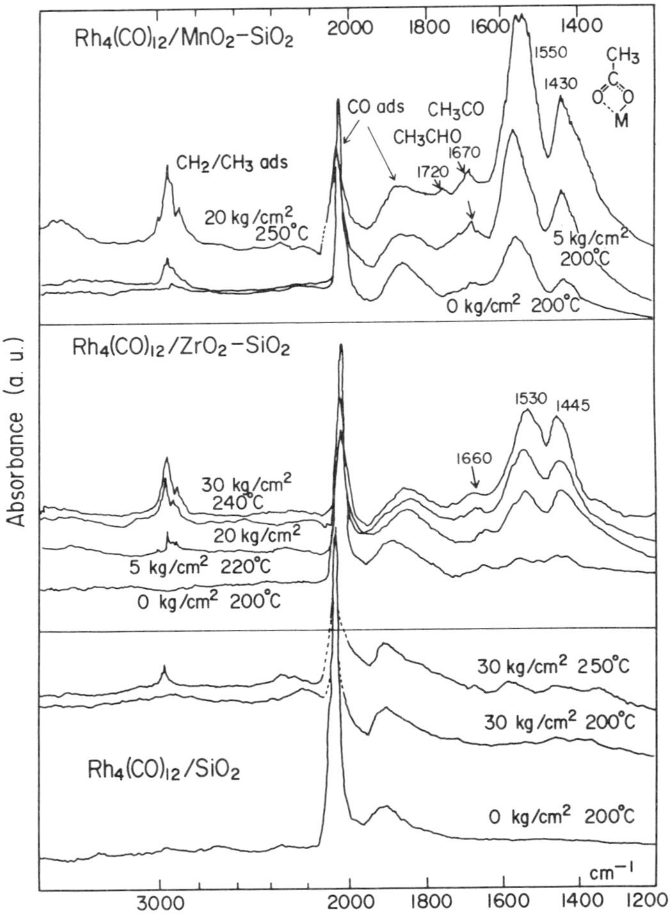

Fig. 23. High-pressure IR spectra of catalysts derived from $Rh_4(CO)_{12}$ impregnated on Mn- and Zr-oxide containing silicas exposed to CO–H_2 mixtures at various flow rates. IR spectrum of (a) $Rh_4(CO)_{12}/MnO_2-SiO_2$ in the presence of $CO/H_2 = 0.5$ at a flow rate of 100 ml min^{-1} (cf. that of $Mn(CH_3COO)_3$ (ν_{CH_3COO} 1561, 1420 cm^{-1})); (b) $Rh_4(CO)_{12}/ZrO_2-SiO_2$ in the presence of $CO/H_2 = 0.5$ at a flow rate of 200 ml min^{-1} (cf. that of $ZrO(CH_3COO)_2$ (ν_{CH_3COO} 1560, 1460 cm^{-1})); and (c) $Rh_4(CO)_{12}/SiO_2$ in the presence of $CO/H_2 = 0.5$ at a flow rate of 80 ml min^{-1}. The spectra were obtained using a high-pressure IR cell with a CaF_2 window in conjunction with a Hitachi 260–50 IR spectioneter.

at elevated pressures (10–30 kg cm^{-2}; CO/H_2 = 0.5) and at 220–250 °C. It is reasonable to assign these to the corresponding acetate species such as 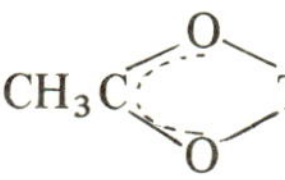 Ti/Zr. In addition weak but distinct IR bands were also observed at ca. 1620–1650 cm^{-1}, possibly due to acetyl species since they compare well with those of acetyl metal complexes, e.g. $Rh(CH_3CO)Cl(I)(PBu_3)_2$ (ν_{CO} 1670 cm^{-1}) and $Rh(CH_3CO)Cl_2(PMe_2Ph)_2$ (ν_{CO} 1655 cm^{-1}). The acetates would be formed from the spillover of acetyl species migrating from Rh to higher valency metal oxides such as MnO_2, TiO_2 and ZrO_2 sites. These metal oxides, which are possibly localized in the vicinity of the Rh metal, considerably enhance CO dissociation catalyzed by Rh, and also contribute toward the stability of the acetyl species thus improving the yield of C_2 oxygenate formed. The larger boundary sites associated with highly dispersed metals prepared from rhodium clusters would be expected to aid the promotion of the oxide counterparts.

Accordingly, C_2-oxygenated carbon products such as ethanol may be considered as arising from a surface species such as acetyl which is formed via CO insertion into surface hydrocarbons $[CH_x]$ (x probably 2 or 3) derived from dissociatively chemisorbed CO on Rh aggregates. The acetyl species may be converted preferentially by hydrogen to ethanol under hydrogen-rich or highly promoted hydrogenation conditions, as shown in Figure 24.

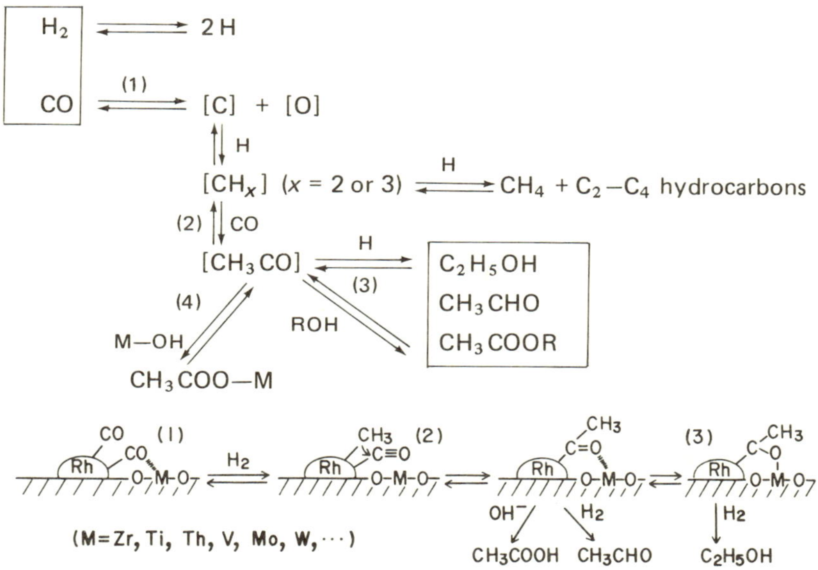

Fig. 24. Proposed mechanism for CO–H_2 conversion to ethanol and other C_2-oxygenated products over Rh catalysts supported on metal oxides.

5.3. BIMETALLIC Rh–Co CLUSTER-DERIVED CATALYSTS IN $CO-H_2$ CONVERSION

Ichikawa [113] has studied the effects of cobalt in metal cluster catalysts on the product distribution in $CO-H_2$ reactions. Typical results obtained using impregnated $Rh_2Co_2(CO)_{12}$ and $RhCo_3(CO)_{12}$ catalysts on ZrO_2, La_2O_3 and ZnO supports are summarized in Figure 25.

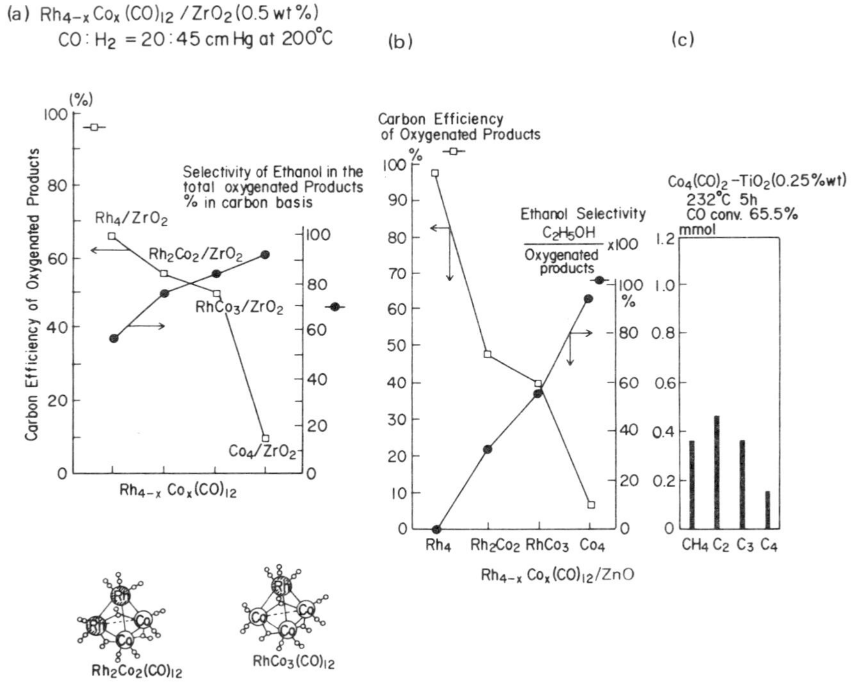

Fig. 25. Product distributions in $CO-H_2$ reaction 1 atm pressure over various cluster-derived catalysts of different metal composition. (a) $Rh_{4-x}Co_x(CO)_{12}$(x = 0–4) (0.5 wt.%) impregnated onto ZrO_2; CO/H_2 = 20:45; temp. 200 °C; (b) $Rh_{4-x}Co_x(CO)_{12}$ (x = 0–4) (0.5 wt.%) impregnated onto ZnO; CO/H_2 = 20:45; temp. ca. 185 °C; and (c) $Co_4(CO)_{12}$ (0.25 wt.%) impregnated onto TiO_2; 65.5% CO conversion after 5 h at 232 °C.

The proportion of ethanol in the oxygenated products increased with cobalt content in the cluster, while the carbon efficiencies of the total oxygenated products decreased. The yields of C_1-C_4 hydrocarbons increased at the expense of the oxygenated products such as methanol and ethanol. Catalysts derived from $Co_4(CO)_{12}$ provide methane and higher hydrocarbons preferentially, in a similar manner to the so-called Fischer–Tropsch catalysts consisting of conventional Fe, Co and Ru metals. Cobalt metal, which catalyzes the dissociative chemisorption of CO, may promote an increase in the concentration of surface hydrocarbon species in the $CO-H_2$ reaction. This could result in an enhancement in the yield of C_2-oxygenated products such as ethanol. With mixed Rh–Co cluster catalysts, both CO conversion and selectivities toward ethanol gradually decreased as the $CO-H_2$ reaction time increased, since cobalt may be eliminated from the bimetallic Rh–Co aggregates associated with the supporting oxides. On the other hand, it was found that conventional Rh–Co catalysts prepared by co-impregnation of $RhCl_3$ and $CoCl_2$ onto ZrO_2 and La_2O_3 supports exhibited poor selectivities (less than 20% in a carbon basis) toward total oxygenated products with preferential formation of C_1-C_4 hydrocarbons, in comparison to the behavior of the corresponding bimetallic cluster-derived catalysts. This suggests that the impregnation of bimetallic Rh–Co carbonyl clusters provides a more uniform surface composition in the resulting metal particles (at higher metal dispersions) than in the conventional counterparts, as has been demonstrated by EXAFS measurements for bimetallic cluster-impregnated catalysts on γ-Al_2O_3 [60]. Conventional Rh–Co supported catalysts exhibit a preferential concentration of cobalt atoms on the catalyst surface rather than rhodium.

5.4. METHANOL SYNTHESIS BY RHODIUM, PLATINUM AND IRIDIUM CLUSTER-IMPREGNATED CATALYSTS

Methanol has been produced catalytically with high selectivities (65–98% on a carbon basis) over catalysts prepared from $[Pt_3(CO)_3(\mu_2\text{-}CO)_3]_n \cdot 2Et_4N$ (n = 2–5) and $Ir_4(CO)_{12}$ impregnated on basic oxides such as MgO, CaO, ZnO, La_2O_3 and Nd_2O_3. Methane, ethanol and CO_2 were also obtained as byproducts in the reaction. The yields of ethanol were moderately dependent on the type of oxide supports employed, although the C_2H_5OH/CH_3OH molar ratio varied over the range 0.02–0.07 with the Pt carbonyl cluster-impregnated catalysts, compared with a value of 0.04–0.06 for the C_2H_6/CH_4 molar ratio. The rates of methanol formation in the steady-state $CO-H_2$ reaction apparently obeyed the empirical equation $r_{CH_3OH} = kP_{H_2}^{1.1}P_{CO}^{0.2} \exp(15 \pm 1 \times 10^3/RT)$ over the $[Pt_9(CO)_{18}] \cdot 2Et_4N/MgO$ catalyst [114].

Induction periods for methanol formation over some of the freshly prepared catalysts were observed, which were substantially affected by the presence of H_2O during catalyst preparation, e.g. during evacuation of the oxide supports, due to the insufficient dring of the solvent and other moisture sources. Addition of H_2O prior to commencing the $CO-H_2$ reaction shortened the induction period and the time necessary to reach a steady-state reaction rate, and improved methanol selectivities. On the other hand, a large amount of CO was taken up reversibly at 100–240 °C over catalysts active toward methanol synthesis, e.g. $[Pt_{15}(CO)_{30}]\cdot 2Et_4N/MgO$, $[Pt_9(CO)_{18}]\cdot 2Et_4N/La_2O_3$ and $Rh_6(CO)_{16}/La_2O_3$, and was evolved over the temperature range 240–280 °C. During such CO uptake and reversible evolution over these catalysts, no appreciable amounts of CO_2 were formed up to a temperature of 270 °C. This CO uptake could be similar to that associated with formate ($HCOO^-$) production from CO and H_2O and/or surface hydroxyl groups over MgO, as previously suggested by Scholten *et al.* [115].

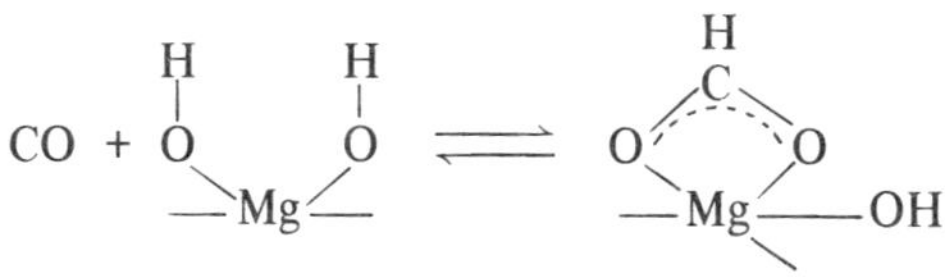

In support of this suggestion, Ichikawa and Schikakura [114] have demonstrated by IR spectroscopic measurements that prolonged contact between the above catalysts and a mixture of CO and H_2 (1:2 volume ratio, 45 cmHg) led to the generation of absorption bands at 1620–1580 and 1400–1360 cm^{-1}, which may possibly be assigned to formate-like species 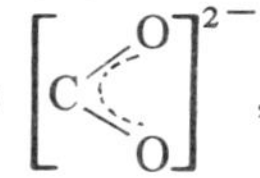, similar to those formed over MgO and ZnO (1572, 1370 cm^{-1}) [116]. IR absorption bands at 1050–1030 cm^{-1}, which are observed during the course of the CO–H_2 reaction at 200 °C over Pt and Rh cluster-impregnated MgO or La_2O_3 catalysts, may be attributed to methoxy and methanol formation.

The attack on coordinated carbon monoxide molecules by nucleophiles such as OH^- and OCH_3^- provides one possible route to their reduction and to the catalysis of the water–gas shift reaction as mentioned previously [22]. Martinengo and Chini [21] have reported that Rh_4, Rh_6 and Ir_4 carbonyl clusters react with OH^- and CH_3O^- to form hydrido-carboxylate complexes:

$$Rh_4(CO)_{12} + CH_3O^- \rightleftharpoons [Rh_4(CO)_{11}C(=O)OCH_3]^-$$

$$Rh_6(CO)_{16} + OH \rightleftharpoons [Rh_6(CO)_{15}C(=O)OH]^- \rightleftharpoons [HRh_6(CO)_{15}]^- + CO_2$$

$$[HRh_6(CO)_{15}]^- \xrightarrow{H^+} H_2$$

$$Ir_4(CO)_{12} + NaOH \rightleftharpoons [HIr_4(CO)_{11}CO_2]^-Na^+ \quad (\nu_{CO_2}\ 1580, 1440, 1380\ cm^{-1})$$

$$[HIr_4(CO)_{11}CO_2]^-Na^+ \rightarrow [Ir_4(CO)_{11}]^- + HCOONa$$

$$[HIr_4(CO)_{11}CO_2]^-Na^+ \rightarrow [HIr_4(CO)_{11}]^- + CO_2$$

In the case of $Ir_4(CO)_{12}$, sodium formate was obtained in aqueous alkaline solution [117].

Accordingly, it is suggested that carbonyl ligands bound to Pt, Rh and Ir aggregates react with adjacent basic OH^- and $O^{2-}(H_2O)$ sites to form formate-like species, such species being stabilized by the countercations of the oxide supports, e.g. Zn^{2+}, Mg^{2+} and La^{3+}. These carboxyl species can be successively reduced by hydrogen to yield formaldehyde and methanol under mild reaction conditions. Ponec *et al.* [118] have recently suggested that promoting metal oxides such as MgO and La_2O_3 stabilize partially oxidized Pd (possibly Pd^I) which may be the active surface species for methanol synthesis during Pd/MgO and Pd/La_2O_3 catalysis [119].

A similar promoting effect has been reported by Naito *et al.* [120] for small alkali cations, e.g. Li^+ and Na^+, as doping agents for conventional Pd/SiO_2 catalysts, which leads to an acceleration of methanol synthesis over such catalysts. Alkali metal cations located at the boundaries of Pd particles activate the oxygen atoms of oxide supports such as SiO_2, TiO_2 and ZrO_2, and stabilize reaction intermediates such as surface formate species. In fact, isotopic exchange rates between $C^{18}O$ and surface oxygen ($^{16}O^{16}O$) were considerably enhanced over Na-doped Pd/SiO_2 catalysts in comparison to those over the non-doped catalyst which exhibits a low activity for methanol formation.

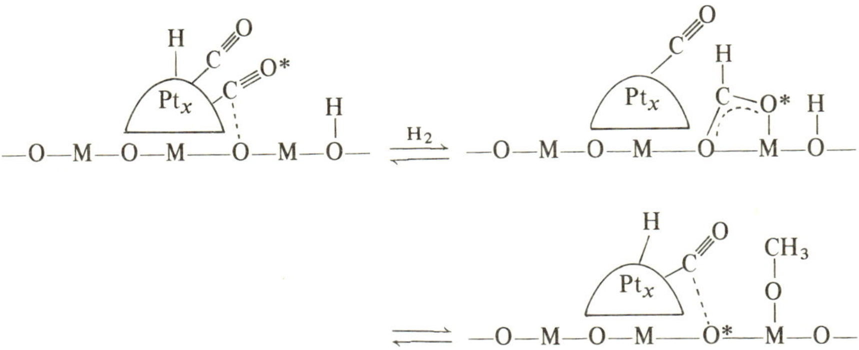

The relative rates of methanol formation in a $CO–H_2$ reaction conducted at 1 atm pressure were markedly dependent upon the kinds of oxide supports and noble metal precursors employed. As shown in Figure 26, Pt carbonyl clusters such as $[Pt_3(CO)_6]_3 \cdot 2Et_4N$ and $[Pt_3(CO)_6]_5 \cdot 2Et_4N$ provide highly active catalysts for methanol synthesis when impregnated on suitable basic oxides such as MgO, La_2O_3, Y_2O_3, Nd_2O_3, CeO_2 (rare earth metal oxides) or ZrO_2 and TiO_2, rather than the conventional Pt salt (H_2PtCl_6) catalysts. Oxides of metals whose electronegativities range from 4.0–14.0 exhibit an appropriate basicity with weak Lewis acid sites, and catalyzed methanol synthesis with higher efficiencies in combination with Pt, Ir and Pd [88, 119]. When Pt carbonyl clusters were impregnated on acidic or strongly basic oxides such as SiO_2, Al_2O_3, WO_3 and Li_2O, the resulting catalysts gave poor or negligible activities for methanol formation, but instead exhibited preferential activities toward methane and high hydrocarbons.

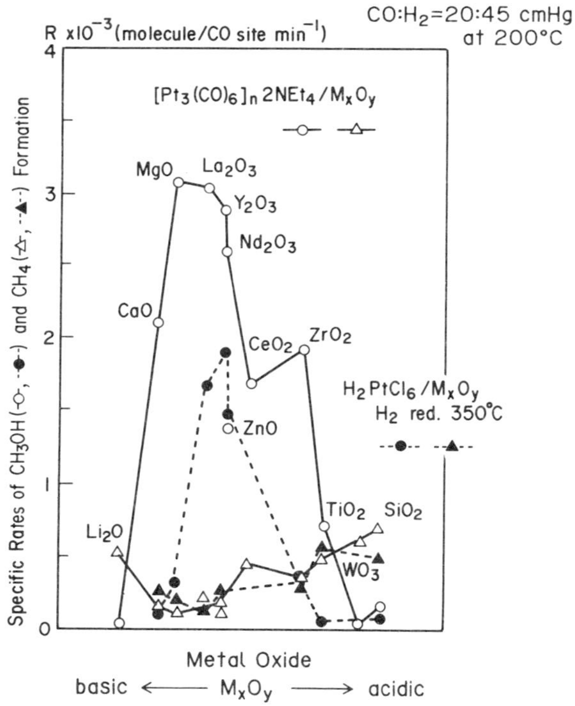

Fig. 26. Rate dependencies and selectivities for methanol and methane synthesis in the CO–H_2 reaction with $[Pt_3(CO)_6]_n \cdot 2Et_4N/M_xO_y$ (data indicated as ○ and △) and H_2PtCl_6/M_xO_y following H_2 reduction at 350 °C (data indicated as ● and ▲) as catalysts. Reaction conditions: CO/H_2 = 20:45; total pressure, 1 atm; temp., 200 °C.

The striking activity differences for methanol formation between Pt carbonyl cluster-derived catalysts and conventional salt-derived catalysts may be associated not only with a higher Pt dispersion for the cluster-derived materials, but also with poisoning by residual Cl^- ions in conventional catalysts blocking the active surface oxygens of the supports and the Pt sites for CO hydrogenation, which are associated with stabilization of reaction intermediates such as the formate species. In contrast, the activities for methane formation were only marginally affected by the supporting metal oxides and Pt precursors, as shown in Figure 26. This suggests that CO dissociation and methane formation proceed irrespective of the nature of the Pt surface, with the specific activities for such catalysts not being influenced to the same extent by the surface properties of the oxide supports.

TABLE XV
Rates of methanol formation in the CO–H_2 reaction over an $Rh_4(CO)_{12}$/MgO catalyst in the presence of varying CO and CO_2 partial pressures[a]

Synthesis gas composition			Product distribution (mmol)[b]			
CO	/ CO_2	/ H_2	CH_3OH	C_2H_5OH	CH_4	H_2O
20 (13.5)[c]	: 0 (0.5)[c]	: 45	1.24	0.010	0.028	0.10
15 (10.5)	: 5 (1.5)	: 45	1.48	0.015	0.029	0.16
10 (9.5)	: 10 (3.5)	: 45	1.65	0.026	0.030	0.39
5 (7.5)	: 15 (5)	: 45	1.19	0.022	0.029	0.82
0 (7.5)	: 20 : (7.5)	: 45	0.90	0.017	0.028	1.75

[a] Reaction conditions: $P_{CO} + P_{CO_2} = 20$ cmHg: $P_{H_2} = 45$ cmHg; temp., 220 °C. The catalyst was prepared by impregnating 0.30 g of $Rh_4(CO)_{12}$ onto 20 g of MgO which had been preheated *in vacuo* at 160 °C
[b] Yields after 23.5 h reaction.
[c] Partial pressures in cmHg after 23.5 h reaction.

It has also been demonstrated that hydrogenation of CO_2 rather than CO occurs catalytically when a mixture of the two gases was passed over Rh or Pt carbonyl cluster-impregnated MgO or La_2O_3 catalysts at 220 °C, to produce methanol and CO in almost equimolar quantities. According to the kinetic data presented for $Rh_4(CO)_{12}$/MgO in Table XV, methanol is produced by the direct hydrogenation of chemisorbed CO_2 not through the successive hydrogenation of the CO evolved during the reaction. The yields of methanol were observed to be greater in the CO_2–H_2 reaction compared to the CO–H_2 reaction under the same reaction conditions. In addition, IR spectroscopic studies have shown that on admitting CO_2 and H_2 onto impregnated $[Pt_{15}(CO)_{30}]{\cdot}2Et_4N$ and $[Pt_9(CO)_{18}]{\cdot}2Et_4N$ catalysts on MgO at 150–200 °C, formate and carbonates were produced (exhibiting a similar IR absorption band distribution to that produced by a mixture of CO and H_2), with the particular band due to the methoxy species at 1040 cm^{-1} being also detected in the products from the prolonged reaction, as shown in Figure 27.

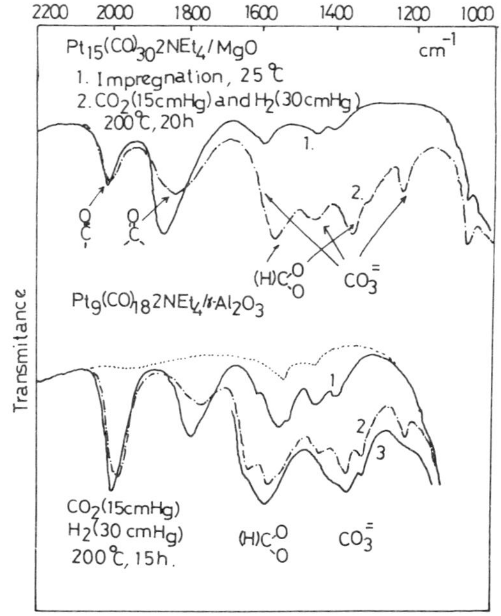

Fig. 27. IR spectra of Pt carbonyl cluster-impregnated catalysts on oxides at 25 °C (curve 1) and after admitting CO_2 (15 cmHg) and H_2 (30 cmHg) at 200 °C for 20 h (curve 2). (a) Spectra arising from $[Pt_{15}(CO)_{30}]\cdot 2Et_4N/MgO$ system; (b) spectra arising from $[Pt_9(CO)_{18}]\cdot 2Et_4N/\gamma\text{-}Al_2O_3$ system.

The $[Pt_9(CO)_{18}]\cdot 2Et_4N/\gamma\text{-}Al_2O_3$ system, catalyst for methanol formation from CO_2 and H_2, exhibited preferential formation of carbonates ($\nu_{CO_3^{2-}}$ 1650, 1470, 1280 cm^{-1}) with formates as minor species. It is suggested that carbon dioxide is readily converted to the formate species by hydrogen at the active sites associated with Pt and the oxide supports, and then successively hydrogenated to methanol and CO (and H_2O) through a reaction mechanism essentially the same as that for $CO–H_2$ conversion into methanol.

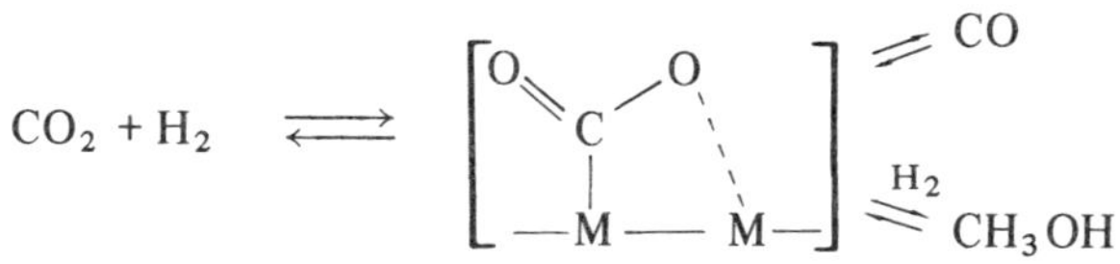

The Research Institute of Catalysis,
Hokkaido University,
Sapporo, 060
Japan

References

1. B. F. G. Johnson (ed.): *Transition Metal Clusters*, Wiley, New York (1980).
2. J. P. Collman, L. S. Hagedies, M. P. CooKe, J. P. Kooke, J. R. Norton, G. Dalcetti and D. N. Marquardt: *J. Am. Chem. Soc.* **94**, 1789 (1972); C. U. Pittman and R. C. Ryan: *Chem. Technol.* 8, 170 (1978).
3. B. C. Gates and J. Lieto: *Chemtech.* 195 (1980).
4. N. D. Parkyns: *Proc. 3rd Int. Congr. Catal. Amsterdam* (*1964*), Vol. 2, p. 914, North-Holland, Amsterdam (1965).
5. R. B. Bjorklund and R. L. Burwell: *J. Colloid Interface Sci.* **70**, 383 (1979). R. L. Burwell and A. Brenner: *J. Mol. Catal.* **1**, 77 (1975/76).
6. R. F. Howe, D. E. Davidson and D. A. Whan: *J. Chem. Soc., Faraday Trans. I* **68**, 2266 (1972).
7. J. R. Anderson and D. M. Mainwaring: *J. Catal.* **35**, 162 (1974).
8. A. K. Smith and J. M. Basset: *J. Mol. Catal.* **2**, 229 (1977).
9. G. C. Smith, T. P. Chojnacki, S. R. Dasgupta, K. Iwatate and K. L. Waters: *Inorg. Chem.* **14**, 1419 (1975).
10. M. Ichikawa: *J. Inst. Pet. Soc.* **19**, 828 (1976); *J. Chem. Soc., Chem. Commun.* 11, 26 (1976); *Hyomen* **19**, 555 (1981); *Chemtech.* 674 (1982).
11. J. H. Sinfelt: *Acc. Chem. Res.* **10**, 15 (1977); *Science* **195**, 64 (1977); R. L. Garten and J. H. Sinfelt: *J. Catal.* **62**, 127 (1980).
12. E. L. Muetterties: *Bull. Soc. Chim. Belg.* **84**, 959 (1975); *Science* **196**, 839 (1977).
13. R. Ugo: *Catal. Rev.* **11**, 225 (1975).
14. P. Chini: *Gazz. Chim. Ital.* **109**, 225 (1979).
15. G. A. Ozin: *Catal. Rev. Sci. Eng.* **16**, 191 (1977).
16. E. Montovani, N. Palladino and A. Zanobi: *J. Mol. Catal.* **3**, 285 (1977/78).
17. G. Gelin, Y. Ben Taarit and C. Naccache: *Stud. Surf. Sci. Catal.* **B15**, 898 (1980).
18. A. Theolier, A. K. Smith, M. Leconte, J. M. Basset, G. M. Zanderigh, R. Pasaro and R. Ugo: *J. Organometal. Chem.* **191**, 1415 (1980).
19. A. K. Smith, F. Hugues, A. Theolier, J. M. Basset, R. Ugo, G. M. Zanderighi, J. L. Bilhou, Bilho-Bougnol and W. F. Graydon: *Inorg. Chem.* **18**, 3105 (1979).
20. M. Ichikawa, K. Sekizawa, K. Shikakura and M. Kawai: *J. Mol. Catal.* **11**, 167 (1981).
21. S. Martinengo and P. Chini: *Gazz. Chim. Ital.* **102**, 344 (1972).
22. (a) A. K. Smith, B. Besson, J. M. Basset, R. Psaro, A. Fusi and R. Ugo: *J. Organometal. Chem.* **192**, C31 (1980); (b) F. Hugues, A. K. Smith, Y. Ben Taarit, J. M. Basset, D. Commereuc and Y. Chauvin: *J. Chem. Soc., Chem. Commun.* 68 (1980).
23. M. Deeba and B. C. Gates: *J. Catal.* **67**, 303 (1981).
24. S. L. T. Anderson, K. L. Waters and R. F. Howe: *J. Catal.*, **69**, 212 (1981).
25. M. Ichikawa: *Chem. Lett.* 335 (1976).
26. J. C. Calabrese, L. F. Dahl, P. Chini, G. Longoni and S. Martinengo: *J. Am. Chem. Soc.* **96**, 2614 (1974).
27. R. P. Eishens, W. A. Pliskin and S. A. Francis: *J. Chem. Phys.* **22**, 1786 (1954); C. W. Garland, R. C. Lord and P. F. Troiano: *J. Phys. Chem.* **69**, 1188 (1965).
28. M. Ichikawa: *J. Chem. Soc., Chem. Commun.* 26 (1976).
29. G. M. Dartigues, A. Chambellan and F. G. Gault: *J. Am. Chem. Soc.* **98**, 856 (1976); J. R. Anderson, R. J. MacDonald and Y. Shimoyama: *J. Catal.* **20**, 147 (1971).
30. O. Zahraa, F. Garin and G. Maire: *Faraday Discuss. Chem. Soc.* **72**, 43 (1981).
31. V. L. Kuznetsov, A. T. Bell and Yu. I. Yermakov: *J. Catal.* **65**, 374 (1980).
32. (a) M. Deeba and B. C. Gates: *J. Catal.* **67**, 303 (1981); (b) M. Deeba, B. J. Streusand, G. L. Schrader and B. C. Gates: *J. Catal.* **69**, 218 (1981); (c) M. Deeba, J. P. Scott,

R. Barth and B. C. Gates: *J. Catal.* **71**, 373 (1981); and (d) P. L. Watson and G. L. Schrader: *J. Mol. Catal.* **9**, 129 (1980).
33. R. Psaro, R. Ugo, G. M. Zanderighi, B. Besson, A. K. Smith and J. M. Basset: *J. Organometal. Chem.* **213**, 215 (1981).
34. P. Gallezot, G. Coudurier, M. Primet and B. Imelik: *Molecular Sieves II, ACS Symp. Ser. No. 40*, p. 144, Am. Chem. Soc., Washington, DC (1977); D. Ballivert-Tkatchenko, G. Coudurier, I. Tkatchenko and D. Figueiredo: *Prepr. Am. Chem. Soc. Div. Pet. Chem.*, (1980).
35. D. Commereuc, Y. Chauvin, F. Hugues, J. M. Basset and D. Olivier: *J. Chem. Soc., Chem. Commun.* 154 (1980); J. R. Katzer, A. W. Sleight, P. Gajardo, J. B. Michel, E. F. Gleason and S. McMillan: *Faraday Discuss. Chem. Soc.* **72**, 121 (1981).
36. G. Collier, D. J. Hunt, S. D. Jackson, R. B. Moyes, I. A. Pickering, P. B. Wells, A. F. Simpson and R. Whyman: *J. Catal.* **80**, 154 (1981).
37. H. Knötzinger and Y. Zao: *J. Catal.* **71**, 337 (1981).
38. J. R. Anderson, P. S. Elmes, R. F. Howe and D. E. Mainwaring: *J. Catal.* **50**, 508 (1977).
39. R. A. Della Betta and M. Selef: *J. Catal.* **48**, 111 (1977).
40. A. Brenner and D. A. Hucul: *Inorg. Chem.* **18**, 2836 (1979).
41. Y. Iwasawa, M. Yamada, S. Ogasawara, Y. Sato and H. Kuroda: *Chem. Lett.* 621 (1983).
42. M. Iwamoto, H. Kusano and S. Shikagawa: *Shokubai* **25**, 410 (1983).
43. A. S. Lisitisyn, V. L. Kuznetsov and Yu. I. Yermakov: *React. Kinet. Catal. Lett.* **14**, 445 (1980).
44. R. Nakamura, A. Oamura, N. Okada and E. Echigoya: *Chem. Lett.* 1463 (1982).
45. Y. Iwasawa, M. Yamada and S. Ogasawara, *Prepr. Catal. Soc. Jpn. Meet.*, 178 (4ED) (1983).
46. A. Brenner and R. L. Burwell, Jr.: *J. Catal.* **52**, 364 (1978).
47. M. Ichikawa: *J. Chem. Soc., Chem. Commun.* 11 (1976); *Shokubai* **17**, 164 (1975).
48. V. W. Day, R. O. Day, J. S. Kristoff, F. J. Hirsekorn and E. L. Muetterties: *J. Am. Chem. Soc.* **97**, 2571 (1975).
49. Y. Kim and K. Seff: *J. Am. Chem. Soc.* **100**, 175 (1978).
50. D. Brodzki, B. Denise and G. Pannetier: *J. Mol. Catal.* **2**, 149 (1977).
51. A. Brenner: *J. Mol. Catal.* **5**, 157 (1979); D. A. Hucul and A. Brenner: *J. Phys. Chem.* **85**, 496 (1981).
52. A. Brenner, D. A. Hucul and S. J. Hardwick: *Inorg. Chem.* **18**, 1478 (1979).
53. J. Smith, R. F. Howe and D. A. Whan: *J. Catal.* **52**, 364 (1978).
54. M. A. Schroeder and M. S. Wrighton: *J. Am. Chem. Soc.* **98**, 551 (1976).
55. A. Brenner and D. A. Hucul: *J. Catal.* **61**, 216 (1980).
56. T. J. Thomas, D. A. Hucul and A. Brenner: *ACS Symp. Ser. No. 192*, p. 249, Am. Chem. Soc., Washington, DC (1982).
57. J. Robertson and G. Webb: *Proc. R. Soc. London, Ser. A* **341**, 384 (1974).
58. A. G. Burden, J. G. J. Martos, R. B. Moyes and P. B. Wells: *Faraday Discuss. Chem. Soc.* **72**, 95 (1981).
59. M. Boudart, A. Aldag, J. E. Benson, N. A. Dougharty and C. G. Harkins: *J. Catal.* **6**, 92 (1966)
60. T. Yokoyama, K. Yamazaki, N. Kosugi, H. Kuroda, M. Ichikawa, and T. Fukushima: *J.C.S., Chem. Commun.* 962 (1984).
61. (a) R. V. Lindsey, C. W. Parchal and Y. G. Stobberg: *Inorg. Chem.* **5**, 109 (1966); (b) Yu. A. Ryndin, Zh. Loran, B. N. Kuznetsov, V. I. Koyalchuk, A. Pentenero and Yu. I. Yermakov: *Kinet. Katal.* 22 (1981).

62. L. Guzz, Z. Schay, K. Matusek, I. Bogyay and G. Steffer: *Stud. Surf. Sci. Catal.*
63. J. P. Barbier and P. Braunstein: *J. Chem. Research* **412**, 5028 (1978).
64. J. H. Sinfelt, G. H. Via and F. W. Lytle: *J. Chem. Phys.* **72**, 4832 (1980); J. H. Sinfelt, G. H. Via, F. Lytle and R. B. Greegor: *J. Chem. Phys.* **75**, 5527 (1981).
65. J. H. Sinfelt, J. L. Carter and D. J. C. Yates: *J. Catal.* **24**, 283 (1972); J. H. Sinfelt: *J. Catal.* **29**, 308 (1973).
66. A. K. Smith, A. Theolier, J. M. Basset. R. Ugo, D. Commereuc and Y. Chauvin: *J. Am. Chem. Soc.* **100**, 2590 (1978).
67. K. Whitmire and D. F. Shriver: *J. Am. Chem. Soc.* **102**, 1456 (1980).
68. D. Ballivert-Tkatchenko and G. Coudurier: *Inorg. Chem.* **18**, 558 (1979).
70. C. Masters, J. A. van der Woude and van Doon: *J. Am. Chem. Soc.* **101**, 1634 (1979).
71. G. C. Demitras and E. L. Muetterties: *J. Am. Chem. Soc.* **99**, 2796 (1977).
72. A. Wong and J. D. Atwood: *ACS Symp. Ser. No. 152*, p. 265, Am. Chem. Soc., Washington, DC (1981); A. Wong, M. Harris and J. D. Atwood: *J. Am. Chem. Soc.* **102**, 4529 (1980).
73. (a) D. Commereuc, Y. Chauvin, F. Hugues, J. M. Basset and D. Oliver: *J. Chem. Soc., Chem. Commun.* 154 (1980) (b) F. Hugues, B. Besson, P. Bussiere, J. A. Dalmon, M. Leconte and J. M. Basset: *ACS Symp. Ser. No. 192*, p. 255, Am. Chem. Soc., Washington, DC (1982).
74. R. C. Brady and R. Pettit: *J. Am. Chem. Soc.* **102**, 6182 (1980); C. E. Summer, Jr., R. E. Riley, R. E. Davis and R. Pettit: *ibid.* **102**, 1752 (1980).
75. F. Hugues, P. Bussiere, J. M. Basset, D. Commereuc, Y. Chauvin, L. Bonneviot and D. Olivier: *Stud. Surf. Sci. Catal.* 7, 418 (1981).
76. D. F. Frenkel and B. C. Gates: *J. Am. Chem. Soc.* **102**, 2478 (1980).
77. (a) P. Centola, G. Terzaghi, R. D. Rosso and I. Pasque: *Chim. Ind. (Milan)* **54**, 775 (1972) (b) T. Yashima, Y. Oshida, M. Ebisawa and H. Hara: *J. Catal.* **36**, 320 (1975); H. Arai, T. Kaneko and T. Kunugi: *Chem. Lett.* 265 (1975).
78 M. Ichikawa: *J. Catal.* **56**, 127 (1979).
79. P. Chini, S. Martinengo and G. Garlashelli: *J. Chem. Soc., Chem. Commun.* 709 (1972).
80. M. Ichikawa: *Shokubai* **20**, 8 (1978).
81. M. Ichikawa: *J. Catal.*, **59**, 67 (1979).
82. (a) M. A. Vannice: *J. Catal.* **40**, 129 (1975); *Catal. Rev. Sci. Eng.* **14**, 153 (1978); (b) G. G. Low and A. T. Bell: *J. Catal.* **57**, 397 (1979).
83. I. Toyoshima and G. A. Somorjai: *Catal. Rev.* **19**, 105 (1979); G. Broden, T. N. Rhodin and C. Brucker: *Surf. Sci.* **59**, 593 (1976); J. A. Rabo, L. F. Elek and J. N. Francis: *Stud. Surf. Sci. Catal.* 7, 490 (1981).
84. J. A. Rabo, A. P. Risch and H. L. Poutma: *J. Catal.* **53**, 295 (1978).
85. M. M. Bhasin, W. J. Bartley, P. C. Ellgen and T. P. Wilson: *J. Catal.* **54**, 120 (1978); Union Carbide Co. Inc.: U.S. Pat. 551,661; 551,660 ( ).
86. M. Ichikawa: *Bull. Chem. Soc. Jpn.* **51**, 2268 (1978); *ibid.* **51**, 2273 (1978).
87. M. Ichikawa: *J. Chem. Soc., Chem. Commun.* 566 (1978).
88. M. Ichikawa and K. Shikakura: *Stud, Surf. Sci. Catal.* 7, 925 (1981).
89. R. L. Pruett and W. F. Walker: U.S. Pat. 3,833,634 (1974); J. S. Bradley: *J. Am. Chem. Soc.* **101**, 7419 (1979); B. D. Dombek, *ibid.* **102**, 6855 (1980).
90. J. M. Rathke and H. M. Feder, *J. Am. Chem. Soc.* **100**, 3623 (1978).
91. Exxon Inc.: U.S. Pat. 242,638 (1982).
92. J. Falbe: *Carbon Monoxide in Organic Synthesis*, Springer-Verlag, Berlin (1970).
93. R. C. Ryan, C. U. Pittman and J. P. O'Conner: *J. Am. Chem. Soc.* **99**, 1986 (1977); C. U. Pittman and R. C. Ryan, *Chemtech.* **8**, 170 (1978).

94. R. M. Laine: *J. Am. Chem. Soc.* **100**, 6541 (1978); G. Brace, G. Sbranna, F. Piacenti and P. Pins: *Chim. Ind. (Milan)* **52**, 1091 (1970).
95. M. Hidai, M. Orisaku, M. Uei, Y. Uchida, K. Yasufuku and H. Yamasaki: *Shokubai* **23**, 90 (1981).
96. J. F. Roth, J. H. C. Craddock, A. Hershman and F. E. Paulik: *Chemtech.* **1**, 600 (1971); Monsanto Co. Inc.: U.S. Pat. 3,769,329 (1973).
97. I. Wender, R. A. Friedel and M. Orchin: *Science* **113**, 206 (1951); I. Wender: *Catal. Rev.* **14**, 97 (1976).
98. J. F. Knifton: *ACS Symp. Ser. No. 152*, p. 226, Am. Chem. Soc., Washington, DC (1981).
99. (a) A. Deluzache, R. Fanseca, G. Gnner and A. Kinnemann, *Erdoel, Kohle, Erdgas, Petrochem. Brennst.-Chem.* **32**, 313 (1979); (b) W. Keim, M. Berger and J. Schlupp: *J. Catal.* **61**, 359 (1980); (c) D. R. Fakey: *J. Am. Chem. Soc.* **103**, 136 (1980).
100. J. L. Vidal and W. E. Walker: *Inorg. Chem.* **19**, 896 (1980); Union Carbide Co. Inc.: U.S. Pat. 3,833,634 (1974); U.S. Pats. 3,878,214, 3,878,290, 3,878,292 (1975).
101. J. L. Vidal, R. C. Schoening and W. E. Walker: *ACS Symp. Ser. No. 155*, p. 61, Am. Chem. Soc., Washington, DC (1981).
102. P. C. Ford, P. Yarrow and H. Cohen: *ACS Symp. Ser. No. 152*, p. 95, Am. Chem. Soc., Washington, DC (1981).
103. C. Ungerman, V. Landis, S. A. Moya, H. Cohen, H. Walker, R. G. Pearson, R. G. Rinker and P. C. Ford: *J. Am. Chem. Soc.* **101**, 5922 (1979); H. Kang, C. H. Mauldin, T. Cole, W. Slegeir, K. Cann and R. Pettit: *ibid.* **99**, 8323 (1977); R. M. Laine: *ibid.* **100**, 6451 (1978).
104. V. A. Zakharov, V. K. Dudchenko, E. A. Paukshtis, L. G. Karakchiev and Yu. I. Yermakov: *J. Mol. Catal.* **2**, 421 (1977); Yu. I. Yermakov and V. A. Zakharov: *Adv. Catal.* **24**, 173 (1975); Yu. I. Yermakov, B. N. Kuznetsov and V. A. Zakharov: *Stud. Surf. Sci. Catal.* **8** (1982).
105. M. Ichikawa and M. Kawai: *Shokubai* **23**, 93 (1981); L. L. Murrell and D. J. C. Yates: *Stud. Surf. Sci. Catal.* **7** 1470 (1981).
106. K. Tanaka, K. Sekizawa, K. Shikakura and M. Ichikawa: *Shokubai* **23**, 96 (1981).
107. D. F. Shriver: *ACS Symp. Ser. No. 152*, p. 1, Am. Chem. Soc., Washington, DC (1981); *Chem. Br.* 482 (1983).
108. S. B. Butts, E. M. Holt, S. H. Strauss, N. W. Alcock, R. E. Stimson and D. F. Shriver: *J. Am. Chem. Soc.* **101**, 5864 (1979); *ibid.* **102**, 5093 (1980).
109. F. Correa, R. Nakamura, R. E. Stimson, R. L. Burwell and D. F. Shriver: *J. Am. Chem. Soc.* **102**, 5112 (1980).
110. A. Alichi, N. J. Nelson, D. Strape and D. E. Shriver: *Inorg. Chem.* **11**, 2976 (1972).
111. C. Keller: *The Chemistry of the Transuranium Elements*, p. 151, Verlag Chemie, Berlin (1971) A. Navrotsky; in *MTP Int. Rev. Sci., Inorg. Chem. Ser.*, (ed. A. G. Sharpe), Vol. 5, Butterworths, London (1975).
112. M. Ichikawa and T. Fukushima; to be published.
113. M. Ichikawa: *Chemtech.* 556 (1982).
114. M. Ichikawa and K. Shikakura: *Shokubai* **21**, 253 (1979).
115. J. J. F. Scholten, P. Mars, P. G. Menon and R. van Hardeveld; *Proc. 3rd Int. Congr Catal. Amsterdam (1964)*, p. 881, North-Holland, Amsterdam (1965).
116. A. Ueno, T. Onishi and K. Tamaru: *Trans. Faraday Soc.* **66**, 756 (1970).
117. S. Martinengo, P. Chini and G. Giorgano: *Gazz. Chim. Ital.* **102**, 330 (1972).
118. J. M. Driessen, E. K. Poels, J. P. Hinderma and V. Ponec: *J. Catal.* **82**, 26 (1983).
119. M. Ichikawa, K. Shikakura and T. Matsumoto: *Nihon Kagaku Kai Shi* 213 (1982).

120. S. Natio, H. Yoshioka, S. Kagami and K. Tamaru: Shokubai **25**, 326 (1983); Y. Kikuzono, S. Kagami, S. Naito, T. Onishi and K. Tamaru: *Faraday Discuss. Chem. Soc.* 72, 135 (1981).
121. A. Fumagell, T. F. Koetzle, F. Takasagawa, P. Chini, S. Martinengo and B. T. Heaton: *J. Am. Chem. Soc.* **102**, 1740 (1980).
122. M. Ichikawa, H. Suzuki and K. Shikakura: *Prepr. 5th Soviet–Japan Catal. Seminar, Tashkent,* 217 (1979).
123. M. Kawai, M. Uda, and M. Ichikawa: *J. Phys. Chem.* 89, 1654 (1985).

A. TAI AND T. HARADA

ASYMMETRICALLY MODIFIED NICKEL CATALYSTS

1. Introduction

The asymmetrically modified nickel catalyst (MNi) is the simplest catalyst for the preparation of optically active compounds. MNi can be prepared easily by soaking activated nickel powder in an aqueous solution of an optically active compound. The unique functions of MNi are those associated with cooperation between an activated metal surface of macromolecular dimensions and the optically active compound implanted on the metal surface in a dispersed monomolecular form. The metal surface functions as a generator of activated hydrogen for hydrogenation while the optically active compound acts as a stereochemical regulator of the reactant molecule. In this respect, studies on MNi catalysts have contributed not only to the development of a significant catalyst for the syntheses of optically active compounds, but also to an understanding of various characteristic phenomena taking place on the surface of nickel catalysts.

Initially in this chapter, some principles of the reactions related to MNi well be introduced briefly. Secondly, a history of the development of high performance MNi catalysts and their synthetic applications will be given, while the final part of the chapter will be devoted to a discussion of the mode of action of MNi.

2. General Considerations in Reactions Relating to MNi

2.1. STEREO-DIFFERENTIATING REACTIONS

A chiral compound produced and consumed in a biological system will consist of a single enantiomer and show optical activity. The same compound synthesized in the laboratory will usually be as an equimolar mixture of two enantiomers and will be optically inactive overall. Studies of so-called asymmetric synthesis have developed from the desire to produce optically active compounds without using biological systems.

The formation of an optically active compound is achieved when a reagent or a catalyst differentiates the stereostructure of a substrate in a reaction and

Y. Iwasawa (ed.), Tailored Metal Catalysts, 265–324.

generates one enantiomer in excess. This process is referred to as a stereo-differentiation process, and the reaction which is involved in this process is called a stereo-differentiating reaction [1]. Based on the concept of stereo-differentiation, the mode of the reaction employed for asymmetric synthesis may be divided into two types as illustrated in Figure 1.

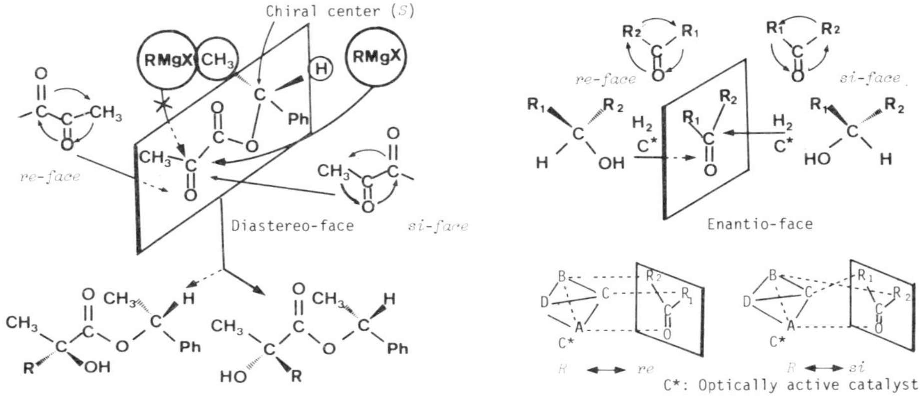

(A) Diastereo-face Differentiating Reaction

(B) Enantio-face Differentiating Reaction

Fig. 1. Mode of reaction employed for asymmetric synthesis. (A) Diastereo-face differentiating reaction; (B) enantio-face differentiating reaction.

As shown in Figure 1(A), when a prochiral carbonyl group is covalently bound to an optically active chiral center, the two faces of the carbonyl group (designated as the *re* and *si* faces [2]) become diastereomeric relative to substituents of the chiral center (for example, CH_3— is close to the *re* face whereas H— is located close to the *si* face). Apparently, the stereo-environments of the two faces are different from each other. These two faces are referred to as diastereo faces. The diastereo faces of a substrate can be recognized by any kind of reagent to give a new chiral center with one enantiomeric form in excess. This type of reaction is called a diastereo-differentiating reaction because the reagent (or catalyst) differentiates one of the diastereo faces of the substrate.

In the reaction shown in Figure 1(A), a Grignard reagent attacks from the least sterically hindered side (*si* face) to give product II in excess. The rule which predicts the configuration of the product in this reaction mode is known as Prelog's rule [3].

The other type of stereo-differentiating reaction is provided by a system which consists of a prochiral substrate and an optically active chiral reagent or catalyst. As shown schematically in Figure 1(B), the *re* and *si* faces of a substrate molecule can interact in different ways with an optically active reagent or catalyst, schematically shown as $R \leftrightarrow re$ and $R \leftrightarrow si$ in the figure. When the reagent can differentiate between these two types of interaction, the product is optically active. Such a reaction is called an enantio-differentiating reaction and the two faces of the substrate are referred to as enantio faces.

The efficiencies of both diastereo- and enantio-differentiating reactions are expessed by the extent to which one stereoisomer is produced in excess of the other as expressed in eqn. (1).

$$\text{Stereoisomer (diastereomer or enantiomer) excess (\%)} = [(A - B)/(A + B)] \times 100 \quad (1)$$

where A and B are the concentrations (weights) of the stereoisomers produced.

The efficiency of an enantio-differentiating reaction can also be expressed in terms of the 'optical yield', when the reaction product is estimated by polarimetry. Other than in special cases, the term 'optical yield' can be used synonymously with enantiomer excess (eqn. (2)).

$$\begin{aligned}\text{Optical yield} &= \frac{[\alpha]_{\mathrm{D}}^{20} \text{ for reaction product}}{[\alpha]_{\mathrm{D}}^{20} \text{ for optically pure sample}} \times 100 \\ &= [([R] - [S])/([R] + [S])] \times 100 \\ &= \text{enantiomeric excess (e.e.\%)} = \text{e.d.a. (\%)} \end{aligned} \quad (2)$$

where $[R]$ and $[S]$ represent the concentrations the (R)- and (S)-isomers, respectively.

The efficiency of an enantio-differentiating reaction depends on how rigorously the reagent or catalyst can differentiate the prochirality of the substrate. To express this kind of ability, the term 'enantio-differentiating ability' (here after abbreviated as e.d.a.) has been introduced. In order to measure e.d.a., the optical yield of the reaction carried out under the stated conditions is employed.

A wide variety of effective diastereo-differentiating reactions and enantio-differentiating reactions has been developed and applied to the preparation of optically active compounds. However, these reactions have a disadvantage in that the reaction requires more than 1 mol of optically active reactants in order to produce 1 mol of optically active products. The reaction is, of course, useful for the small-scale production of various optically active compounds in the laboratory, but cannot provide an efficient procedure for large-scale industrial production. Thus, more attention has been focused on enantio-differentiating reactions with optically active catalysts, since these reactions can produce large amounts of optically active materials from a catalytic amount of an optically active substrate, and in addition the mode of this reaction is similar to that of an enzymatic reaction. In spite of the importance of these reactions, to date the number of well-established reaction systems amount to less than 10 owing to the lack of fundamental information about the stereo-differentiating procedure. Of well-documented reaction systems, the most spectacular success has been achieved in enantio-differentiating hydrogenations [4]. The MNi system is one which has contributed to this achievement [5].

2.2. CATALYSTS FOR ENANTIO-DIFFERENTIATING HYDROGENATION REACTIONS

The most intensively studied and most effective catalyst systems for enantio-differentiating hydrogenation are MNi [5] and an optically active rhodium phosphine complex catalyst [Rh—P] [6]. The mode of action of these two catalysts is in striking contrast to each other, but despite this fact they have both played an important role in the great advance of enantio-differentiating hydrogenation reactions.

MNi is a heterogeneous metal catalyst and exhibits a high performance in the enantio-differentiating hydrogenation of certain kinds of prochiral ketones to secondary alcohols (eqn. (3)), but gives poor results in the hydrogenation of prochiral olefins. On the other hand, Rh—P is a homogeneous catalyst and functions well in the enantio-differentiating hydrogenation of certain kinds of prochiral olefins to saturated compounds (eqn. (4)), but usually gives poor chemical and optical yields in the hydrogenation of prochiral ketones.

$$R{-}\underset{\underset{O}{\|}}{C}{-}CH_2{-}COOCH_3 \xrightarrow[\text{MNi}]{H_2} R{-}\overset{\overset{H}{|}}{\underset{\underset{OH}{|}}{C}}{-}CH_2{-}COOCH_3 \qquad (3)$$

$$\begin{matrix} R \\ H \end{matrix}\!\!>\!C{=}C\!<\!\!\begin{matrix} COOH \\ NHCOR \end{matrix} \xrightarrow[\text{Rh--P}]{H_2} R{-}CH_2{-}\overset{\overset{COOH}{|}}{\underset{\underset{NHCOR}{|}}{C}}{-}H \qquad (4)$$

where Rh–P depicts, for example,

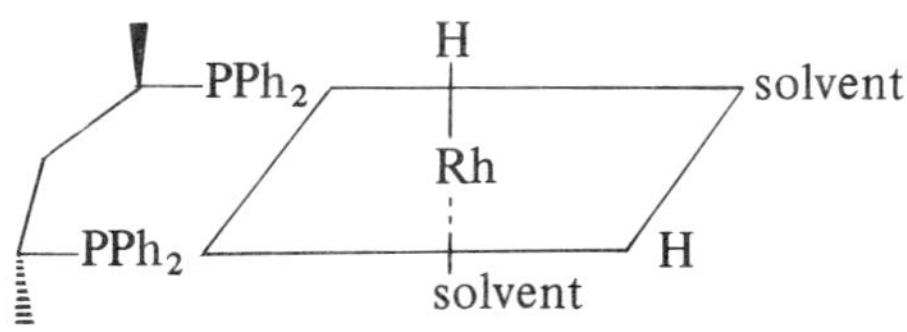

The starting material for the preparation of MNi is a conventional nickel catalyst, of the type widely utilized in various branches of chemical industry. Thus, once a highly efficient catalyst has been developed, it is immediately applicable to large-scale industrial production. Recently developed MNi catalysts (TA–NaBr–MRNi), comprised of tartaric acid (TA), NaBr and Raney nickel (MRNi), have been shown to give 90% optical yield and a quantitative chemical yield in the hydrogenation of β-ketoesters and β-diketones. They are already being utilized in the commercial-scale production of optically active β-hydroxy acid and β-glycol [7].

It is well documented that Rh—P is an excellent catalyst for the preparation of β-amino acid in almost 100% optical yield. However, this system still has problems to be solved before it can provide an industrial catalyst. In addition to the price and recycling problems associated with Rh, the development of an economical preparation of phosphine ligands also poses a difficulty. The recently developed ligand 'SKEWPHOS' [8] has, however, the advantage that it can be produced on an industrial scale, since it can be readily prepared from optically active 2,4-pentanediol as shown in eqn. (5) [9], the starting materials, (2*R*,4*R*)- and (2*S*,4*S*)-pentanediol already being produced on a large scale by the enantio-differentiating hydrogenation of acetylacetone over MNi [7, 10].

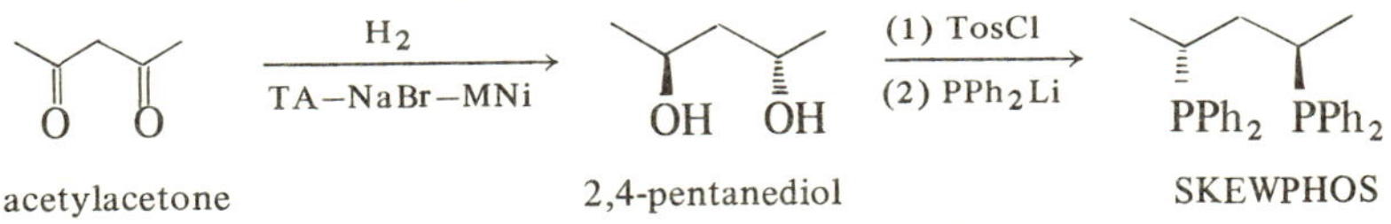

(5)

Such cooperation in the use of heterogeneous and homogeneous catalysts could open up the way to widely applicable systems for enantio-differentiating hydrogenation.

3. MNi as a Practical Enantio-differentiating Hydrogenation Catalyst

3.1. DEVELOPMENT AND IMPROVEMENT OF CATALYST AND REACTION CONDITIONS

3.1.1. *The discovery of MNi*

The first heterogeneous catalyst for enantio-face differentiating hydrogenation was the silk palladium catalyst developed by Akabori and coworkers. This catalyst is prepared by the hydrogenolysis of $PdCl_2$ impregnated into silk. The catalyst functioned very well in the type of reaction shown in eqn. (6) and gave a 35.5% optical yield when a particular kind of silk was used as a support [11].

However, when the source of silk was changed, no reproducible optical yields were obtained.

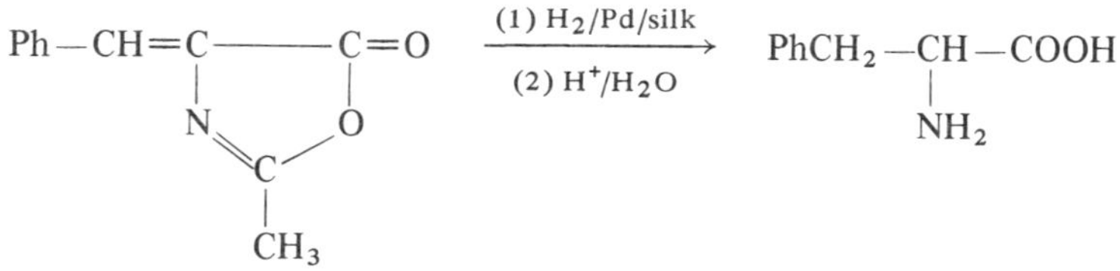

(6)

During investigations on this silk palladium catalyst, Izumi *et al.* noticed that the e.d.a. of silk palladium arose not from the chiral macromolecular template provided by silk as Akabori and coworkers had proposed, but from the microscopic asymmetric environment provided by the optically active amino acids adsorbed on metallic palladium. This consideration was extended to the preparation of a Raney nickel catalyst (RNi) whose surface had been modified with optically active materials (MRNi). The first MRNi was prepared by soaking a Raney nickel catalyst in a boiling aqueous solution of monosodium L-glutamate (Glu–MRNi) [12].

It was sheer luck that methyl acetoacetate (MAA) was employed as a substrate; even today MAA is still employed as one of the best substrates for the MNi system. The optical yield of the reaction reached 15% and as a result was sufficently high to provide motivation for further intensive investigations of this system.

Even before Izumi's discovery, Isoda *et al.* had attempted the enantio-differentiating hydrogenation of diethyl α-keto glutarate with Raney nickel or amino acid treated Raney nickel [13], but no further exploitation of this system had been made by Isoda's group. Immediately after the discovery by Izumi *et al.*, Klabunovskii and coworkers started systematic studies of this catalyst [14], and indeed almost all the early work on MNi was carried out by the groups led by Izumi and Klabunovskii.

As illustrated in Scheme 1, the preparation of MNi is simple and no special reagents or expensive materials are required. However, the variables affecting the performance of the catalyst are considerable in number and too many to be optimized readily. For this reason, it took more than 15 years to increase the e.d.a. of the catalyst from 15% to 90%.

3.1.2. *Modification variables*

In early work, the modification variables were extensively investigated using MAA as the substrate and RNi as the starting catalyst. The e.d.a. of MNi was expressed in terms of the optical purity of the methyl 3-hydroxybutyrate (MHB)

Source of Nickel Catalyst
Raney alloy, nickel oxide and other nickel salts

↓ ← Preparation variables

Activated Nickel Catalyst
RNi, HNi, DNi, Ni/SiO$_2$, etc.

Modification Procedure
Activated catalyst soaked in modifying solution for stated period at stated temperature. The resulting MNi catalyst washed with water, methanol and solvent to be used for hydrogenation

Type of Modifying Reagent
Hydroxy acid, or amino acid and supplementary modifying reagent such as NaBr

↓ ← Modification variables
Modifying temperature, pH concentration of solution, modifying period, etc.

Modified Nickel Catalyst

Hydrogenation Procedure

Substrate, H_2 and Solvent

↓ ← Reaction variables
Type of solvent, hydrogen pressure, reaction temperature, etc.

Optically Active Product

Scheme 1.

obtained. The number of the modifying reagents examined in this early work amounted to over 100 and included amino acids and their derivatives, peptides, amino alcohols, and hydroxy acids and their derivatives. The details of this work have been reviewed [15] and an extensive list of references is given there.

In general, α-amino acids or α-hydroxy acids were found to give favorable results. Of the commerically available optically active compounds, the best five were as follows: tartaric aicd (TA) > malic acid (MA) > glutamic acid (Glu) > aspartic acid (Asp) = valine (Val).

The stereochemistry of reactions carried out with catalysts prepared from the above modifying reagents is illustrated in Figure 2. In general, (*S*)-α-amino acid–MNi gave (–)-(*R*)-MHB in excess, whereas (*S*)-α-hydroxy acid–MNi gave (+)-(*S*)-MHB in excess.

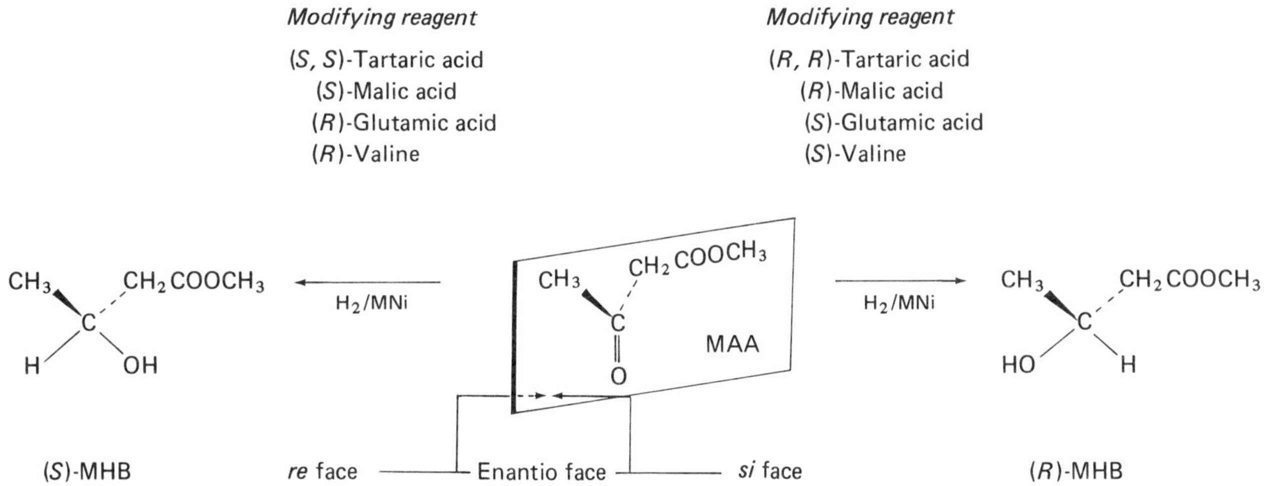

Fig. 2. Stereochemistry of the hydrogenation of MAA over various MRNi catalysts (from H. Ozaki, A. Tai and Y. Izumi: *Chem. Lett.* 635 (1974)).

On modify the conditions, the pH and temperature of the modifying solution were first found to strongly affect the e.d.a. of the resulting catalyst. For this reason extensive studies have been carried out on these variables [15]. However, an explanation for the results has only been provided recently. The major reason for this complexity is now recognized as the effect of unstable aluminum contaminants on the RNi empolyed. Details of this topic are discussed below in Sections 4.3.2 and 4.3.3.

As a result of investigations of modifying variables for RNi, TA and threo-(−)-methyl tartaric acid were found to be the best modifying reagents. Modification of RNi with a slightly acidic solution of TA and an aqueous NaOH solution at ca. 100 °C gave MRNi with the highest e.d.a. When the modifying pH was adjusted through the use of an alkaline hydroxide other than NaOH such as LiOH, KOH, RbOH and NH_4OH, the resulting TA–MNi exhibited a very low e.d.a. [18]. Although the reason has not yet been clarified, stereochemical factors relating to the cation, such as its ion radius, may affect the mode of adsorption of TA onto the catalyst.

3.1.3. *Preparation variables of the metal catalyst*

Modified catalysts can be prepared from various activated metal catalysts other than nickel. Klabunovskii *et al.* have examined the e.d.a.'s of various TA modified metal catalysts (Raney-Co, -Fe, -Cu and -Ru) in the hydrogenation of ethyl acetoacetate [19] and found that the highest e.d.a. was obtained through the use of a nickel catalyst. Murakami *et al.* obtained similar results in the hydrogenation of 4-hydroxy-2-butanone [20].

The investigation of the preparation variables of nickel catalyst brought about a great improvement in e.d.a. Representative nickel catalysts employed for the TA modification and the e.d.a.'s of the resulting MNi catalysts are listed in Table I. The activated Ni catalysts prepared by reaction in the liquid phase showed high hydrogenation activity but the e.d.a.'s of the MNi's prepared from them were generally low (entries 1–12 in Table I). Of these, RNi gave the best results.

It is well documented that various types of RNi may be obtained by leaching Ni–Al alloys under different conditions. The catalyst prepared at a low leaching temperature (RNiL) showed higher hydrogenation activity and aluminum content than the catalyst prepared at a high leaching temperature (RNiH). Upon TA modification, RNiH gave a higher e.d.a. than RNiL. The same tendency has also been reported by Gross and Rys [29], who found that the most active RNi (W-6-RNi) gave the lowest e.d.a. and the least active RNi (W-1-RNi) gave the highest e.d.a. The preparative conditions for W-6-RNi and W-1-RNi were similar to those for RNiL and RNiH, respectively, listed in Table I. These results

TABLE I
The enantio-differentiating abilities of TA–MNi catalysts prepared from various nickel catalysts

No.	Source of activated Ni	Modifying conditions pH	Modifying conditions Temp. (°C)	Reaction conditions Amount of MNi (g)	MAA (ml)	Solvent (ml)	AcOH (ml)	Temp. (°C)	H_2 pressure (atm)	Catalysts e.d.a. (%)	Ref.
1	Ni–Al/NaOH (RNiL)[b]	5.1	0	3.6	17.5	–	–	70	1	25.6	[25]
2	Ni–Al/NaOH (RNiH)[a]	5.0	100	2.5	10	–	–	60	1	60–70	[26]
3	Ni–Al/NaOH (RNiH)	5.0	0	0.6	17.5	–	–	60	90	24	[17]
4	Ni–Al/NaOH (RNiH)	5.0	100	0.6	17.5	–	–	60	90	44	[16]
5	Ni–Al/NaOH (RNiL)	5.0	0	0.8	11.5	MP[d] (23)	0.2	100	90	35	[21]
6	Ni–Al/NaOH (RNiH)	5.0	0	0.8	11.5	MP (23)	0.2	100	90	40	[21]
7	Ni–Al/NaOH (RNiH)	5.0	100	0.8	11.5	MP (23)	0.2	100	90	52	[16]
8	Ni–Mg/tartaric acid	5.0	100	0.6	17.5	–	–	60	90	6	[16]
9	Ni–Si/NaOH	5.0	100	0.6	17.5	–	–	60	90	44	[16]
10	Urushibara Ni–Al	4.1	85	1.0	11.5	THF[e] (23)	0.2	100	90	4	[16]
11	$Ni(OAc)_2/NaBH_4$ (BNi)	5.1	83	1.0	10	EA[f] (10)	–	60	10	3.0	[22]
12	$Ni(OAc)_2/NaH_2PO_2$ (PNi)	5.1	83	1.0	10	EA (10)	–	60	10	1.5	[22]
13	$Ni(CO)_4$/heat	4.1	85	0.6	11.5	THF (23)	0.2	85	90	58–75	[16]
14	$Ni(OOCH)_2$/pyrolysis (DNi)	5.0	100	0.6	17.5	–	–	100	90	55	[16]
15	$Ni(OOCH)_2$/pyrolysis (DNi)	5.1	83	3.0	10	EA (10)	–	60	10	42	[22]
16	NiO/H_2 (HNi)	4.1	85	0.8	11.5	THF (23)	0.2	100	90	65–82	[16]
17	NiO/H_2 (HNi)	5.1	83	3.0	10.0	EA (10)	–	60	10	56	[22]
18	Ni/Kg[c] (Shimalite) 1:1	4.1	85	0.6	11.5	THF (23)	0.2	85	90	37–54	[16]
19	Ni/Kg (Yoneyama Yakuhin Co.) 1:1	5.1	83	0.88	10.0	EA (10)	–	60	10	29–48	[23]
20	Ni/Pd/Kg (Kokusan Chemical Co.) 1:0.01:1	4.3	85	1.15	22.0	THF (44)	HCO_2H (0.04)	100	85	91	[24]
21	Ni/Pd/Kg (Kokusan Chemical Co.) 1:0.01:1	4.1	20	1.15	40.0	–	–	85–108	60–80	52	[34]
22	Ni/Kg (Kokusan Chemical Co.) 1:14.1	4.1	85	1.15	31.0	–	–	120	60–80	50	[24]
23	Ni/Al_2O_3 (ICN Woelm Lab., Inc.) 1:1	4.1	85	0.6	17.5	THF (23)	0.2	120	90	68	[16]
24	Ni/SiO_2 (Nakarai Chemicals Co.) 1:1	5.1	83	0.88	10.0	EA (10)	–	60	10	56	[23]
25	Ni/SiO_2 (wide pore, Royal Dutch Shell Lab.) (Ni, 12.5 wt. %)	5.0	100	1.0	5.0	MeOH (5)	–	–	1	40	[28]

[a] RNiH: leaching temp., 85 °C; ageing temp., 100 °C.
[b] RNiL: leaching temp., 20 °C; ageing temp., 75 °C.
[c] Kg = Kieselguhr.
[d] MP = methyl propionate.
[e] THF = tetrahydrofuran.
[f] EA = ethyl acetate.

indicate that a nickel catalyst which is least contaminated with Al or other metal may provide an MNi with the high e.d.a. In fact, the use of aluminum-free nickel (metals) or supported nickel catalysts prepared by the dry process resulted in high e.d.a.'s as listed in Table I (entries 13–25). Thus, an e.d.a. of almost 60% was attained with TA–MNi prepared from HNi (catalyst prepared by the hydrogenolysis of NiO), Ni/SiO_2 or Ni–Pd/kieselguhr (catalyst prepared by the hydrogenolysis of supported metal oxide on kieselguhr; Ni/Pd/kieselguhr = 1.0.01:1) as shown in Table I (Entries 17, 21 and 24).

3.1.4. *Reaction variables*

The optical yield of the reaction is a function of the e.d.a. of the catalyst, as well as reaction variables such as hydrogen pressure, reaction temperature, type of solvent and additives employed. High-pressure liquid-phase hydrogenations in an autoclave are not usually carried out at constant pressure, but even when the intial pressure was changed from 80 to 120 kg cm^{-2} no detectable change in the optical yield was observed. Thus, the optical yield is expected to be insensitive to changes in H_2 pressure. In fact, Klabunovskii *et al.* have reported that the optical yield is not affected by variations in the H_2 pressure from 25 to 100 kg cm^{-2} when MRNi is employed [30]. Over the range 1 to 30 kg cm^{-2} H_2 pressure, however, complicated changes in the optical yield have been reported [32].

Changing the reaction temperature from 60 °C to 100 °C in high-pressure liquid-phase hydrogenations also did not strongly affect the optical yield. On the other hand, when the reaction was conducted at 1 atm H_2 pressure complicated changes in the optical yield occurred as the reaction temperature was changed [25, 32].

The solvent and additives employed in the reaction system have a significant influence upon the optical yield. In the early work with TA–MRNi, it was reported that the use of a polar protic solvent such as methanol, ethanol or isopropanol increased the optical yield of the reaction [33]. However, recent studies by Orito *et al.* with Ni/Pd/kieselguhr have shown that the use of an aprotic semipolar solvent such as THF or ethyl acetate results in high optical yields, whereas methanol or ethanol give poor results [34]. We have also confirmed the effectiveness of THF and methyl propionate. Of the solvents we have examined to date, the best for the hydrogenation of MAA has been found to be methyl propionate [35]. Pronounced effects of additives such as water, carboxylic acid and unsaturated organic compounds were noted by Izumi and coworkers [36], but the significance of these additives in enhancing the optical yield was only recognized later by Orito and coworkers [24]. The optical yield of the reaction is also increased appreciably by the addition of carboxylic acids such as acetic acid, benzoic acid and succinic acid.

As shown in Table I (entries 16 and 20), optical yields greater than 80% may be achieved in the hydrogenation of MAA over TA–MHNi or TA–MNi/Pd/kieselguhr in the presence of THF as a solvent and acetic acid as an additive. The MHNi catalysts gave high optical yields, but their hydrogenation activities were not as high as that of MRNi since they required 15–22 h to complete the reaction at 100 °C. Consequently, the chemical yields with these catalysts did not exceed 85% and some polymeric impurities were also formed. At this stage, the development of an ideal catalyst having a high e.d.a. as well as a high hydrogenation activity still remained a challenging problem.

3.1.5. *Supplementary modifying reagents*

MRNiH gave an optical yield of 44% and a quantitative chemical yield when the hydrogenation of MAA was carried out in mehtyl propionate containing acetic acid (Table I, entry 7). The reaction was completed within 6 h at 60 °C. For this reason efforts have been made to further improve the e.d.a. of MRNi without sacrificing its hydrogenation activity. During the course of studying the modification variables of MRNi, it was frequently noted that the presence of Cl^- or SO_4^{2-} ions as contaminants in the water used for modification increased the e.d.a. of the resulting catalyst. Hence an examination of the effects of various inorganic salts was carried out, and as shown in Table II the effect of NaBr is outstanding. Thus RNi modified with a solution containing TA and NaBr (TA–NaBr–MRNi) showed the same e.d.a. as the of TA–Ni/Pd/Kg without any loss of the high hydrogenation activity of the original RNi [16, 21, 35]. To find out the optimum preparation conditions for this catalyst, the effect of the concentration of NaBr in the modifying solution (g $(100\ \text{ml})^{-1}$) on the e.d.a. of the resulting TA–NaBr–MRNi system was investigated. The optimum concentration of NaBr was found to be 8–10 g in 100 ml of the modifying solution containing 1 g of TA at a modifying pH of 3.2. The TA–NaBr–MRNi system prepared under optimum modification conditions showed an e.d.a. of 83% for the hydrogenation of MAA.

A further improvement of e.d.a. was obtained when the modification procedure was repeated. Thus, RNi treated three times with the modifying solution under the optimum modifying conditions exhibited an e.d.a. of 88%. This successive modification procedure was also found to be effective in eliminating aluminum contaminants from the surface of the catalyst.

TABLE II[a]
Enantio-face differentiating hydrogenation of MAA over TA–inorganic salt–MRNi[b]

Amt. of inorganic salt in modifying solution (g)	$[\alpha]_D^{20}$ of product (°)	e.d.a. (%)
none	– 8.99	39
NaBr (10)	–19.07	83
NaF (3)	–13.95	61
NaCl (10)	–16.55	72
NaI (5 × 10^{-4})	–11.74	51
Na_2SO_4 (10)	–12.95	56
$NaH_2PO_4 \cdot 2H_2O$ (1)	– 8.19	36
$NaNO_3$ (0.1)	–12.17	53
$NiBr_2$ (1)	–14.36	63
$LiBr \cdot H_2O$ (10)	–14.25	62

[a] After T. Harada, M. Imaida, H. Ozaki, A. Tai and Y. Izumi: *Bull. Chem. Soc. Jpn.* **54**, 2323 (1981).
[b] Reaction conditions: amount of RNi, 0.8 g. Modifying solution contained TA (1 g) and stated amount of inorganic salt. Modification was undertaken at pH 3.2 and 100 °C for 1 h.

3.1.6. *Preparation of TA–NaBr–MRNi and its use for the hydrogenation of MAA*

This section is devoted to the description of a typical preparation of TA–NaBr–MRNi and its use for the hydrogenation of MAA.

Preparation of RNi

The Ni–Al alloy (Ni/Al = 42:58, Kawaken Fine Chemical Co., 19 g) was added as portions to a solution consisting of 45 g of NaOH dissolved in 200 ml of water. After the vigorous reaction had ceased, the resulting suspension of RNi was maintained at 100 °C for 1 h. The catalyst thus obtained was washed 15 times with 300 ml portions of deionized water.

Modification

TA (10 g) and NaBr (100 g) were dissolved in deionized water (1000 ml) and the pH of the resulting solution adjusted to 3.2 by the addition of 1 M NaOH. The solution was heated in boiling water, and then poured upon the RNi prepared previously one portion. After maintaining the mixture at 100 °C for 1 h with occasional shaking, the supernatant liquid was removed by decantation and the residue washed successively with one 100 ml portion of deionized water, two 500 ml portions of methanol and one 250 ml portion of the solvent to be used for hydrogenation (methyl propionate).

Hydrogenation

Into an autoclave (1000 ml capacity) was introduced 115 ml of MAA, 2 ml of acetic acid and the catalyst suspended in 230 ml of methyl propionate. After evacuation, the autoclave was charged with H_2 to a pressure of 95 kg cm^{-1}. The mixture in the autoclave was kept at 100 °C with shaking until no further consumption of hydrogen was observed. The autoclave was then cooled to room temperature and the H_2 atmosphere removed. After separation of the catalyst by decantation, the reaction product was distilled under reduced pressure to give MHB (b.p., 68–70 °C/20 mmHg) of 82–85% optical purity in quantitative yield.

3.2. APPLICATION OF MNi FOR THE PREPARATION OF OPTICALLY PURE MATERIALS

Various types of prochiral compounds including those containing C=C, C=N [37] and C=O groups have been subjected to hydrogenation over MNi. Of the substrates so far examined, certain types of carbonyl compounds gave excellent results but others gave poor results. In this section, the performance of TA–NaBr–MNi in the enantio-differentiating hydrogenation of appropriate substrates will be mentioned together with a supplementary procedure which allows for the imperfect nature of enantio-differentiating hydrogenation and provides optically pure compounds from the hydrogenation products.

3.2.1. *Hydrogenation of β-ketoester*

MAA has been most widely investigated as a substrate in hydrogenations involving the MNi system. Two homologs of MAA are also excellent substrates for MNi. In Table III (entry 1), the results of the hydrogenation of various esters of acetoacetic acid over TA–NaBr–MRNi are listed [38]. With the exception of MAA, all higher esters gave an optical yield of ca. 88%. For the preparation of optically active 3-hydroxybutyric acid, the use of ethyl acetoacetate rather than MAA is recommended.

The results of the hydrogenation of the other homologs of MAA are also summerized in Table III (entry 2). Again, the higher homologs gave better results than MAA. In all cases, (*R,R*)-TA–NaBr–MRNi gave (*R*)-isomers and (*S,S*)-TA–NaBr–MRNi gave (*S*)-isomers in excess.

The isolation of optically pure 3-hydroxyalkanoic acid from the reaction product (optical purity >85%) was achieved by the preferential crystallization of its dicyclohexyl ammonium salt [39]. This facile method could make a considerable contribution to synthetic chemistry, since optically pure 3-hydroxyalkanoic acid is important not only as a starting material for natural products containing the moiety of 3-hydroxyalkanoic acid [40], but also as a chiral pool in synthetic chemistry of optically active secondary hydroxyl groups.

TABLE III

Hydrogenation of β-ketoesters over (*R*,*R*)-TA–NaBr–MRNi

Entry No.	β-Ketoester	Optical yield (%)	Configuration of product
1	$CH_3C(=O)CH_2COOR^1$		
	$R^1 = CH_3$	83.0–85.0	*R*
	C_2H_5	87.9	*R*
	n-C_3H_7	87.6	*R*
	i-C_3H_7	87.7	*R*
	n-C_4H_9	88.0	*R*
	i-C_4H_9	88.1	*R*
	n-C_8H_{17}	87.7	*R*
2	$R^2C(=O)CH_2COOCH_3$		
	$R^2 = CH_3$	83.0–85.0	*R*
	C_2H_5	86.0	*R*
	n-C_3H_7	87.0	*R*
	i-C_3H_7	87.5	*R*
	n-C_4H_9	86.7	*R*
	i-C_4H_9	85.5	*R*
	n-C_8H_{17}	87.0	*R*

3.2.2. *Hydrogenation of β-diketones*

Acetylacetone is also a very good substrate for the enantio-differentiating hydrogenation of TA–NaBr–MRNi [41]. The reaction proceeds in a stepwise fashion as shown in eqn. (7):

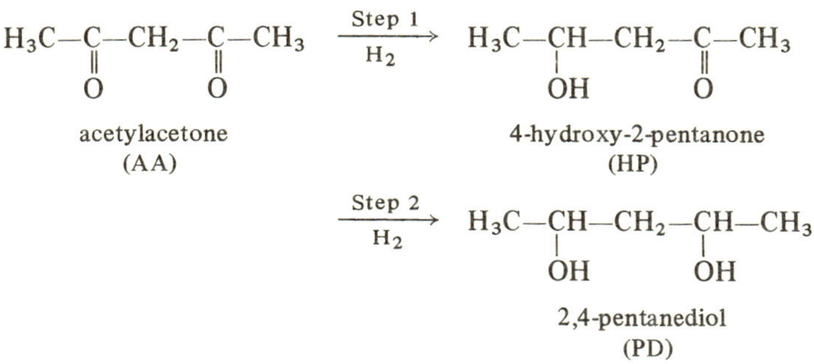

(7)

Step 1 is the usual enantio-differentiating reaction which gives 4-hydroxy-2-pentanone (HP) in 74% optical yield (Scheme 2). HP, which is a substrate of step 2, is a chiral compound and hence the conversion of HP to 2,4-pentanediol(PD) is a diastereo-face differentiating reaction. When an optically active catalyst is employed in this reaction, the enantio-differentiating reaction takes place simultaneously. The results of the hydrogenations of (*R*)- and (*S*)-HP over (*R,R*)-TA–NaBr–MRNi are shown in Scheme 2.*

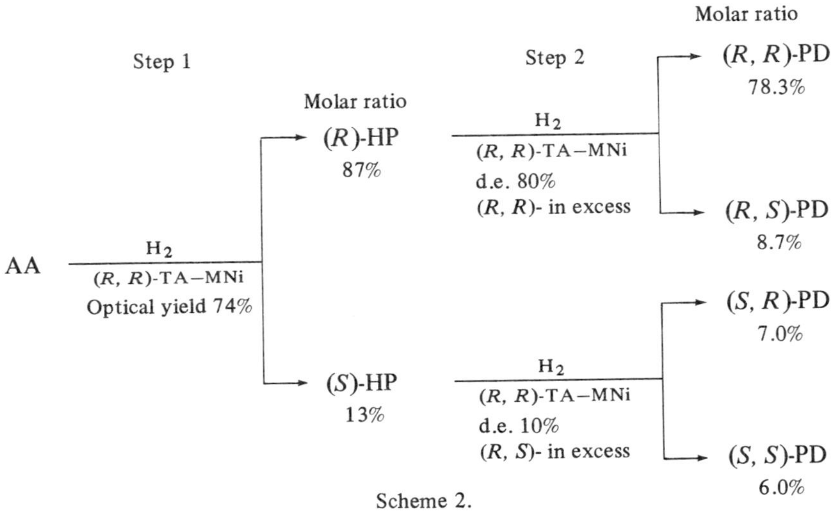

Scheme 2.

(*R*)-HP was converted predominantly to (*R,R*)-PD whereas (*S*)-HP was converted to an almost equimolar mixture of (*S,S*)- and (*S,R*)-PD. Thus, the diastereo-differentiation favors the formation of the (*R,R*)-diastereomer, i.e. (*R*)-HP favors the formation of the *R*-configuration. In the hydrogenation of (*R*)-HP over (*R,R*)-TA–NaBr–MRNi, both enantio- and diastereo-differentiations function in the same direction. As a result, (*R,R*)-PD is produced in large excess. On the other hand, in the reaction of (*S*)-HP over (*R,R*)-TA–NaBr–MRNi, the directions of the two stereo-differentiations work counter to each other to give an equimolar mixutre of diastereomers. A material balance in the direct hydrogenation of AA to PD in Scheme 2 has been estimated from the data shown in Table IV (entries 2, 4 and 5). The good agreement between the experimental and calculated values clearly shows that the reaction proceeds exactly in the manner illustrated in Scheme 2.

* This scheme is taken from A. Tai, K. Ito and T. Harada: *Bull Soc. Chem. Jpn.* 54, 223 (1981).

TABLE IV
Hydrogenation of acetylacetone and related compounds over TA–NaBr–MRNi

<table>
<tr><th rowspan="3">Entry No.</th><th rowspan="3">Catalyst</th><th rowspan="3">Mole ratio H_2/substrate</th><th rowspan="3">Substrate</th><th colspan="4">Product (mol %)</th><th colspan="2">Optical purity (%)</th></tr>
<tr><th rowspan="2">AA</th><th rowspan="2">HP</th><th colspan="2">PD</th><th rowspan="2">HP</th><th rowspan="2">(R*,R*)-PD</th></tr>
<tr><th>R*,R*</th><th>R*,S*</th></tr>
<tr><td>1</td><td>RNi</td><td>1.1</td><td>AA</td><td>22</td><td>34</td><td colspan="2">44</td><td>–</td><td>–</td></tr>
<tr><td>2</td><td>(R,R)-TA–NaBr–MRNi</td><td>1.1</td><td>AA</td><td>0</td><td>91</td><td colspan="2">9</td><td>74 (R)</td><td>–</td></tr>
<tr><td>3</td><td>(S,S)-TA–NaBr–MRNi</td><td>1.1</td><td>AA</td><td>0</td><td>90</td><td colspan="2">10</td><td>73 (S)</td><td>–</td></tr>
<tr><td>4</td><td>(R,R)-TA–NaBr–MRNi</td><td>1</td><td>(R)-HP</td><td>–</td><td>–</td><td>90</td><td>10</td><td>–</td><td>–</td></tr>
<tr><td>5</td><td>(R,R)-TA–NaBr–MRNi</td><td>1</td><td>(S)-HP</td><td>–</td><td>–</td><td>45</td><td>55</td><td>–</td><td>–</td></tr>
<tr><td>6</td><td>(R,R)-TA–NaBr–MRNi</td><td>0.55</td><td>racemic-HP</td><td>–</td><td>45</td><td>44</td><td>11</td><td>30 (S)</td><td>48(R,R)</td></tr>
<tr><td>7</td><td>(R,R)-TA–NaBr–MRNi</td><td>2</td><td>AA</td><td>–</td><td>–</td><td>87
(84)[a]</td><td>13
(16)[a]</td><td>–
–</td><td>86 (R,R)
(86)[a]</td></tr>
<tr><td>8</td><td>(R,R)-TA–NaBr–MRNi</td><td>1.7</td><td>AA</td><td>–</td><td>30</td><td>65</td><td>5</td><td>–</td><td>98 (R,R)</td></tr>
</table>

[a] Calculated values based on data given in Scheme 2.

The data in Table IV (entry 6) illustrate another important feature of this reaction. When the hydrogenation of racemic HP over (*R,R*)-TA–NaBr–MRNi was stopped halfway through the reaction, the recovered HP was the (*S*)-isomer in excess, while the product PD was present as the (*R,R*)-isomer in excess. In other words, enantiomer differentiation (kinetic resolution) of (*R*)- and (*S*)-HP took place with (*R,R*)-TA–NaBr–MRNi. (*R*)-HP was preferentially hydrogenated at an early stage in the reaction to give (*R,R*)-PD in large excess. Taking into account the material balance of the reaction in Scheme 2, the optical purity of the (*R,R*)-PD produced should attain a high value if the reaction is discontinued at an appropriate conversion. As expected, when the reaction was stopped at 85% conversion, the resulting (*R,R*)-PD showed 98% optical purity.

In conclusion, it is obvious that the hydrogenation of AA to PD over TA–NaBr–MRNi provides an example of an ideal stereo-differentiating reaction, i.e. all three types of stereo-differentiating reactions, enantio face, enantiomer and diastereo face, are involved in increasing the optical purity of the final reaction product.

The major product, (*R,R*)-PD, is readily separated from (*R,S*)-PD by recrystallization of the reaction product from ether at –10 °C. The (*R,R*)-PD thus isolated was found to be optically pure already [42]. The use of (*R,R*)-TA–NaBr-MRNi gave (*R,R*)-PD and of (*S,S*)-TA–NaBr–MRNi gave (*S,S*)-PD. The average yield of optically pure PD isolated was 65% based on AA. The usefulness of PD as a starting material for phosphine ligands in the Rh–P system has been already mentioned in Section 2.2.

The higher homologs of AA can also be converted to optically pure β-diols by the same method as that deseribed for PD. Optically pure diols were obtained in yields of 20–60% from the corresponding β-diketones [43].

3.2.3. *Hydrogenation of miscellaneous carbonyl compounds*

Some analogs of β-ketoesters are hydrogenated with high optical yield over TA–NaBr–MRNi. Some typical results are listed in Table V. The hydrogenation of β-ketosulfone proceeded smoothly in giving β-hydroxysulfone in high optical yield (entries 1, 2 and 3). The optically pure materials were easily obtained by recrystallization of the products [44].

4-Hydroxy-2-butanone and its methyl ether were also found to be good substrates for TA–NaBr–MRNi (entries 4 and 5). The resulting 1,3-butanediol may also be purified to the optically pure material by recrystallization of its bis (phenyl urethane) [20]. TA–NaBr–MRNi was also found to be an excellent catalyst for the enantio-differentiating hydrogenation of methyl alkyl ketones (entries 6–10) [45]. The optical yield obtained is the highest value known to

TABLE V
Enantio-face differentiating hydrogenation of various ketones

Entry No.	Catalysts / Solvent	Substrate		Product $[\alpha]_D^{20}$ (°)	Optical yield (%)	configuration	Optically pure material $[\alpha]_D^{20}$ (°)
	(*R,R*)-TA–NaBr–MRNi / THF/AcOH	$RCOCH_2SO_2CH_3$	⟶	$RCH(OH)CH_2SO_2CH_3$			
1		R = CH_3CH_2		−22.9 (c 5, EtOH)	71.0[a]	–	– 33.9[b]
2		$CH_3(CH_2)_4$		−13.7 (c 5, EtOH)	68.0[a]	–	– 19.7[b]
3		$CH_3(CH_2)_7$		−10.5 (c 5, EtOH)	67.0	*R*	– 15.6[b]
	(*R,R*)-TA–NaBr–MRNi / THF/AcOH	$CH_3COCH_2CH_2OR$	⟶	$CH_3CH(OH)CH_2CH_2OR$			
4		R = H		−20.8 (c 10, EtOH)	70.2	*R*	– 29.6[c]
5		CH_3		−20.6 (c 10, EtOH)	68.0[a]	*R*	–
	(*R,R*)-TA–NaBr–MRNi / THF/pivalic acid	CH_3COR	⟶	$CH_3CH(OH)R$			
6		R = CH_2CH_3		+6.79 (neat)	48.9	*S*	–
7		$(CH_2)_3CH_3$		+7.62 (neat)	65.9	*S*	–
8		$(CH_2)_5CH_3$		+6.40 (neat)	65.6	*S*	–
9		$CH(CH_3)_2$		+3.04 (neat)	62.6	*S*	–
10		$C(CH_3)_3$		+5.72 (neat)	74.2	*S*	–
	cinchonidine–MPt / benzene	$RCOCO_2CH_3$	⟶	$RCH(OH)CO_2CH_3$			
11		R = CH_3		+6.66 (neat)	80.8	*R*	+ 8.2
12		Ph		−107.6 (c 4, $CHCl_3$)	83.9	*R*	−128.5

[a] Determined by NMR spectroscopy in the presence of a chiral shift reagent.
[b] Obtained by preferential crystallization from ethyl acetate or ether.
[c] Obtained as phenyl urethane by preferential crystallization from ethanol.

date for catalytic hydrogenations or for the chemical reduction of simple alkyl ketones including diastereo- and enantio-differentiating reactions. The reaction was strongly affected by the presence of carboxylic acids as additives. The use of a large amount of pivalic acid gave excellent results in the hydrogenation of 2-alkanone. The highest optical yield (74%) was obtained in the hydrogenation of methyl *t*-butyl ketone [46]. The hydrogenation of α-ketoesters over MNi invariably gave poor results (Section 4.4). The recent development of the cinchonidine-modified platinum catalyst (cinchonidine–MPt) by Orito and coworkers has overcome the disadvantage of MNi and leads to favorable results in the hydrogenations of methyl benzoylformate and methyl pyruvate. The results are listed in Table V (entries 11 and 12) [47, 48].

4. MNi, its Structure and Mode of Action

4.1. ENANTIO-DIFFERENTIATING AND HYDROGENATION STEPS IN THE REACTION PATH

The MNi catalyst is a dual function catalyst. Not only does it catalyze the hydrogenation of the substrate but it also differentiates one of the enantio faces of the substrate to yield one enantiomer in excess. From a kinetic viewpoint, enantio-differentiating hydrogenation may be considered a competitive reaction giving (*R*)- and (*S*)-isomers at different rates. The Curtin–Hammet principle states that the ratio of products obtained in a competitive reaction depends on the difference between the activation free energies of the competitive reaction paths (i.e. $G_R^{\ddagger} - G_S^{\ddagger}$ in Figure 3(B)). This kinetic description of enantio-differentiation rests on the assumption that differentiation takes place in concert with the rate-determining step for the reaction, and that the ratio of (*R*)- and (*S*)-isomers depends on the difference between the true activation energies, $\Delta G^{\ddagger}$ (Figure 3(A)). However, this assumption is not always correct, especially when a multistep reaction is carried out over a heterogeneous catalyst. In this case, the difference between the activation energies of the competitive reaction paths does not necessarily originate from the difference between the true activation energies. As shown in Figure 3(A), the difference between the free energies at the adsorption step could bring about the apparent energy difference at the transition state ($G_R^{\ddagger} - G_S^{\ddagger} = \Delta\Delta G^{\ddagger} + \Delta\,(G_{ad})$).

In order to understand the mechanism of enantio-differentiation of MAA over MNi, it is essential to explore the relationship between the enantio-differentiating and hydrogenation steps during the course of reaction.

As for liquid-phase hydrogenation, two different approaches, i.e. a comparative study of the rate of hydrogenation and e.d.a. of MRNi together with a study using a stereochemical probe, have been carried out by our group [25, 49,

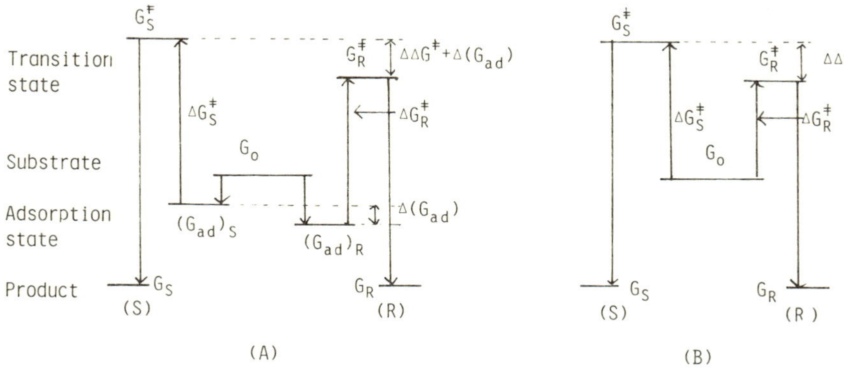

Fig. 3. Schematic illustration of the energy diagram for a stereo-differentiating reaction (from Y. Izumi, A. Tai and T. Harada: *Kagaku Sosetsu* 142 (1981)).

55]. The rate study of hydrogenation in the liquid phase was carried out at 1 atm H_2 pressure, and examples of Arrhenius plots with respect to the rate of hydrogenation of MAA over various types of MRNi catalysts are shown in Figure 4.

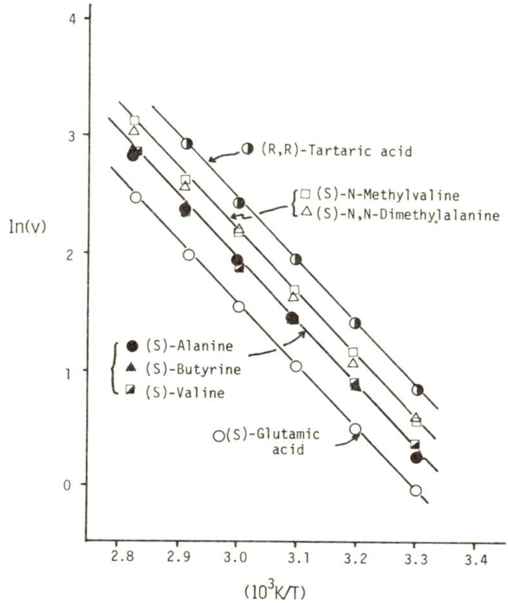

Fig. 4. Arrhenius plots of the rate of hydrogenation of MAA over various MRNi catalysts (from T. Harada, Y. Hiraki, Y. Izumi, J. Muraoka, H. Ozaki and A. Tai: *Proc. 6th Int. Congr. Catal., London (1976)*, p. 1024, The Chemical Society, London (1977)).

All the plots are parallel to each other with those corresponding to modifying reagents of a similar structure overlapping. These results strongly suggest that the mechanism of the rate-determining process is not influenced by the type of modifying reagent employed on the catalyst. The values of the apparent activation energies were 44.2 kJ mol^{-1} in all cases. A plot of the reciprocal of the initial rate ($1/v$) *versus* the reciprocal of the concentration of the substrate ($1/[MAA]$) was found to be linear and hence the rate law may be expressed as:

$$v = \frac{\beta[\text{MAA}]}{1 + \alpha[\text{MAA}]} \tag{8}$$

From the rate, the order with respect to [MAA] was found to be 0.2–0.3 and a linear relationship existed between the rate and the amount of catalyst employed up to 15 times that used under standard reaction conditions. These facts indicate that supplies of MAA and H_2 to the active area of the catalyst are sufficient under the reaction conditions employed. The value of the apparent activation energy suggests that the rate-determining step in the hydrogenation process is a surface reaction between the adsorbed MAA and H_2. The experimental results may be explained satisfactorily by the following process:

$$\text{MAA} + \text{C} \begin{array}{l} \overset{K_R}{\rightleftharpoons} [\text{MAA*C}]_R \xrightarrow[\text{H*}]{k_R} (R)\text{-MHB} \\ \underset{K_S}{\rightleftharpoons} [\text{MAA*C}]_S \xrightarrow[\text{H*}]{k_S} (S)\text{-MHB} \end{array} \tag{9}$$

where K_R and K_S are the adsorption equilibrium constants for the hydrogenation of MAA to (*R*)- and (*S*)-MHB, respectively, k_R and k_S are the rate constants for the formation of (*R*)- and (*S*)-MHB, respectively, C is the adsorption site for MAA on the MRNi, $[MAA^*C]_R$ and $[MAA^*C]_S$ are the concentrations of adsorbed MAA on the catalyst which are hydrogenated to (*R*)- and (*S*)-MHB, respectively, and H* is the activated hydrogen associated with the catalyst.

A comparison of the rate and optical yield has also been undertaken using TA–MRNi and malic acid (MA)–MRNi. The overall rate of hydrogenation ($v = v_R + v_S$) and the rates of formation of (*R*)- and (*S*)-MHB (v_R and v_S) were estimated graphically from the time course of the hydrogenation and optical yield, respectively, and are illustrated in Figure 5. As shown in this figure, the overall rates, v_{TA} and v_{MA}, are identical to each other at all times during the reaction, i.e. all the kinetic parameters, including the decay of catalytic activity, must be the same for TA–MRNi and MA–MRNi. In contrast, $(e.d.a.)_{TA}$ and $(e.d.a.)_{MA}$ differ from each other.

The rates and e.d.a.'s for various MRNi catalysts determined by the above procedure are listed in Table VI together with the amounts of modifying reagent

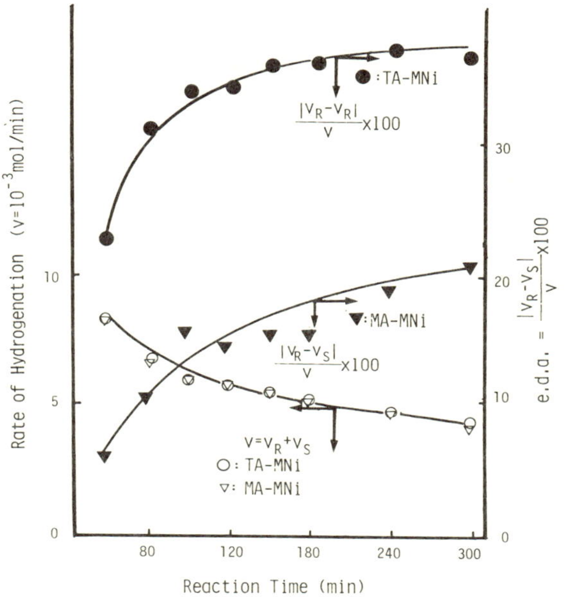

Fig. 5. Time dependencies of the rate and optical yield during the hydrogenation of MAA over TA–MRNi and MA–MRNi.

TABLE VI
Effect of modifying reagent on the rate and optical yield during the hydrogenation of MAA over various MRNi catalysts

Modifying reagent	Optical yield (%)	Rate of hydrogenation of MAA (mmol h^{-1})	Amt. of modifying reagent on catalyst (mmol (g catalyst)$^{-1}$)
TA	26.2	13.1	0.12
MA	19.5	13.1	0.12
Ala	0	9.8	0.18
Val	13.3	10.2	0.18
Leu	4.3	10.3	0.19
Ileu	8.1	9.8	–
Glu	11.6	4.3	0.22
Oru	6.5	3.3	–
Lys	6.4	3.1	0.27

adsorbed on the catalyst. In this case no systematic relationships are apparent for the rates and e.d.a.'s. Furthermore, no Arrhenius-type correlations could be obtained between the reciprocal of the reaction temperature ($1/T$) and the logarithm of the ratio of enantiomers produced ($\ln([R]/[S])$) [25, 32]. These results strongly suggest that the reaction steps involving enantio-differentiation and hydrogenation are independent of each other, with the enantio-differentiation step being not rate-determining, i.e. the rate constants k_R and k_S in eqn. (9) are not involved in enantio-differentiation ($k_R = k_S$). When the rate and the amount of absorbed modifying reagent listed in Table VI are plotted against each other, however, a linear relationship is obtained. The data in Table VI also show that MRNi catalysts prepared from modifying reagents which are homologs of each other (the same class of MRNi) contain the same amounts of modifying reagent, and lead to the same rates of the reaction but different e.d.a.'s. This suggests that the modifying reagent changes the number of adsorption site available for MAA, although the existence of the modifying reagent on the catalyst does not affect the equilibrium constant (K) between MAA and the vacant adsorption sites on the catalyst. Since all the various homologous modifying reagents were adsorbed onto the catalyst in the same amount, the same adsorption site for MAA(C) and the same adsorption constant (K) would be expected to apply for the same class of MRNi (e.g. TA–MRNi and MA–MRNi). Thus, the total amount of adsorbed MAA, i.e. $[\mathrm{MAA^*C}]_R + [\mathrm{MAA^*C}]_S$, is the same for the same class of MRNi since only the the mole fractions of the adosrbed species, γ_R and γ_S ($\gamma_R + \gamma_S = 1$), are determined by the intrinsic e.d.a. of the modifying reagent. The adsorption equilibrium constants K_R and K_S, in eqn. (9) are related to each other by the expression $K_R + K_S = (\gamma_R + \gamma_S) K = K$.

The rate and e.d.a. of MRNi respectirely, may be written as

$$v_R = \frac{kK_R\,[\mathrm{MAA}]\,[\mathrm{C}]\,[\mathrm{H^*}]}{1 + (K_R + K_S)[\mathrm{MAA}]} \tag{10}$$

$$v_S = \frac{kK_S\,[\mathrm{MAA}]\,[\mathrm{C}]\,[\mathrm{H^*}]}{1 + (K_R + K_S)[\mathrm{MAA}]} \tag{11}$$

from which

$$v = \frac{k(K_R + K_S)[\mathrm{MAA}]\,[\mathrm{C}]\,[\mathrm{H^*}]}{1 + (K_R + K_S)[\mathrm{MAA}]} = \frac{kK\,[\mathrm{MAA}]\,[\mathrm{C}]\,[\mathrm{H^*}]}{1 + K\,[\mathrm{MAA}]} \tag{12}$$

and

$$\text{e.d.a.} = \frac{(v_R - v_S)}{v} \times 100 = \frac{(K_R - K_S)}{K} \times 100 = (\gamma_R - \gamma_S) \times 100 \tag{13}$$

In these equations, k is the rate for the overall hydrogenation process.

Equations (12) and (13) when applied to the experimental results from these various catalysts indicate that the same class of MRNi's gave the same rate but different e.d.a.'s, with the rate and e.d.a. being independent for each MRNi. From this it was logically proposed that the ratio of enantiomers formed depends principally on the term γ_R/γ_S, and that the enantio-differentiating step occurs during adsorption which occurs before the rate-determining step.

Kinetic studies of the gas-phase reaction over MDNi have been carried out by Yasumori and coworkers [50–52]. They have developed the rate law shown in eqn. (14) and have proposed that the rate-determining step in the reaction is that of addition of hydrogen to the substrate. On the basis of kinetic parameters listed in Table VII, the results of MAA deuteration and deuterium exchange with MHB, it was shown that enantio-differentiation occurred during adsoprtion which took place prior to the rate-determining step.

$$v = k \left(\frac{K_H P_H}{1 + K_H P_H}\right) \left(\frac{K_{MAA} P_{MAA}}{1 + K_{MAA} P_{MAA}}\right) \qquad (14)$$

Yasumori and coworkers also compared TA- and MA–MDNi and respect to their hydrogenation activities and e.d.a.'s, and showed that the results of the gas-phase reaction were comparable with those for the liquid-phase reaction mentioned above.

TABLE VII
Kuietic parameters for the hydrogenation of MAA in the gas phase over various modified MNi catalysts

Catalyst	Apparent activation energy ($kJ\ mol^{-1}$)	Reaction order with respect to MAA	Reaction order with respect to H_2	e.d.a. (%)	Relative reaction rate	Reference
DNi	44.3 ± 2.5	1.0 ± 0.1	0.0 ± 0.1	–	1.00	[50]
TA–MDNi[a]	43.9 ± 2.0	0.8 ± 0.1	0.0 ± 0.1	17.8 ± 1.0	–	[51]
TA–MDNi[b]	44 ± 2	0.8 ± 0.1	0.0 ± 0.1	48	0.22	[52]
MA–MDNi	–	–	–	32	0.27	[52]
HNi/SiO_2	61 ± 5	0.40 ± 0.05	0.20 ± 0.03	–	–	[53]
$TA–MHNi/SiO_2$	61 ± 5	0.40 ± 0.05	0.20 ± 0.02	≤ 15	–	[53]

[a] TA modification carried out in air.
[b] TA modification carried out *in vacuo*.

Kinetic studies of the gas-phase hydrogenation of MAA have also been carried out by Sachtler's group [28, 53]. They used low loadings of Ni on wide-pore silica (Ni content, 0.5 wt.%) as a catalyst and determined the activation energies for hydrogenation over TA-modified and unmodified catalysts under conditions such that only a surface reaction takes place and pore diffusion is completely excluded (high space velocity (8400 h^{-1}) and low MAA/H_2 ratio (10^{-3}). As shown by the second entry in Table VII, the e.d.a. of the modified catalyst is very low. For both the modified and unmodified catalysts (entries 5 and 6), the activation energies for the heterogeneous reaction (61 kJ mol^{-1} in both cases) are somewhat higher than that for the liquid-phase reaction (~44 kJ mol^{-1}). These authors have suggested that the low activation energy in the liquid-phase reaction would originate from participation of a diffusion-controlled process associated with H_2.

Nitta and her coworkers have recently carried out a kinetic study on the high-pressure liquid-phase hydrogenation of MAA over TA–MNi/SiO_2 [54]. They modified eqns. (10) and (11) to the forms shown in eqns. (15) and (16), using the assumption that enantio-differentiating reaction sites (E = e.d. site) and non-enantio-differentiating reaction sites (N = nod-e.d. site) exist on the catalyst (see Section 4.3.1), and that the former give the optically pure product while the latter give a racemic product; hydrogen and MAA are adsorbed competitively on the same catalyst site under a high pressure of hydrogen.

$$v_E = \frac{\alpha^2 k_E K_H C_H (K_E)_{MAA} C_{MAA}}{(1 + K_H C_H + (K_E)_{MAA} C_{MAA})^2} \tag{15}$$

$$v_N = \frac{(1-\alpha)^2 k_N K_H C_H (K_N)_{MAA} C_{MAA}}{(1 + K_H C_H + (K_N)_{MAA} C_{MAA})^2} \tag{16}$$

where K is the adsorption equilibrium constant, k is the rate constant, E is an e.d. site, N is a non-e.d. site, C_{MAA} is the concentration of MAA and α is the fraction of e.d. sites on the catalyst.

The optical purity of the product, P, may be expressed as

$$P = \frac{v_E}{v_E + v_N}$$

$$= 1 \Big/ \left[1 + \left(\frac{1}{\alpha} - 1\right)^2 \times \frac{k_N (K_N)_{MAA} (1 + K_H C_H + (K_E) C_{MAA})^2}{k_E (K_E)_{MAA} (1 + K_H C_H + (K_N)_{MAA} C_{MAA})^2}\right] \tag{17}$$

The experimental data, the relationship between P and C_{MAA} and that between P and H_2 pressure, are well simulated by eqn. (17) provided that $k_N = k_E$, $\alpha = 0.5$, $(K_N)_{MAA}/(K_E)_{MAA} = 2.5$, $K_H = 0.2$ (kg $cm^{-2})^{-1}$, and $(K_E)_{MAA} = 0.07$ (vol.%)$^{-1}$. The fact that such good simulation is obtained when k_E is put equal to k_N suggests that enantio-differentiation takes place during the adsorption step. Although P changes little with changes in the H_2 pressure, the competitive adosrption of H_2 and MAA does provide a plausible mechanism for the high-pressure liquid-phase hydrogenation.

Kinetic studies support our suggestion that enantio-differentiation takes place during the adsorption step but have not provided direct evidence for interaction between the substrate and the modifying reagent in this step. Methyl 2-methyl-3-oxobutyrate (MMOB) should provide a suitable substrate for studying such interaction since the configuration at the C-2 position in this compound is readily interconvertible on contact with MNi. However, once MMOB has been changed to methyl 3-hydroxy-2-methylbutyrate (MHMB) through hydrogenation of its carbonyl group, no further epimerization can occur at the C-2 position. Hence, the change in the stereochemistry of the C-2 position during hydrogenation should provide an internal probe for the detection of the stereo-differentiating step [55].

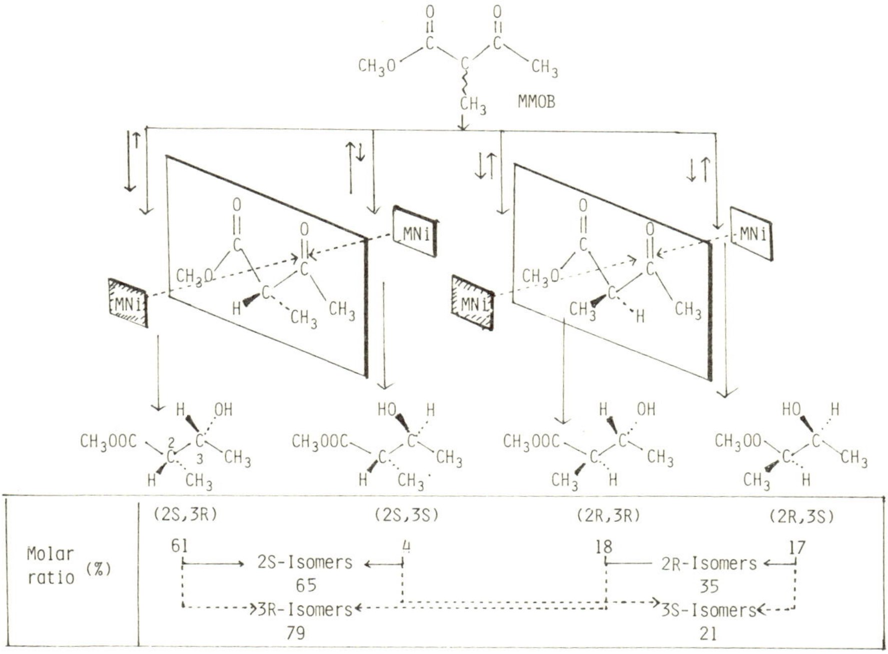

Fig. 6. Enantio-differentiating hydrogenation of methyl 2-methyl-3-oxobutyrate (MMOB) to methyl 3-hydroxy-2-methylbutyrate (MMOB).

The results of studies of the hydrogenation of racemic-MMOB over TA–MHNi are summarized in Figure 6. The excess formation of (2*S*)-isomers (2*S*,3*R* + 2*S*,3*S*) rather than (2*R*)-isomers (2*R*,3*R* + 2*R*,3*S*) despite the use of racemic-

MMOB clearly indicates interaction between MMOB and the modifying reagent leading to (2*S*)-isomers in excess on the catalyst surface at the adsorption step. Since the extent of formation of (3*R*)-isomers (2*S*,3*R* + 2*R*,3*R*) is greater than that of (3*S*)-isomers (2*S*,3*S* + 2*R*,3*S*), the carbonyl group of MMOB must be hydrogenation in the same enantio-differentiating manner as that of MAA. From the similarities between the structure of MAA and that of MMOB, it is obvious that not only MMOB but also MAA and its analogs interact significantly with the modifying reagent during adsorption step.

That the MNi catalyst which gives the greater excess of (3*R*)-isomers also provide a greater excess of the (2*S*,3*R*)-isomer indicates that the degree of enantio-face differentiation at the C-3 position and the change of configuration at the C-2 position are governed by the same factor, i.e. the ability of the modifying reagent to interact with MMOB. The configurations at the C-2 and C-3 atoms are expected to be determined simultaneously through formation of MMOB–modifier–Ni complexes as will be discussed in Section 4.4.

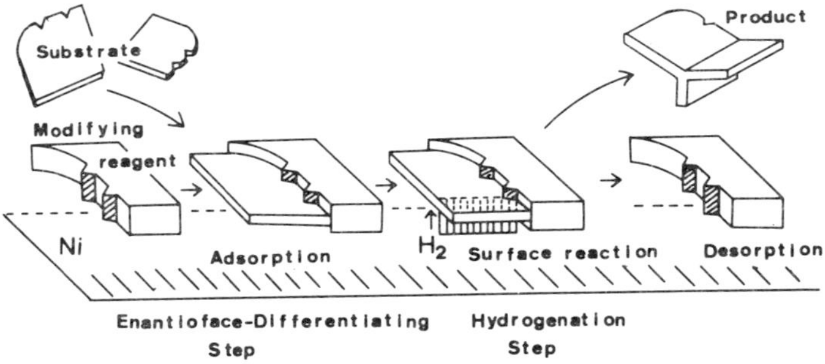

Fig. 7. Schematic illustration of the mechanism of enantio-differentiating hydrogenation over MRNi (from Y. Izumi, A. Tai and T. Harada: *Kagaku Sosetsu* 142 (1981).

From the above results of kinetic and stereochemical studies, the reaction process involved in enantio-differentiating hydrogenation over an MNi catalyst may be illustrated schematically as in Figure 7.

4.2. MODE OF ADSORPTION OF MODIFYING REAGENT ON NICKEL CATALYSTS

4.2.1. *α-Hydroxy acids*

Detailed investigations of the effects of pH on the amount of the modifying reagent adsorbed on the catalyst (abbreviated hereafter as a.m.a.) and on the e.d.a. of the catalyst have provided important information regarding the nature of the adsorbed species under the catalyst conditions normally employed [56, 57].

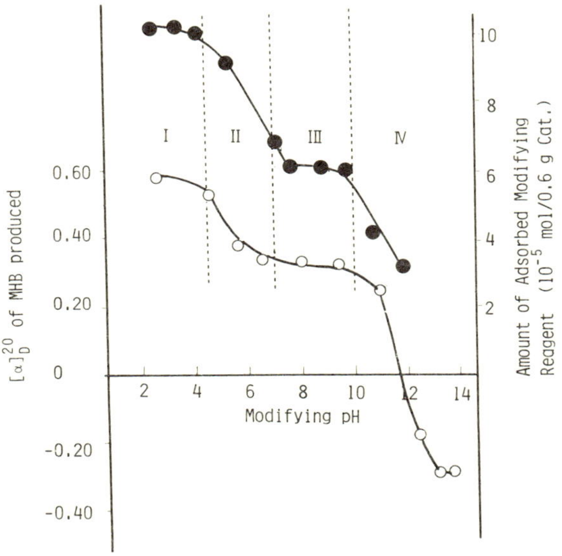

Fig. 8. Effect of pH on the amount of (*S*)-2-hydroxy-3-phenylpropionic acid (HPA) adsorbed on MRNi (•) and on the e.d.a. of (*S*)-HPA–MRNi (○) (from T. Harada: *Bull. Chem. Soc. Jpn.* **48**, 3236 (1975)).

Figure 8 shows the relationship between a.m.a. and the pH, and between e.d.a. and pH, when (*S*)-2-hydorxy-3-phenylpropionic acid (HPA) and RNi were employed as the modifying reagent and catalyst, respectively. The profiles of the two plots are very similar to each other and indicate that both e.d.a. and a.m.a. are governed by a common pH-dependent factor. Over the pH region 4–12, the plots are also closely related to the titration curve for HPA with NaOH. This strongly suggests that both e.d.a. and a.m.a. depend on the type of species of the modifying reagent adsorbed, i.e. whether the modifying reagent is adsorbed as a free acid or as its sodium salt.

When the pH was less than 4.5, some of the metallic nickel was attacked by the modifying reagent and entered the solution as Ni^{2+} ions. For this reason, the modifying reagent was expected to exist at the catalyst surface as its nickel salt rather than as the free acid. It is also to be expected that after treatement at pH values less than 4.5, the surface is no longer the same as that of the original catalyst. Such a modification of the surface as well as changes in the species of modifying reagent adsorbed are also indicated by the profiles of the plots shown in Figure 8.

As indicated in the figure, the profiles of the plots may be subdivided into four pH regions. Over region I, the modifying reagent is adsorbed onto the acid-corroded metal surface as Ni–$(HPA)_2$. Over region III, the adsorbed species is expected to be Na–HPA, a suggestion confirmed by the fact that when the catalyst is modified by the addition of a solution of potassium glycolate (pH 6.1), the molar ratio of glycolate to potassium ion adsorbed onto the resulting catalyst was exactly 1:1 [57]. Since no corrosion of the Ni was observed, the original surface of the catalyst was not changed significantly by modification under these conditions. Over region II, co-existence of the two species found in regions I and III is expected. Over region IV, the large excess of NaOH present may compete with Na–HPA during the adsorption process and could to affect the mode of adsorption of Na–HPA significantly. However, further details are still awaited for this region.

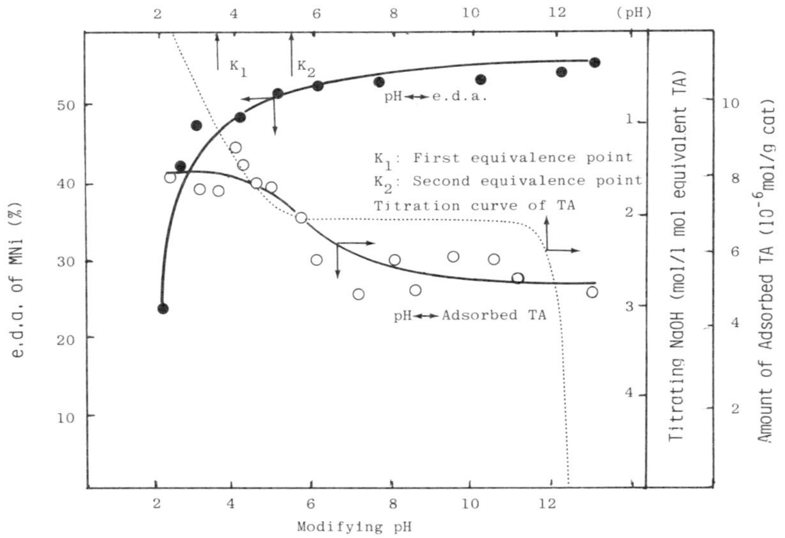

Fig. 9. Effect of the pH on the amount of TA adsorbed (○) and the e.d.a. of TA–MHNi (●) relative to the titration curve for TA with NaOH (.......) (from T. Harada, M. Imaida, H. Ozaki, A. Tai and Y. Izumi: *Bull. Soc. Chem. Jpn.* 54, 2323 (1981).

The plots of e.d.a. *versus* pH and a.m.a. *versus* pH for TA on RNi exhibit some complex features. However, a simple relationship based on the concept derived from the HPA study has recently been obtained through the use of HNi as shown in Figure 9 [16]. The pK_{a1} and pK_{a2} values for tartaric acid are very close to each other (2.9 and 4.2) so that the titration curve does not exhibit a distinct first titration point but rather a gentle change of slope with pH until the second titration point is reached.

The plot of a.m.a. *versus* pH shows two levels of a.m.a., i.e. a high value over the pH region 2–4.5 and a low value over the pH region above 6. The low level a.m.a. value arose when two carboxyl groups of TA were completely neutralized with NaOH. Hence the adsorbed species in this region should be TA–Na_2. The a.m.a. value begins to decrease at pH 4.5 (near the pK_{a2} value), and for this reason it is obvious that the modifying solution must contain tartaric acid or sodium hydrogen tartarate in order for the high level of a.m.a. to be attained. If the nickel corrosion over in this pH region is taken into account, the adsorbed species giving a high a.m.a. value is either TA–Ni or $(TA–Na)_2Ni$.

In contrast to the variation of a.m.a., the e.d.a. changes sharply near the pK_{a1} value. The e.d.a. value was extremely low at pH <2.5, increased sharply to the 50% level at pH ~ 3 (corresponding to pK_{a1}) and increased gently to a plateau at pH >4.5 (pK_{a2}).

The results clearly show that the favored adsorption species generating a high e.d.a. value for the catalyst is not nickel tartarate (TA–Ni) nor nickel hydrogen tartarate; the adsorption of TA accompanied by one or two sodium ions as counterions is the process responsible for a high e.d.a. value for TA–MNi. As mentioned in Section 3.1, conventional TA–MNi or TA–NaBr–MNi catalysts are prepared at pH values in the range 3.2–5.0. Hence, the adsorbed species on these catalysts must be $(TABa)_2Ni$ and $TANa_2$.

A similar conclusion has been arrived at by Hoek and Sachtler [58] from a study of the TA–MNi/SiO_2 system. These workers have pointed out that the presence of Ni^{2+} and Na^{+} ions in the modifying solution lead to a high e.d.a. value for MNi, and they propose that TA–Ni (adsorbed over the pH range 4.5–5.5) or TA–Ni/NaOH (in a molar ratio of 0.5) were the most favorable species for producing a catalyst with a high e.d.a. value.

XPS studies of the modified surface of DNi with respect to C(1s), O(1s), Ni($2p_{3/2}$) and Ni(valence band) have been carried out by Yasumori and coworkers [51, 52, 59]. Figure 10(A) shows the C(1s) spectrum for (a) TA adsorbed onto DNi at pH 5, (b) that of TA–Ni, (c) that of TA–Na_2 and (d) of TA alone. Two peaks with the same intensity (III and II) have been assigned to the carbon atom of the carboxy group and that in the α-position, respectively. A small peak at ca. 285 eV (I) is the carbon peak of a contaminant. The binding energies of peaks III and II for adsorbed TA (288.7 and 286.7 eV) are much closer to those for TA–Ni (288.7 and 286.7 eV) and TA–Na_2 (288.6 and 286.6 eV) than those for free TA (289.7 and 287,3 eV). As shown in Figure 10(B), the O(1s) peak for TA adsorbed on Ni (a) appeared at ca. 533 eV. The peak for TA–Ni (b) was broad and appeared at a lower binding energy to that of adsorbed TA whereas the peak for free TA (d) appeared at a higher binding

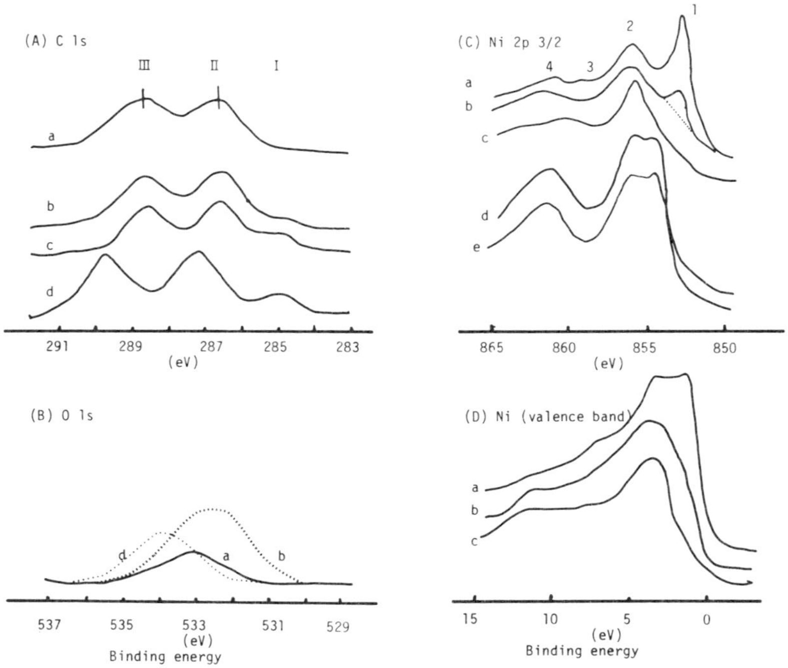

Fig. 10. X-ray photoelection spectra of the modified surface of DNi. (A) C (1s) spectra: (a) difference spectrum between unmodified Ni and TA-modified Ni (pH 5); (b) nickel tartarate; (c) sodium tartarate; (d) tartaric acid. (After Y. Inoue, K. Okabe and I. Yasumori: *Bull. Chem. Soc. Jpn.* **54**, 613 (1981).) (B) O (1s) spectra: (a) difference spectrum between unmodified Ni and TA-modified Ni (pH 5); (b) nickel tartarate; (d) tartaric acid. (After Y. Inoue, K. Okabe and I. Yasumori: *Bull. Chem. Soc. Jpn.* **54**, 613 (1981).) (C) Ni ($2p_{3/2}$) spectra: (a) TA-modified Ni; (b) unmodified Ni; (c) nickel tartarate; (d) NiO treated with tartaric acid; (e) NiO. (After I. Yasumori: *Pure Appl. Chem.* **50**, 971 (1978).) (D) Ni (valence band): (a) TA-modified Ni; (b) unmodified Ni; (c) nickel tartarate. (After Y. Inoue and I. Yasumori: unpublished results.)

energy to that of adsorbed TA. The broad O(1s) peak of for TA–Ni has been explained in terms of the participation of OH groups as well as COOH groups in the formation of the nickel complex. On the basis the that the C(1s) spectrum for TA adsorbed on Ni is similar to that for TA–Ni and TA–Na_2, while the O(1s) spectrum of TA adsorbed on Ni differs from that of TA–Ni, Yasumori *et al.* have suggested that the hydroxyl groups of adsorbed TA undergo little interaction with the surface Ni atoms.

Figure 10(C) shows the Ni($2p_{3/2}$) spectrum (a) for the surface of TA–MNi, (b) that of unmodified Ni, (c) that of TA–Ni, (d) of NiO treated with TA, and (e) of NiO. The spectrum of the modified nickel surface (TA–MNi) exhibits a characteristic peak at 852.7 eV (peak 1) which is assigned to the peak for the pure nickel surface. Peaks 2 and 4 corresponds to the peaks for NiO. The small peak at 858.5 eV (peak 3) is a satellite peak for the nickel surface. When the modified surface was exposed to air, peak 1 remained unchanged. In the spectra of TA–Ni (c) and of NiO treated with TA (d), no peak appeared near the position of peak 1. The spectrum of unmodified nickel (b) showed a small peak at a higher binding energy than peak 1 (853.1 eV) which disappeared when the surface was exposed to air.

Figure 10(D) shows the Ni(valence band) spectra for (a) TA–MNi, (b) unmodified Ni and (c) TA–Ni. TA–MNi gave two peaks at binding energies 1.1 eV and 3.2 eV greater than the Fermi level. The former peak corresponds to the d-band for metallic nickel while the latter is similar to the d-band for nickel oxide. A large peak corresponding to nickel oxide appeared in the spectrum of unmodified Ni, but the peak corresponding to metallic nickel was negligible.

It follows from these results that a large part of the nickel surface modified with TA is maintained in the metallic state and protected from oxidation. The protective effect of the modifying reagent has also been demonstrated by the resistance of MRNi to corrosion with acetylacetone [60].

Hatta and Suetaka have compared the IR reflectance spectrum for a thin film of HPA deposited on a nickel surface with the conventional absorption spectrum for HPA [61]. Since the oscillating radiation vector is polarized normal to the metal surface in the reflectance method, the difference between the band intensities in the reflectance and absorption spectra is related to the direction of the molecule's transition moments. Therefore, knowing the crystal structure of HPA, the orientation of the molecule on the surface can be evaluated.

In the reflectance spectrum, the C=O stretching mode (1725 cm^{-1}) appears with a high intensity but the C—O stretching and C—O—H in-plane deformation absorption bands (1245 and 1197 cm^{-1}) are very weak while the carnoxylic O—H stretching absorption (2920 and 2570 cm^{-1}) has disappeared completely. In contrast, absorption bands originating from the alcoholic O—H groups appear unchanged. These results indicate that the plane which includes the C=O and C—O—H units of the carboxy group is arranged perpendicular to the nickel metal surface whereas the alcoholic O–H groups is not in the same plane as that which includes the carboxy group.

The crystal conformation of HPA has not been determined directly but is expected to be similar to that of the closely related compound 3-(*p*-bromophenyl)-2-hydroxypropionic acid whose conformation is as shown in Figure 11(a). On the basis of this conformation, the authors have proposed the molecular orientation shown in Figure 11(b) for a thin film of (*S*)-2-hydorxy-3-phenylpropionic acid (HPA) on a nickel metal surface.

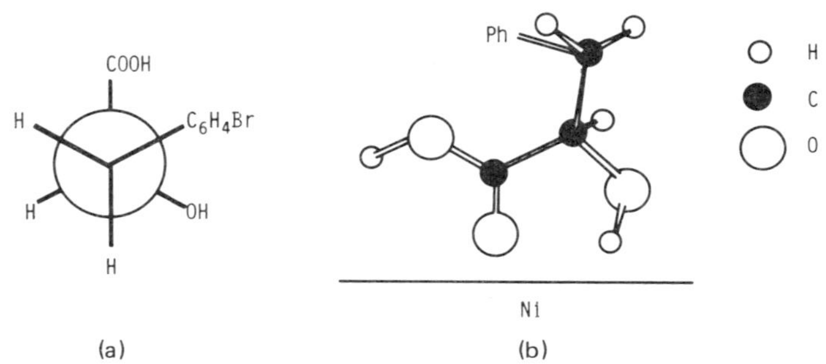

Fig. 11. (a) The molecular conformation of (*S*)-3-(*p*-bromophenyl)-2-hydroxypropionic acid as vew alone the C_α–C_β bond. (b) The molecular orientation in a thin film of (*S*)-2-hydroxy-3-phenylpropionic acid on a nickel metal surface. The phenyl group has been omitted from this figure for the sake of clarity. (From A. Hatta and W. Suetake: *Bull. Soc. Chem. Jpn.* 48, 2428 (1975).)

The reflectance spectrum for such a film does not provide direct information about the first layer of absorbed HPA. However, the orientation of the surface species is expected to determine the direction of crystal growth, and from this viewpoint the mode of adsorption shown in Figure 11(b) is considered to provide the most plausible orientation for HPA on a surface. The results of studies on such adsorbed species suggest that the carboxy group of HPA shown in Figure 11(b) could be replaced by the carboxylate group of HPA in the first layer.

The arrangement of an adsorbed species can also be deduced from its coverage of the metal surface. Fish and Ollis have determined the amount of TA adsorbed onto a nickel electrode by an electrochemical method and have estimated the coverage as 5.3×10^{-6} mol m^{-2} at a pH of 4 [62]. From the amount of TA on TA–MHNi prepared at a pH of 4 and at 0 °C, and the surface area of HNi as determined by the BET method, we have estimated the amount of TA–MHNi adsorbed to be 2.7×10^{-6} mol m^{-2}.

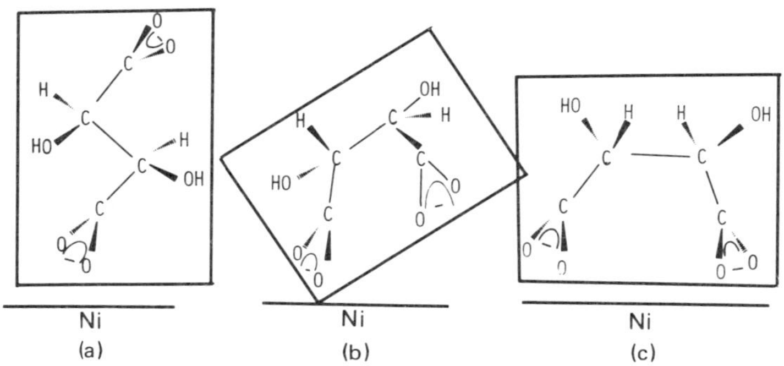

Fig. 12. Possible modes of adsorption of TA onto a nickel catalyst.

The maximum amounts of TA adsorbed as evaluated for the three different adsorption modes shown in Figures 12(a), (b) and (c) are 8.1×10^{-6}, 5.1×10^{-6} and 3.3×10^{-6} mol m^{-2}, respectively. When the observed value of the amount adsorbed and the calculated adsorption area for MAA are compared, monolayer adsorption in the mode shown in Figure 12(b) is more plausible than in the vertical (Figure 12(a)) or completely flat modes (Figure 12(c)).

4.2.2. *α-Amino acids*

The adsorption mode of α-amino acids has been investigated in the main by IR spectral methods. Thus, Groenewegen and Sachtler [63] prepared a nickel catalyst disc supported on silica and modified by means of glycine and alanine introduced in the vapor phase. Such modification proceeded readily with evolution of hydrogen gas. The IR spectrum of the catalyst modified by vapor phase glycine was quite similar to that of nickel glycinate. Characteristic absorption bands corresponding to NH_2 stretching (3500, 3300 cm^{-1}) and CO_2 stretching (1595, 1405 cm^{-1}) modes were observed but no absorptions corresponding to the NH_3^+ zwitterionic form of glycine were noted.

Thus, modification of nickel metal with glycine resulted in the formation of an Ni–glycine complex on the surface of the catalyst with the generation of hydrogen. The same result was also obtained through vapor-phase modification with alanine. The mode of adsorption proposed by these authors is shown in Figure 13(B).

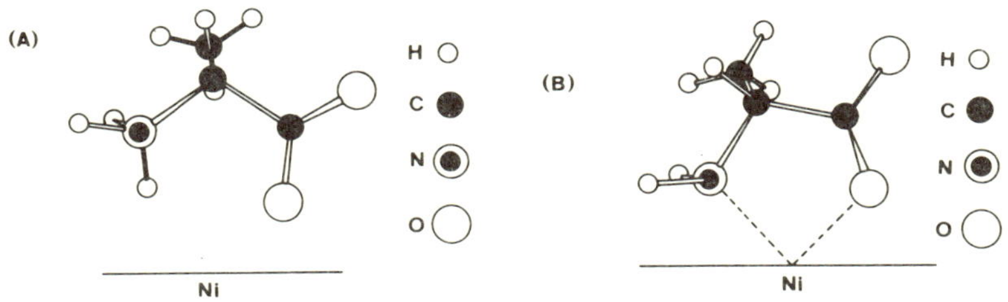

Fig. 13. Proposed modes of adsorption of α-amino acid onto a nickel catalyst (from A. Hatta and W. Suetaka: *Bull. Chem. Soc. Jpn.* 48, 3441 (1975).)

XPS studies of aspartic acid and alanine on DNi support the above conclusion [59]. These studies showed that the N(1s) spectra exhibit a single peak at the same binding energy as that of $[Ni(NH_3)_6]Cl_2$ but no peaks corresponding to NH_3^+. The C(1s) spectra of the carboxy group appeared in the same position as peak III of adsorbed tartaric acid shown in Figure 10(A). Hence, the adsorbed forms of the amino acids studied should be as their nickel chelates as depicted in Figure 13(B).

Epitaxially grown films of amino acids on nickel surfaces have also been studied by high sensitivity IR reflectance spectroscopic methods. Suetaka *et al.* have prepared a Raney Ni plate by leaching an Al/Ni alloy loaded onto a nickel plate and modifying it with alanine or phenylalanine at a pH of 5 and 100 °C [64]. The reflectance spectrum of the modified plate showed characteristic absorptions corresponding to the NH_2 stretching mode but no absorptions corresponding to NH_3^+ were observed. Hence, the alanine was not adsorbed onto the surface as the free acid (or zwitterion) but as its nickel complex. The comparison of the reflectance spectrum of alanine on a nickel surface with the absorption spectrum of the Ni–alanine complex has shown that the NH_2 scissoring mode (1570 cm^{-1}) disappeared, the asymmetric CO_2^- stretching mode (1610 cm^{-1}) diminished slightly, while the symmetric CO_2^- stretching mode (1400 cm^{-1}) was diminished significantly. This suggests that the plane of the CO_2^- group is inclined in a vertical direction to the surface to some extent.

The results may be interpreted by visualizing the adsorption of alanine in the orientation shown in Figure 13(B). In this orientation, the plane of the NH_2 group is parallel to the surface so that its scissoring mode becomes parallel to the direction of the transition moment which is inactive in the reflectance method. Such a mode of adsorption is in good agreement with the results of Sachtler. When alanine or phenylalanine is deposited onto a pure nickel metal surface at 0 °C, the epitaxially grown film consists of the zwitterionic form of the amino acid. The orientation, however, is not significantly different from that of the Ni complex (Figure 13(A)). The similar orientation of the chelate to that of the zwitterion in the epitaxially grown crystal strongly suggests that the chelate molecules in the first layer of the film can act as good nuclei for epitaxial growth of the zwitterion. Thus, even in the above case, the strong possibility exists that the first amino acid layer on the nickel surface is composed of the nickel chelate.

Tanabe and Uyeda have investigated the arrangement of epitaxially grown amino acids on a nickel plane, using an electron diffraction method [65]. They crystallized amino acids onto (001) nickel films prepared by vacuum condensation using a KCl cleavage plane as the substrate, and measured the diffraction pattern of the amino acid crystallites together with that of the nickel substrate on which the crystallites were grown. Figure 14(A) is a schematic diagram of the electron diffraction pattern of (001) nickel and (*S*)-phenylalanine. The most preferred positions for mutual orientation have been determined by a statistical treatment of the angular distribution of the appearance frequency. The result is shown as the histogram in Figure 14(B) from which three preferred orientations may be assumed for phenylalanine. The schematic drawing of these lattice orientations on a (001) nickel film is shown in Figure 14(C). The

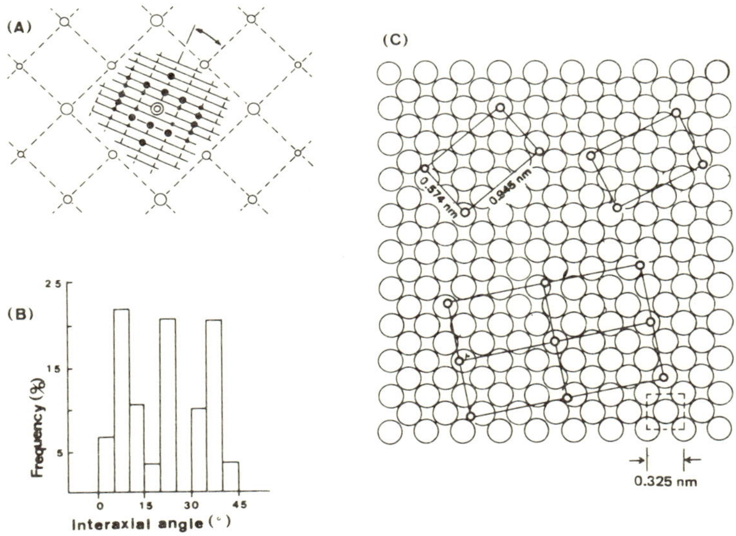

Fig. 14. (A)Schematic diagram of the electron diffraction pattern of (001) nickel film and an overgrown (*S*)-phenylalanine layer. (B) Distribution of the angular position of orientation between adjacent crystallites in the layer. (C) The lattice orientation of (*S*)-phenylalanine crystallites on a (001) nickel film. (From K. Tanabe and N. Uyeda: *Bull. Inst. Chem. Reas. Kyoto Univ.* 52, 616 (1974).)

lattice orientations of alanine and glutamic acid on a (001) nickel film have also been determined. Tanabe and Uyeda have that some differences exist between the angular distribution patterns for the glutamic acid–(001) Ni systems prepared at 100 °C and 0 °C. These differences may affect the e.d.a. of MNi when the relationship between the temperature and the e.d.a. of the resulting catalyst is taken into account. However, further detailed information has not, as yet, been obtained.

By means of an investigation employing IR, XPS and electron diffraction methods, it has been shown that amino acids adsorbed onto the catalyst are arranged in the form shown in Figure 13(B) in each crystallite, with the walls consisting of NH_2, α-C and CO_2^- units. Interaction of the adsorbed amino acid with adsorbed MAA presumably takes place at these walls.

4.3. ENANTIO-DIFFERENTIATING AND NON-ENANTIO-DIFFERENTIATING SITES ON MNi

4.3.1. *Surface condition of catalyst and proportion of enantio-differentiating sites*

As has been mentioned briefly in Section 3.1.3, the e.d.a. of MNi varies widely depending on the variables employed during the preparation of the nickel catalyst, even when the same modifying conditions and modifying reagent were employed. From these results, we have suggested that two types of reaction sites exist; an enantio-differentiating reaction site (e.d. site) where an optically active product is produced with the aid of the intrinsic e.d.a. of the adsorbed modifying reagent, and a non-enantio-differentiating site (non-e.d. site) where a racemic product is produced. According to this hypothesis, the e.d.a. of the catalyst can be expressed as follows:

$$\text{e.d.a.} = iE/(E + N) \tag{18}$$

where i is the intrinsic e.d.a. of the modifying reagent, E is the extent of participation of the e.d. site in the reaction, and N is the extent of participation of the non-e.d. site in the reaction. The hypothesis is also well compatible with the results of kinetic studies.

The use of activated nickel metals prepared by liquid-phase reactions resulted in a low e.d.a. compared to catalysts prepared by gas-phase reactions (Table I, Section 3.1). Generally, the catalyst prepared in the liquid phase consists of well-dispersed fine metal particles and has a high hydrogenation activity. It contains substantial amounts of contaminants such as Al, Mg, Si, Zn, B and P. On the other hand, the catalyst obtained by gas-phase reaction is almost free from contaminants and consists of ordered large metal particles with low hydrogenation activities. From these observations, two factors, i.e. impurities on the metal surface and the size of the ordered structure of the metal surface, are apparently capable of controlling the proportion of e.d. to non-e.d. sites.

In the case of RNi, contamination with aluminum is assumed to be a key factor in controlling the e.d.a. of the modified catalyst. Table VIII lists the methods of preparation of RNi catalysts with various aluminum contents and gives a summary of the e.d.a.'s of the resulting TA–MRNi catalysts. Modification was carried out under mild conditions (pH 5, 0 °C) in order to avoid corrosion of the catalyst surface.

As shown (Table VIII, entries 1 and 2), a decrease in the aluminum content resulted in an increase in the e.d.a. of the catalyst. It is well known that metallic aluminum and certain aluminum–nickel alloys ($NiAl$, Ni_3Al) which are difficult to dissolve with NaOH, can be easily dissolved by treatment with an aqueous

TABLE VIII[a]

Preparation and modifying conditions of RNi and e.d.a. of the resulting MRNi catalysts

No.	Catalyst	Preparation conditions	Al content (%)	Modifying conditions pH	Modifying conditions Temp. (°C)	e.d.a. of TA–MRNi catalyst
1	RNiL	Ni/Al (42:58) alloy was leached with 20% aq. NaOH at low temperature (20 °C) and aged at 75 °C	6	5.0	0	35
2	RNiH	Ni/Al (42:58) alloy was leached with 20% aq. NaOH at high temperature (80–100 °C) and aged at 100 °C	5	5.0	0	40
3	RNiH		–	4.1	100	44
4	RNi–A (GA)	RNiH was treated with 1% aq. solution of GA (pH 3.2 adjusted with NaOH) at 100 °C for 1 h. The resulting catalyst was treated with 1 M NaOH at 100 °C for 1 h and then washed with water to remove GA completely.	3	5.0	0	72
5	RNi–A (TA)	RNiH was treated with 1% aq. solution of TA (pH 3.2) at 100 °C for 1 h. The resulting catalyst was washed with water.	2	5.0	0	62

[a] After T. Harada, A. Tai, M. Yamamots, H. Ozaki and Y. Izumi: *Proc. 7th Int. Congr. Catal. Tokyo* (*1980*), p. 364 Elsevier/Kodansha, (1981).

solution of an α-hydroxy acid. Selective elimination of contaminating aluminum from RNi was expected to take place during TA modification at low pH (Table VIII, entry 3).

Selective elimination of Al from RNi may be achieved efficiently by treatment with glycolic acid (GA) or TA in acidic solution at high temperatures. Thus, from a conventional RNiH catalyst with 6% Al content, treatment with GA (RNi–A(GA)) or TA (RNi–A(TA)) yielded catalysts with Al content less than 3%. Modification with TA at pH 5 and 0 °C, gave TA–MRNi catalysts with excellent e.d.a.'s (Table VIII, entries 4 and 5).

Since elimination of Al accompanies corrosion of Ni, the surface state of the resulting RNi–A catalyst is no longer the same as that of the original RNi. In the original RNi, the nickel which co-exists with aluminum compounds has a more disordered structure than the nickel which is present as the pure metal. Thus, elimination of aluminum by treatment with GA is accompanied by the selective elimination of nickel co-existing with aluminum compounds to yield a pure nickel surface.

The e.d.a. of MHNi, an aluminum-free catalyst, was also found to be strongly affected by the source of NiO. Two types of NiO are available from commercial sources. One is a clear green NiO prepared by the direct oxidation of purified nickel metal powder. The other is a grayish green NiO prepared by the decomposition of a nickel salt (e.g., $NiCO_3$). Activated nickels prepared from clear green NiO are referred to below as HNi-I while those prepared from grayish green NiO are referred to as HNi-II.

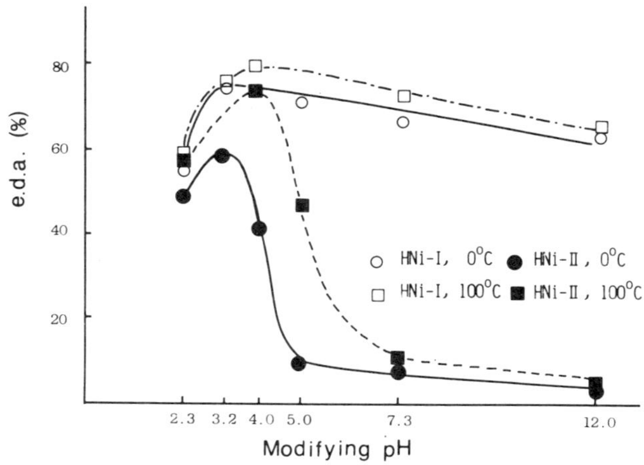

Fig. 15. Relationship between pH and the enantio-differentiating ability (e.d.a.) of TA–MHNi-I and TA–MHNi-II catalysts. The temperatures employed during modification of the various catalysts were as follows: ○ HNi-I at 0 °C; □ HNi-I at 100 °C; ● HNi-II at 0 °C; and ■ HNi-II at 100 °C. (From T. Harada, Y. Imachi, A. Tai and Y. Izumi: *Metal– Support and Metal–Additive Effects in Catalysts*, p. 377, Elsevier, Amsterdam (1982).)

The dependence of e.d.a. on the pH and temperature employed during modification is shown in Figure 15 [66]. When modified with TA at pH 3.2 and 100 °C, the e.d.a.'s of the resulting TA–MHNi catalysts were the same whereas a large difference in e.d.a. was observed when they were modified at pH 7.3 and 100 °C. A large difference was also observed when they were modified at pH 7.3 and 0 °C. The pH dependence of MHNi–II suggests that acid treatment is also effective in improving HNi. In Table IX the effects of TA treatemnt on the e.d.a.'s of MHNi are compared with respect to HNi-I and HNi-II. It is evident that HNi-II contains some non-e.d. sites which can be removed by acid treatment or TA modification at pH 3.2 and 100 °C.

TABLE IX
Comparison of activated nickel catalysts HNi-I and HNi-II

Catalyst	Modifying conditions pH	Temp. (°C)	e.d.a. (%)	Mean crystallite size (D_c) (nm)	Mean particle size (10^3nm)
HNi-I	unmodified		–	32	6.6
	7.3	0	71	32	7.4
	3.2	0	71	–	–
	3.2	100	75	34	10.0
HNi-IA[a]	7.3	0	81	–	–
	3.2	0	77	–	–
HNi-II	unmodified		–	25	3.8
	7.3	0	10	25	4.5
	3.2	0	60	–	–
	3.2	100	77	25	3.8
HNi-IIA[a]	7.3	0	59	–	–
	3.2	0	77	–	–

[a] HNi-I or HNi-II treated with TA at pH 3.2 and 100 °C for 1 h in advance of TA modification.

Mean particle sizes as determined by the resistance change principle (Coulter counter) and mean crystallite sizes (D_c) as determined by X-ray diffraction (XRD) for the two types of catalysts are also listed in Table IX. The mean particle size and D_c of HNi-I are larger than those of HNi-II, although the values for both catalysts remain unchanged even after TA modification at pH 3.2 and 100 °C. Inspection of these HNi catalysts and TA–MHNi catalysts by scanning electron microscopy confirms that the apparent shapes and sizes of HNi-I and -II are different from each another and remain unchanged under the modification conditions listed above [66].

Since acid treatment and high-temperature modification of HNi result in little or no change in the apparent shape, particle size or crystallite size of the catalyst, it follows that such treatment must influence the structure of a thin surface layer of the catalyst. The surface and adjacent substratum are expected to be composed of both a crystalline and an amorphous component. It is reasonable to assume that the amorphous component is corroded by acidic and high-temperature modification to a larger extent than the crystalline component. Taking into account the observation that TA–MHNi-II (modified at pH 3.2 and 100 °C) exhibited a higher e.d.a. than TA–MHNi-II (modified at pH 7.3 and 0 °C), it follows that the amorphous and crystalline components most probably correspond to the non-e.d. and e.d. sites, respectively. Hence, acidic treatment and high-temperature modification must contribute to the removal of non-e.d. sites.

Nitta and coworkers [22] have investigated the relationship between the e.d.a.'s of TA–MRNi, –MPNi, –MBNi, –MDNi and –MHNi and the Al, P and B surface contents of these catalysts as determined by XPS. RNi, PNi and BNi gave MNi's with lower e.d.a.'s as the surface content of Al, P and B, respectively, increased, although the effect of an increase in surface P or B content on the e.d.a. of MPNi or MBNi was much greater than the corresponding decrease in e.d.a. induced by Al for MRNi. These workers also found that refluxing the catalyst in a concentrated solution of NaOH eliminated the P, B and Al contaminants from the catalyst and increased its crystallite size. As shown in Figure 16, there is a relationship between the e.d.a. and D_C, the mean crystallite size, as determined by XRD. Thus, the e.d.a. of MNi increased with an increase in D_C until a constant value was attained for a mean crystallite size of 20 nm and above. On this basis, the authors have proposed that the size of ensembles of regularly arranged nickel atoms on the catalyst surface is a factor governing the e.d.a. of MNi.

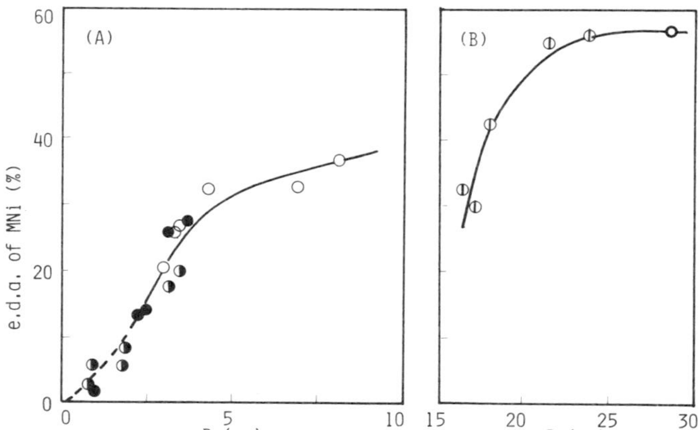

Fig. 16. The dependence of the e.d.a.'s of various catalysts on the mean crystallite size (D_C) of nickel present: ○, RNi; ◑, Ni-B; ●, Ni-P; ⦶, D-Ni; ◎, HNi. (From Y. Nitta, T. Imanaka and S. Teranishi *Bull. Chem. Soc. Jpn.* 54, 980 (1981).)

Nitta and coworkers have also carefully investigated the relationship between the preparation variables and the e.d.a. of the resulting catalyst and shown that the e.d.a. of a supported catalyst is affected by the mean crystallite size in a similar manner to an unsupported catalyst [23]. Thus, in the preparation of a silica-supported catalyst, the use of $Ni(NO_3)_2$ and Na_2CO_3 led to a catalyst with a large D_c and a high e.d.a. Other combinations of reagents, such $Ni(NO_3)_2$/NaOH or $Ni(OAc)_2/Na_2CO_3$ resulted in a small D_c and a poor e.d.a. The amounts of nickel loaded onto the support also affected the e.d.a. of the catalyst; the latter was found to increase with an increase in the Ni loading and reached a value of ca. 60% at a Ni/SiO_2 = 1:1 loading. Increasing the Ni loading also resulted in an increase in the D_c value.

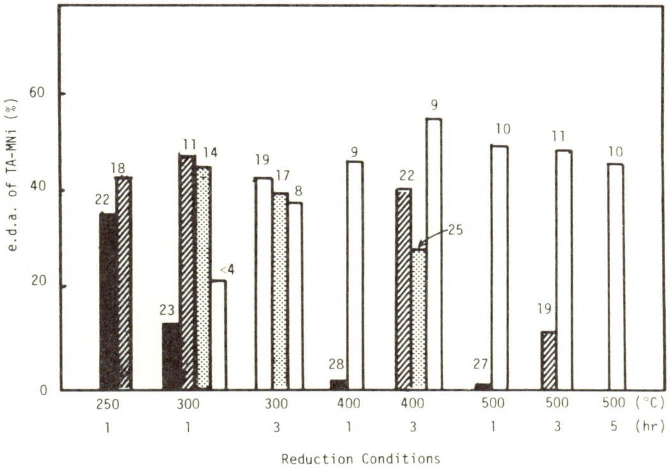

Fig. 17. The influence of the reduction conditions on the e.d.a. of some nickel-supported catalysts: ▆, $Ni-TiO_2$; ▨, $Ni-Al_2O_3$; ⬚, Ni–Kg; □, $Ni-SiO_2$. The figure on each histogram denotes the mean crystallite size (D_c) for the nickel metal in each catalyst. (From Y. Nitta, F. Sekine, T. Imanaka and S. Teranishi *J. Catal.* 74, 382 (1982).)

The influence of the conditions employed for reducing NiO on various supprts (Ni/support = 1:1) on the e.d.a.'s of the resulting catalysts are shown in Figure 17, together with the D_c for each catalyst. With modified Ni/SiO_2, the relationship between D_c and e.d.a. is similar to that for the unsupported catalyst. Ni/SiO_2 prepared by reduction at 300 °C for 1 h had a small D_c and gave MNi's with a poor e.d.a., whereas Ni/SiO_2 catalysts prepared under severe reduction conditions had moderate D_c values and gave MNi's with high e.d.a.'s. However, with other supported catalysts such as Ni/Al_2O_3, Ni/TiO_2 and Ni/kieselguhr (Kg), no systematic relationship between D_c and e.d.a. was observed. In these cases, an increase in the temperature employed for reducing the catalysts resulted in a large decrease in e.d.a. for the modified catalyst despite an increase in D_c.

Nitta *et al.* have explained these somewhat unexpected results on the following basis. The nickel–support interaction of Ni/Al_2O_3 or Ni/TiO_2 has been reported to be stronger than that of Ni/SiO_2 [70]. The surface areas of Al_2O_3 (253 $m^2\ g^{-1}$), TiO_2 (73 $m^2\ g^{-1}$) and kieselguhr (5.7 $m^2\ g^{-1}$) were appreciably less than that of SiO_2 (600 $m^2\ g^{-1}$). With a high nickel loading (Ni/support = 1:1), a homogeneous loading of the nickel salt on these supports cannot be achieved except for the case of SiO_2. However, some of the nickel salt could bind strongly with the support even under highly dispersed conditions, while the other parts of the deposited nickel salt would only interact weakly with the support. At a low reduction temperature, the strongly bound salt would remain as such since the strong nickel–support interaction would retard the reduction process, whereas the weakly bound salt would be readily reduced to nickel crystallites of an intermediate size of almost the same magnitude as that of HNi. Thus, modification of the resulting catalyst gave MNi's with high e.d.a.'s. High-temperature reduction, on the other hand, led to the strongly bound salt being reduced to dispersed nickel crystallites of a small size with high hydrogenation activity, while intermediate size nickel crystallites produced from the weakly bound salt underwent sintering and changed to large crystallites which exhibited no hydrogenation activity. Thus the active sites of the resulting Ni/Al_2O_3, Ni/TiO_2 or Ni/Kg catalysts are expected to consist of small nickel crystallites despite their apparent large D_c values. Hence modification of the resulting catalysts would not be expected to give MNi's with a high e.d.a.

It is apparent from the above investigations that a surface which consists of amorphous or small nickel crystallites functions as a non-e.d. site with a high hydrogenation activity, while a surface containing nickel crystallites of intermediate size functions as an e.d. site exhibiting reasonable hydrogenation activity.

4.3.2. *Control of non-enantio-face differentiating sites by partial poisoning*

Characterization of the e.d. site and methods for preparing catalysts containing a large number of e.d. sites have been discussed in the previous section. However, perfect control of the surface conditions of the catalyst is virtually impossible, and as a result an alternative approach toward controlling the ratio of non-e.d. to e.d. sites on the catalyst is required.

During the course of a kinetic study [49], it has been shown that the hydrogenation activity of a catalyst expressed as the rate of hydrogenation, v, decreased with an increase in conversion, whereas the e.d.a. of the catalyst evaluated as $(v_R - v_S)/v$ increased (see Figure 5, Section 4.1.). Since the relationship $v \propto [\mathrm{MAA}]^{0.2-0.3}$ applies under the reaction conditions employed, the decrease in rate is mainly caused by deactivation of part of the catalyst rather than by changes in the MAA concentration. This suggests that there are non-e.d. sites on the catalyst which are liable to lose their hydrogenation activities during the reaction. This has motivated us to investigate the partial poisoning of MNi and has led to the discovery that NaBr acts as an effective supplementary modifying reagent.

As mentioned in Section 3.1.5, a considerable increase in e.d.a. is observed when a nickel catalyst is modified with a solution containing NaBr and TA. Details of the manner in which NaBr functions have been obtained from studies of the relation between the amount of NaBr adsorbed and the e.d.a. of the catalyst [21].

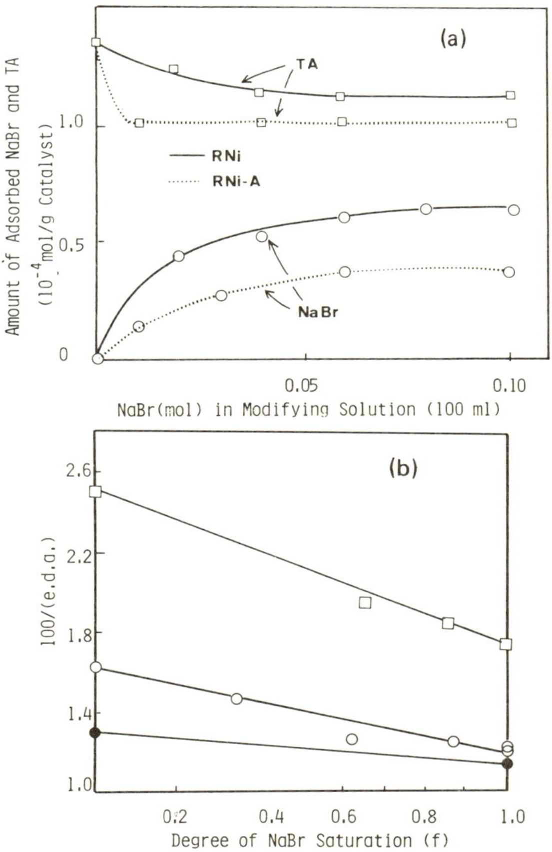

Fig. 18. (a) The relationship between the concentration of NaBr in the modifying solution and the amount of TA or NaBr adsorbed onto RNi and RNi–A. (b) The dependence of 100/e.d.a. on the degree of NaBr saturation of the catalyst. □, RNi; ○, RNi–A (TA); •, HNi. (From T. Harada, A. Tai, M. Yamanoto, H. Ozaki and Y. Izumi: *Proc. 7th Int. Congr. Catal. Tokyo (1980)*, p. 364, Elsevier/Kodansha, (1981).)

When RNi and RNi–A are modified with a mixture of TA and NaBr, a distinct relationship is observed between the amount of each component adsorbed onto the catalyst and the amount of NaBr in the modifying solution (Figure 18(a)). Thus, the amount of TA decreased steeply as the amount of NaBr increased until a constant lower value was attained. In contrast, the amount of NaBr adsorbed on the catalyst gradually increased with increasing amounts of NaBr in the modifying solution until a constant upper satursted value was attained. The e.d.a. of the catalyst increased as the amount of NaBr adsorbed onto the catalyst increased.

The shape of the plot in Figure 18(a) indicates that part of the catalyst preferentially adsorbs TA while another part preferentially adsorbs NaBr, and that the former and the latter may provide the sites for e.d. and non-e.d. reactions, respectively. Since the amount of adsorbed TA is constant under the experimental conditions, it is reasonable to assume that the amount of optically active MHB produced does not change as the amount of NaBr adsorbed on the catalyst increases and that the increase of the e.d.a. brought about by an increase in the amount of NaBr adsorbed results from a decrease in production of racemic MHB from non-e.d. sites, through deactivation of these sites by the poisoning effect of NaBr. If deactivation of non-e.d. sites is proportional to the degree of saturation of NaBr, eqn. (18) may be rewitter as:

$$\text{e.d.a.} = \frac{i \times E}{E + NA + NB(1 - f)} \tag{19}$$

Where NA is the extent of participation of non-e.d. sites which cannot be blocked by NaBr, NB is the extent of participation of non-e.d. sites which can be blocked by NaBr and

$$f = \text{degree of NaBr saturation}$$

$$= \frac{\text{amount of NaBr adsorbed onto catalyst}}{\text{amount of NaBr present on the catalyst at saturation}} \tag{20}$$

Rearranging eqn. (19), we obtain:

$$\frac{1}{\text{e.d.a.}} = \frac{1}{i}\left(\frac{E + NA + NB}{E} - \frac{NB}{E} \times f\right) \tag{21}$$

Plotting the degree of NaBr saturation of the catalyst against the reciprocal of the e.d.a. for the catalyst led to linear relationships of the type depicted in Figure 18(b). This result is completely compatible with the relationships proposed in eqns. (19) and (21).

The validity of eqn. (19) is also demonstrated by the data listed in Table X. Although the e.d.a.'s of the catalysts were different from each other, the e.d.a. values of NaBr-treated catalysts were in all cases 1.4 times higher than those

TABLE X
Influence of NaBr on the e.d.a. of MRNi catalysts

Modifying reagent[a]	Catalyst e.d.a. in absence of NaBr $(\text{e.d.a.})_1$ (%)	Catalyst e.d.a. in presence of NaBr $(\text{e.d.a.})_2$ (%)	$\frac{(\text{e.d.a.})_1}{(\text{e.d.a.})_2}$
(*S*)-valine	11.8	15.9	1.37
(*S*)-malic acid	20.3	28.4	1.39
(*R*,*R*)-tartaric acid	40.0	56.0	1.40

[a] Catalysts modified at pH 5 and 0 °C.

of catalysts which had not been treated with NaBr, i.e. the increase in e.d.a. on NaBr treatment is solely dependent on the ratio of e.d. to non-e.d. sites and independent of the intrinsic e.d.a. of the modifying reagent.

From eqn. (19), the proportions of E, NA and NB may be expressed as:

$$E(\%) = \frac{E \times 100}{E + NA + NB} = \frac{(\text{e.d.a.})_{f=0}}{i} \times 100$$

$$NA(\%) = \frac{NA \times 100}{E + NA + NB} = \left(\frac{(\text{e.d.a.})_{f=0}}{(\text{e.d.a.})_{f=1}} - \frac{(\text{e.d.a.})_{f=0}}{i}\right) \times 100 \qquad (22)$$

$$NB(\%) = \frac{NB \times 100}{E + NA + NB} = \left(1 - \frac{(\text{e.d.a.})_{f=0}}{(\text{e.d.a.})_{f=1}}\right) \times 100$$

As shown in eqn. (22), E and NA are both dependent on i and cannot be determined independently, while NB may be evaluated independently. For simplicity, the proportions of E, NA and NB calculated on the basis that i is 100% for TA are shown in Figure 19 for RNi, RNi–A and HNi, respectively.

Comparision of Figures 19(a) and (b) suggests that acid treatment is quite effective in decreasing the value of NA but not so effective in decreasing NB. The two kinds of non-e.d. sites are expected to be quite different from each other. Thus, cooperation between the acid and NaBr treatments is an indispensable factor in obtaining a catalyst with a high e.d.a.. This fact also demonstrates the existence of non-e.d. sites which cannot be blocked by NaBr, although the NA value cannot be obtained unequivocally since the value of i is unknown. The results also indicate that there is still room for improvement in catalyst e.d.a.'s; the development of a new method for removing non-e.d. sites which cannot be blocked by NaBr is an improtant problem for the future.

Although each portion indicated in the pie charts shown in Figure 19 is not directly related to the absolute surface area, nevertheless such charts are quite useful for evaluating the state of the catalyst surface under working conditions.

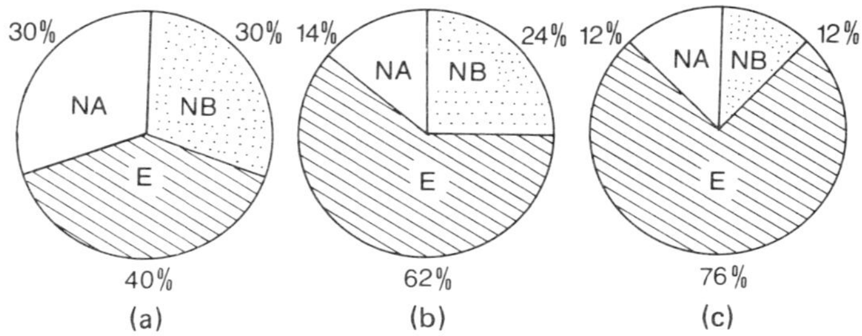

Fig. 19. Participation of enantio-differentiating (e.d.) and non-enantio-differentiating (non-e.d.) sites in the behavior of various catalysts. In this figure, *E* is the extent of participation of e.d. sites, *NA* the extent of participation of non-e.d. sites which cannot be blocked by NaBr and *NB* the extent of participation of non-e.d. sites which can be blocked by NaBr. (a) corresponds to RNi, (b) to RNi–A and (c) to HNi. (From T. Harada, A. Tai, M. Yamamoto, H. Ozaki and Y. Izumi: *Proc. 7th Int. Congr. Catal. Tokyo* (*1980*), p. 364. Elsevier/Kodansha, (1981).)

Studies of the partial poisoning of MNi with amine in the gas phase [67] and in the liquid phase [30] and with thiophene in the liquid phase [68] have also been carried out. The first treatment resulted in some increase in e.d.a., but the second and third treatments led to a large decrease in e.d.a. without a significant decrease in the rate of hydrogenation.

While studying the effect of amine on MNi, we have found that the treatment of TA—NaBr—MNi with a trace amine effectively improved durability of the catalyst. The amine treated catalyst showed a high e.d.a. even after 10 times of repeated uses, while the untreated catalyst lost its original e.d.a. only after 3 times of repeated use. These results are suggestive that amine may interact with TA on the e.d. site rather than the non e.d. site [73].

4.4. MODE OF ENANTIO-FACE DIFFERENTIATION ON MNi

Various attempts to reveal the mode of enantio-differentiation have been undertaken since the discovery of MNi.

At an early stage, Groenewegen and Sachtler proposed a simple model to illustrate the stereochemistry of the hydrogenation process, as shown in Figure 20 [69]. In this model, the modifying reagent and MAA are adsorbed onto two adjacent nickel atoms arranged perpendicularly to the nickel surface. Of the four possible arrangements, that of lowest energy for the α-amino acid was chosen as in Figure 20(A) and that of lowest energy for the α-hydroxy acid as in Figure 20(B). For the α-amino acid, the most favorable arrangement is that which is the least sterically hindered. For the α-hydroxy acid, that stabilized by a hydrogen bond is selected. Hydrogen addition to the adsorbed MAA takes place in the most favorable steric state as shown in Figures 20(C) and (D). The C=O group of the perpendicularly adsorbed MAA is located furthest

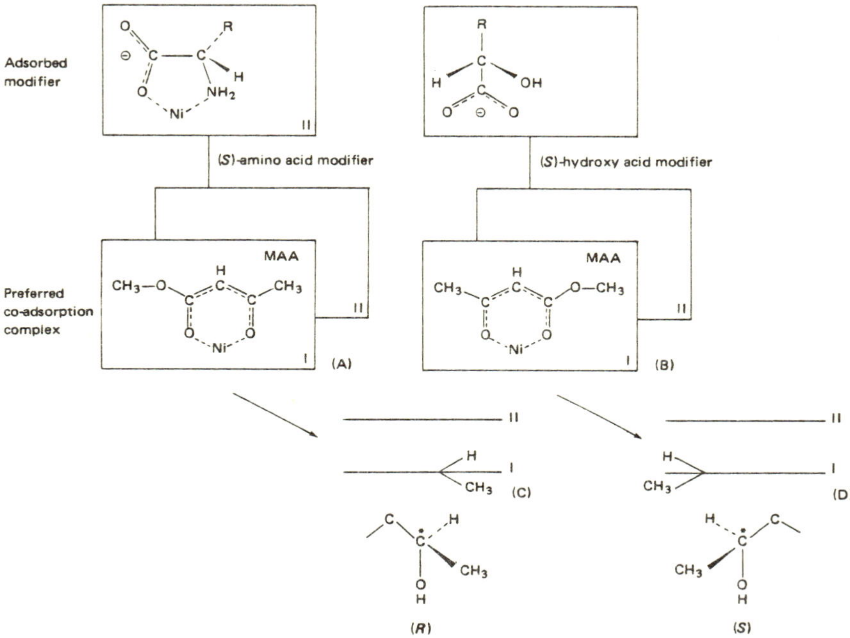

Fig. 20. Schematic representation of enantio-differentiating hydrogenation of MAA with α-amino acid–MRNi and α-hydroxy acid–MRNi, respectively. (From J. A. Groenewegen and W. M. H. Sachtler: *Proc. 6rh Int. Congr. Catal. London (1976)*, p. 1014, The Chemical Society, London (1977).)

away from the catalyst surface, but despite this fact the authors proposed that hydorgenation would proceed through active hydrogen supplied by vertical hydrogen spillover from the catalyst surface. This assumption is not realistic, however; furthermore, this model does not accord with the results of subsequent investigations. Consequently, Hoek and Sachtler [58] have proposed a revised mechanism for hydrogenation over TA–MRNi [58], although in this mechanism they did not mention what factor governed the stereochemistry of the product.

That enantio-differentiation takes place at the adsorption stage independent of the rate-determining hydrogenation reaction provides a promising approach to the elucidation of the mode of differentiation. Such an approach requires no insight into the transition state involved in hydrogenation, which in any case is difficult to speculate on. Enantio-differentiation at the adsorption stage is expected to be governed by only two factors; the mode in which the modifying reagent is adsorbed and interaction between the substrate and the modifying reagent to give a reactant similar to a co-adsorbed species. On this basis, we have been able to present a model for the enantio-face differentiation of MAA and related compounds over an α-hydroxy acid–MRNi [38, 55].

TABLE XI[a]

Enantio-differentiating ability of TA–MRNi catalysts and their analogs deteriaimed via the hydrogenation of MAA

Entry No.	Modifying reagent	Formula	e.d.a. (%)	Product configuration
1	(*R,R*)-TA	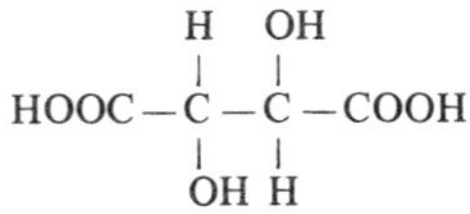	83	*R*
2	(2*R*,3*R*)-2-methyl-TA	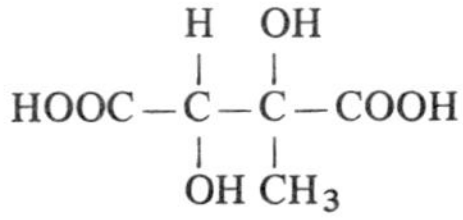	75	*R*
3	(*R,R*)-O-benzoyl-TA	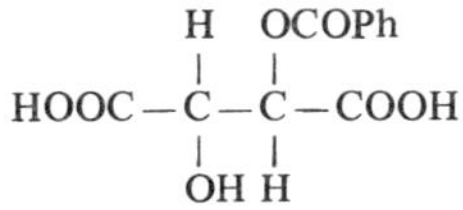	65	*R*
4	(*R,R*)-O-methyl-TA	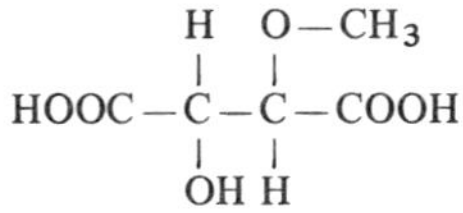	68	*R*
5	(*S*)-malic acid(MA)	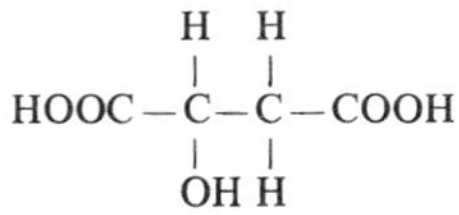	61	*S*
6	(2*S*,3*R*)-2,3-dihydroxybutyric acid	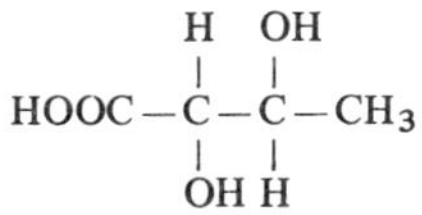	1.2	*S*
7	(*S*)-glyceric acid	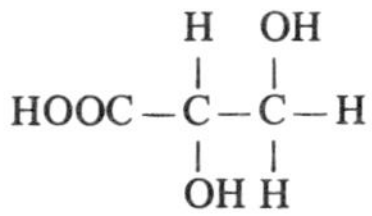	0.0	–
8	(*R,R*)-O,O′-dibenzoyl-TA	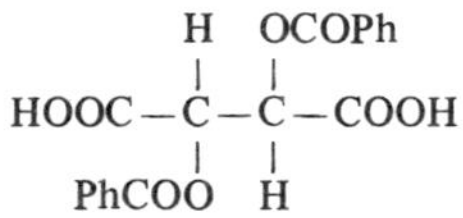	8	*R*
9	(*R,R*)-O,O′-dimethyl-TA	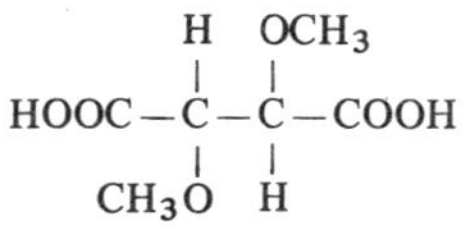	0.2	*R*

[a] From A. Tai, T. Harada, Y. Hiraki, S. Murakami and Y. Izumi: *Bull. Chem. Soc. Jpn.* **56**, 1414 (1983).

In Table XI., the e.d.a.'s of MRNi catalysts prepared from TA or its analogs and NaBr are summarized. Of the compounds examined, only TA and 2-mythyl-TA gave more than 75% e.d.a. in the hydrogenation of MAA (entries 1 and 2). When one of the hydroxyl groups of TA was substituted with a hydrogen, benzoyloxy or methoxy group, an appreciable decrease in e.d.a. was observed (entries 3, 4 and 5). The e.d.a.'s for this class of compounds were around 65%. Substitution of one of the carboxy groups of TA with H or a methyl group resulted in a complete loss of e.d.a. (entries 6 and 7), even if the compound possessed vicinal hydorxyl groups. This drastic decrease in e.d.a. was also observed when both hydroxyl groups of TA were acylated or alkylated (entries 8 and 9). These results clearly show that two hydroxyl groups and two carboxy groups in TA are absolutely vital as functional groups for the effective enantio-face differentiation of MAA.

In Table XII, the results of the hydrogenation of various prochiral ketones over TA–NaBr–MRNi are summarized. For the hydrogenation series of methyl esters of α-, β-, γ- and δ-keto acids (**1a–4a**) over TA–NaBr–MRNi, an excellent optical yield was obtained only with the β-ketoester (entry 2). The results suggest that **2a** can undergo a close-fitting interaction with TA whereas the other substrates (**1a**, **3a**, **4a** and **5a**) cannot do so, for reasons which are discussed below.

As has been mentioned in Section 3.2, two homologs of **2a**, i.e. alkyl acetoacetates and methyl alkanoylacetates, are smoothly hydrogenated over TA–NaBr–MRNi with an uniformly excellent optical yield (Table III). Acetylacetone (AA) and its homologs, in which one of the carbonyl groups takes the place of the ester carbonyl group of MAA, have been shown to be good substrates (Table IV). Compounds in which the ester group of **2a** was substituted with other polar functional group such as SO_2CH_3, CH_2OH and CH_2OCH_3 (Table V, entries 1–5) have also been shown to be favorable substrates for enantio-face differentiating hydrogenation with TA–NaBr–MRNi. The hydrogenation with TA–NaBr–MRNi. The hydrogenation of these substrates over (*R*,*R*)-TA–NaBr–MRNi proceeded with reasonable optical yield to give the (*R*)-isomers in excess.

From these results, it follows that an essential requirement for a good substrate is the presence of an oxygen atom bound to the β-atom of the prochiral carbonyl group to be hydrogenated.

In principle, a rigorous enantio-face differentiating reaction over MRNi should be obtained when the pre-adsorbed modifying reagent is strongly attached to the nickel surface, offering uniform chiral binding sites for the prochiral substrate which approaches this surface. The adsorbed substrates are oriented on the nickel surface through one side of their enantiofaces via a close-fitting interaction between the binding sites of the modifying reagent and the substrate.

TABLE XII
Enantio-face differentiating hydrogenation of prochiral ketones with TA–MaBr–MRNi

Entry No.	Substrate	Product(s)	Optical yield (%)	Configuration
1	methyl pyruvate (**1a**) $CH_3-C(=O)-C(=O)-O-CH_3$	$CH_3-CH(OH)-C(=O)-O-CH_3$ (**1b**)	2.0	*R*
2	MAA (**2a**) $CH_3-C(=O)-CH_2-C(=O)-O-CH_3$	$CH_3-CH(OH)-CH_2-C(=O)-O-CH_3$ (**2b**)	86	*R*
3[a]	methyl 4-oxopentanoate (**3a**) $CH_3-C(=O)-(CH_2)_2-C(=O)-O-CH_3$	$CH_3-CH(OH)-(CH_2)_2-C(=O)-O-CH_3$ (**3b**) and $CH_3-CH(O)-(CH_2)_2-C{=}O$ (**3c**)	38	*R*
4[a]	methyl 5-oxohexanoate (**4a**) $CH_3-C(=O)-(CH_2)_3-C(=O)-O-CH_3$	$CH_3-CH(OH)-(CH_2)_3-C(=O)-O-CH_3$ (**4b**) and $CH_3-CH(O)-(CH_2)_3-C{=}O$ (**4c**)	0	–
5	2-nexanone (**5a**) $CH_3-C(=O)-(CH_2)_3-CH_3$	$CH_3-CH(OH)-(CH_2)_3-CH_3$ (**5b**)	28	*S*

[a] The reaction product was a mixture of hydroxy acid ester and lactone. The optical yield was determined after all the products had been converted to lactone.

According to studies of the mode of adsorption of a modifying reagent onto a nickel surface (Section 4.2), an α-hydroxy carboxylic acid is adsorbed onto the nickel surface solely through the carboxy group which is dissociated into the carboxylate ion, while the hydroxyl group of the acid is essentially free from the surface metal. In order to implant a modifying reagent in an immobile form onto the surface of a catalyst, interaction at at least two sites is required between the modifying reagent and the surface of the catalyst. Experimental results show that modifying reagents with a high e.d.a. are limited to α-hydroxy- and α,α′-dihydroxyl-dicarboxylic acids. Thus, the presence of two carboxyl groups is expected to be one of the necessary conditions for a good modifying reagent.

The poor e.d.a.'s noted for O,O′-dibenzoyl- and O,O′-dimethyl-TA indicate that no effective interaction leading to efficient enantio-face differentiation takes place between the modifying reagent and MAA because of the absence of free hydroxyl groups on the modifying reagent. Although α-hydroxy-dicarboxylic acids (malic acid, O-benzoyl-TA and O-methyl-TA) gave moderate e.d.a.'s, an e.d.a. of more than 75% was obtained only through the use of α,α′-dihydorxy-dicarboxylic acids. It is evident that two hydroxyl groups in a modifying reagent play an indispensable role in the efficient enantio-face differentiation of MAA and its analogs. The two hydroxyl groups in TA are expected to function as hydrogen bond donors, whereas the two carbonyl groups of MAA are hydrogen bond acceptors. In this regard, an attractive force through the formation of hydrogen bonds, rather than a sterically repulsive force as exemplified by Prelog and Cram's rule, is expected to play a key role in enantio-differentiation with MRNi.

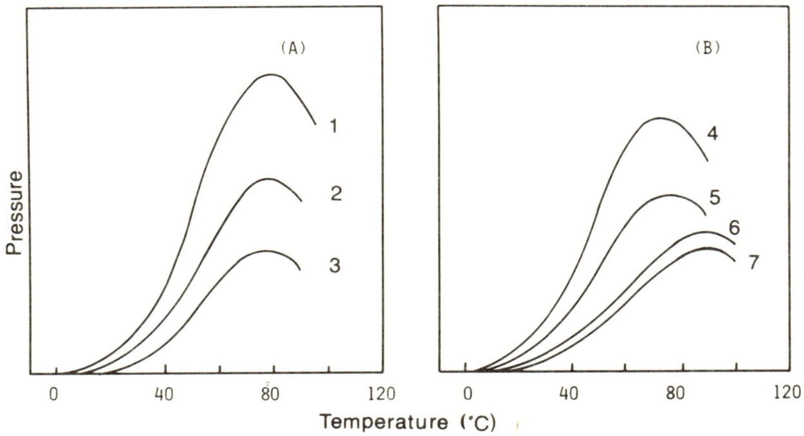

Fig. 21. Thermal desorption spectra of MAA adsorbed onto TA–MDNi and unmodified DNi. Plots of the partial pressure *versus* temperature for (A) the unmodified catalyst and (B) the TA-modified catalyst. (From I. Yasumori: *Pure Appl. Chem.* 50, 971 (1978).)

Yasumori [67] has compared the thermal desorption spectra of MAA adsorbed onto TA–MDNi and unmodified DNi, and drawn the conclusion that an attractive interaction exists between TA and MAA on the catalyst. In the case of MAA on unmodified DNi, the temperature at which the peak maximum occurs does not change with a change in the amount of MAA adsorbed over the range $1.2–3.8 \times 10^{17}$ molecules m^{-2} (Figure 21(A)). With TA–MNi, the position of the peak maximum is shifted upward by 15 °C but remains unchanged until the adsorption reaches 1.7×10^{17} molecule m^{-2}. For amounts of MAA above this value, the position of the peak maximum is the same as that for the unmodified catalyst (Figure 21(B)). These results suggest that the adsorbed MAA is stabilized to a certain extent by TA on the catalyst. Thus, the presence of attractive forces between TA and MAA on the catalyst is proven physicochemically.

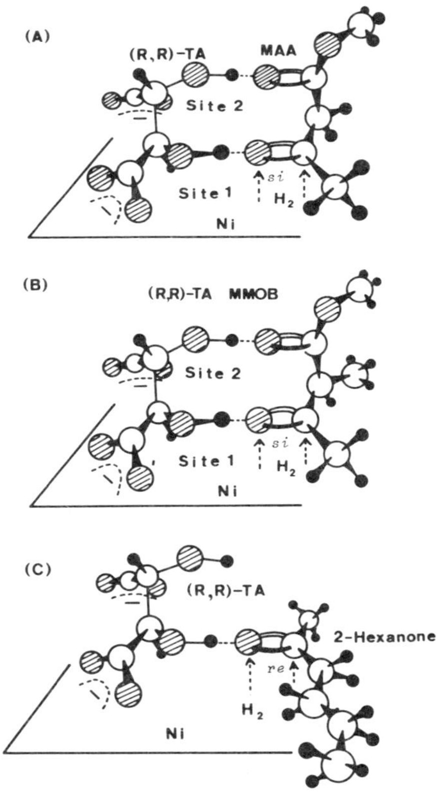

Fig. 22. Schematic representation of the interaction between the substrate and TA. The various substrates depicted are (A) MAA, (B) MMOB and (C) 2-hexanone. (From A, Tai, T. Harada, Y. Hiraki, S. Murakami and Y. Izumi: *Bull Chem. Soc. Jpn.* 56, 1414 (1983).)

On the basis of the hydrogen-bonding ability of the two hydroxyl groups in TA and the two carbonyl groups in MAA, together with the stereochemistry of the hydrogenation reaction of MAA over (*R,R*)-TA–MNi to give (*R*)-MHB, we have proposed the various modes of co-adsorbed MAA and TA on the catalyst depicted in Figure 22(A). As shown in the figure, the two hydroxyl groups of TA are arranged at a distance ca. 0.25 nm from each other and function as hydrogen bond donors. Thus, it may be predicted that not only MAA but also compounds having two hydrogen bond acceptors arranged ca. 0.25 nm apart should be hydrogenated over TA–NaBr–MRNi with high optical yield and with the same stereochemistry. The good predictability of our stereochemical model is well proven by the results summarized in Table III, IV and V [38].

The hydrogenation of methyl 2-methyl-3-oxobutyrate (MMOB) over (*R,R*)-TA–MHNi gives the (2*S*,3*R*)-isomer of methyl 3-hydroxy-2-methylbutyrate (MHMB) in large excess as mentioned in Section 4.1 [55]. These results can also be well explained by a similar model. The interaction between (*R,R*)-TA and MMOB, which gives (2*S*,3*R*)-MHMB, provides the best mode for the involvement of the catalyst as illustrated in Figure 22(B). In this case, the carbonyl group to be hydrogenated faces the catalyst with its *si* face in the same disposition as for MAA with the configuration at C-2 becoming *S* so that the mehtyl group this position is remote from the catalyst giving a less sterically hindered interaction. The formation of a large excess of the (2*S*)-isomer despite the use of a racemic substrate indicates that the substrate which originally has the 2*R*-configuration is converted to the 2*S*-configuration in order to provide a favorable complex from strong attractive hydrogen bonding.

The results quoted in Table XII are also quite well explained by our model. The α-ketoester (**1a**), which exists principally in the *anti*-conformation, cannot interact with TA via two hydrogen bonds. Even in the *syn*-conformation its structure does not allow a close-fitting interaction with TA. Thus, hydrogenation of **1a** leads to extremely poor optical yields.

With the γ-ketoester (**3a**), the distance between the two carbonyl groups is somewhat greater than that between the hydroxyl groups in TA. For this reason, a close-fitting interaction leading to high optical yields is not possible. However, the flexible nature of the carbon chain allows the substrate to achieve a two-site interaction with TA to a certain extent. Thus, hydrogenation over (*R,R*)-TA–NaBr–MRNi gives (*R*)-(**3b**) in excess with a moderate optical yield.

The flexible $CH_3O_2CCH_2CH_2CH_2$ group in the δ-ketoester (**4a**) behaves either as a hydrogen bond acceptor through its ester carbonyl group or simply as a hydrophobic alkyl chain. The virtually negligible optical yield in the reaction of **4a** arises from by existence of two counteracting adsorption modes. One mode is the same as that of **2a** or **3a** and leads to the product expected from *si*-face attack. The other mode is that caused by repulsion of the hydrophobic chain by the hydroxyl group of TA in the manner discussed below.

With the dialkyl ketone (**5a**), the adsorption mode is determined by the single carbonyl group and the hydrophobic alkyl group. The proximity of the hydroxyl groups of TA on the catalyst is expected to lead to the formation of a highly hydrophilic region. When the carbonyl group of **5a** interacts with one of the hydroxyl groups of TA (site 1), the other hydroxyl group of TA (site 2) should displace the very hydrophobic hydrocarbon residue of the substrate to one side. Consequently, the mode of adsorption is as in Figure 22(C), leading to *re*-face attack and giving the (*S*)-isomer in excess, as is found experimentally for **5a**. This type of differentiation arises from the existence of a repulsive force and only single site interaction between the substrate and the modifying reagent. Hence, the optical yield should be less than that of the β-ketoester.

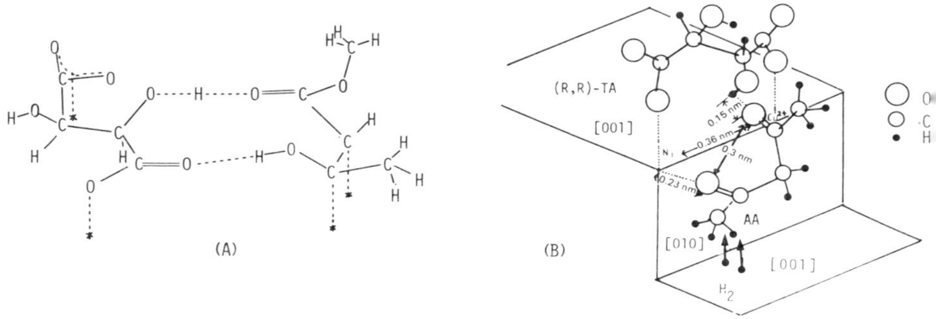

Fig. 23. (A) Interaction mode involved in the enantio-differentiating hydrogenation of MAA over TA–MNi, as proposed by Yasumori and coworkers [52]. (B) Interaction mode involved in the enantio-differentiation hydrogenation of AA over TA–MNi/Cu catalyst, as proposed by Klabunovskii and coworkers [71]

An alternative model for the interaction between MAA and TA has been proposed by Yasumori and coworkers as shown in Figure 23(A) [51, 52]. Klabunovskii *et al.* have also proposed an interaction mode for the reaction between AA and TA on a Ni/Cu alloy catalyst as shown in Figure 23(B) [71]. In both cases, the function of the second hydroxyl group in TA is not taken into account. Hence, a difference in the e.d.a. between TA and other α-hydroxy acids is not contemplated.

With amino acid–MNi's, the importance of the interaction between polar groups in the modifying reagent and the substrate during the enantio-differentiation process has been demonstrated [72]. Thus, in the hydrogenation of α-, β-, γ- and δ-ketoesters over α-amino acid–MNi, the maximum optical yield was obtained with the β-ketoester (Figure 24(A)). On the other hand, hydrogenation with β-amino acid–MNi gave the best results with the γ-ketoester (Figure 24(B)). For this reason, the correspondence of the distance between the NH_2 and

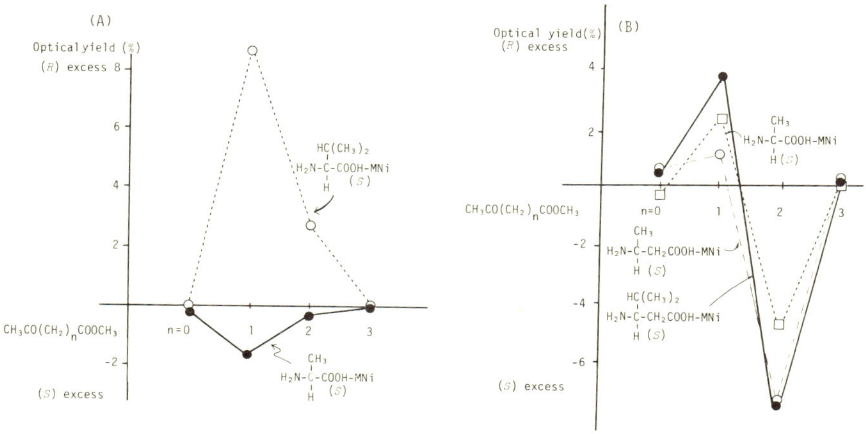

Fig. 24. Relationship between the substrate and the modifying reagent in determining the optical yield in enantio-differentiating hydrogenation over an amino acid–MRNi catalyst. (A) The hydrogenation of $CH_3CO(CH_2)_nCOOCH_3$ over α-amino acid–MRNi ($n = 0–3$). (B) The hydrogenation of $CH_3CO(CH_2)_nCOOCH_3$ over β-amino acid–MRNi ($n = 0–3$). (From S. Okamoto, T. Harada and T. Tai: *Bull. Chem. Soc. Jpn.* **52**, 2670 (1979).)

COOH groups in the modifying reagent with that between the CO and CO_2CH_3 groups in the substrate is expected to play an important role in the enantio-differentiation process. However, sufficient data to allow a mode of interaction to be proposed have not yet been obtained.

From CD spectral studies it has been shown that AA in a complex of the type [AA] [(*S*)- or (*R*)-amino acid] M, when M = Cu, Ni or Co, should adopt an induced puckered conformation of the λ- or δ-type depending on the configuration of the amino acid and the type of metal involved [72]. Metals giving the δ-conformation, e.g. [AA] [(*S*)-Tyr] M, where M = Ni or Co, gave (*S*)-4-hydroxy-2-pentanone (HP), whereas those giving the λ-conformation, e.g. [AA] [(*S*)-Tyr] M, where M = Cu, gave (*R*)-HP, when hydrogenation was conducted over the corresponding Raney metal catalyst modified with (*S*)-Tyr. Similar results were also obtained in the case of ethyl acetoacetate. However, the e.d.a. of the catalyst was not sufficiently high to allow a detailed discussion of the differentiation mode. Refinement and accumulation of data are necessary to develop this model further.

5. Conclusions

It is a striking fact that a simple compound such as tartaric acid when adsorbed onto a nickel catalyst can recognize one of the enantio faces of a substrate with more than 90% accuracy. The excellent performance of the MNi catalyst clearly demonstrates that the reaction proceeding on the solid catalyst can be controlled by the presence of a well-designed substance implanted on the surface.

At present, however, the technique of 'modification' is imperfect and our knowledge of the surface of the solid insufficient. Future advances in this field should provide useful information not only for the design of effective catalyst systems for synthetic chemistry but also for an understanding of the various phenomena occurring on solid catalyst surfaces.

Acknowledgement

We are indebted to Professor Dr. Y. Izumi and Dr. H. Ozaki of Osaka University for reading the original manuscript and for a number of helpful suggestions.

We thank Mrs. T. Kobayashi and Mrs. A. Kobatake for their assistance in the preparation of the manuscript.

Institute for Protein Research,
Osaka University,
Yamadaoka,
Suita-shi
Osaka 565, Japan

References

1. Y. Izumi and A. Tai: *Stereo-differentiating Reactions*, Chap. 4, Kodansha, Tokyo, and Academic Press, New York (1971).
2. R. K. Hanson: *J. Am. Chem. Soc.* 88, 2731 (1966).
3. V. Prelog: *Helv. Chim. Acta* 36, 1178 (1953).
4. J. D. Morrison and H. S. Mosher: *Asymmetric Organic Reactions*, Prentice Hall, Englefield Cliffs, N.J. (1971); Y. Izumi and A. Tai: *Stereo-differentiating Reactions*, Chap. 5, Kodansha, Tokyo, and Academic Press, New York (1971); H. B. Kagan and J. C. Fiand: *Top. Stereochem.* **10**, 175 (1978); P. Valentine, Jr. and J. W. Scott: *Synthesis* 329 (1978); B. Bosnish and M. D. Fryzuk: *Top. Stereochem.* **12**, 119 (1981).
5. Y. Izumi, A. Tai and T. Harada: in *Shokubai Sekkei* (*Catalyst Design*), (*Kagaku Sosetsu, No. 34*), p. 142, Japan Scientific Societies Press, Tokyo (1982).
6. B. R. James: *Adv. Organometal. Chem.* **17**, 319 (1979); B. Bosnish and M. D. Fryzuk: *Top. Stereochem.* **12**, 119 (1981).
7. Catalogue of Wako Pure Chemical Industries (Osaka), 22nd Ed. p. 471 and 547.
8. P. A. MacNeil, N. K. Robert and B. Bosnish: *J. Am. Chem. Soc.* **103**, 2273 (1981).
9. J. Bakos, I. Toth and L. Marko: *J. Org. Chem.* 46, 5427 (1981).
10. K. Ito, T. Harada and A. Tai: *Bull. Chem. Soc. Jpn.* **53**, 3367 (1980).
11. S. Akabori, Y. Izumi, Y. Fujii and S. Sakurai: *Nature* (*London*) **178**, 323 (1956).
12. Y. Izumi, M. Imaida, H. Fukawa and S. Akabori: *Bull. Chem. Soc. Jpn.* **36**, 21 (1963).

13. T. Isoda, A. Ichikawa and T. Shimamoto: *J. Sci. Res. Inst. Tokyo* 34, 134 (1958); *ibid.* 34, 143 (1958).
14. E. I. Klabunovskii and Y. L. Petrov: *Doklady Akad. Nauk SSSR* 162, 853 (1965).
15. Y. Izumi: *Angew. Chem., Int. Ed. Engl.* 10, 871 (1971); E. I. Klabunovskii and A. A. Vedenyapin: *Asymmetricheskiy Kataliz, Hidrogenizatsya no metalakh*, Izdatel'stvo (Nauka), Moscow (1980); M. J. Fish and D. F. Ollis: *Catal. Rev. Sci. Eng.* 18, 259 (1978).
16. T. Harada, M. Yamamoto, S. Onaka, M. Imaida, H. Ozaki, A. Tai and Y. Izumi: *Bull. Chem. Soc. Jpn.* 54, 2323 (1981).
17. S. Tatsumi: *Bull. Chem. Soc. Jpn.* 41, 408 (1968).
18. T. Tanabe, K. Okuda and Y. Izumi: *Bull. Chem. Soc. Jpn.* 46, 514 (1973).
19. E. I. Klabunovskii, V. I. Neupokoev and Yu. I Petrov: *Izv. Akad. Nauk SSSR, Ser. Khim.* 2829 (1970); E. I. Klabunovskii, N. P. Sokolva, A. A. Vendenyapin and Yu. M. Talanov: *ibid.* 1803 (1971); E. I. Klabunovskii, N. P. Sokolova, A. A. Vendenyapin, Yu. M. Talanov, N. D. Zubreva, V. P. Polyakova and N. V. Gorina: *ibid.* 2361 (1972).
20. S. Murakami, T. Harada and A. Tai: *Bull. Chem. Soc. Jpn.* 53, 1356 (1980).
21. T. Harada, A. Tai, M. Yamamoto, H. Ozaki and Y. Izumi: *Proc. 7th. Int. Congr. Catal. Tokyo (1980)*, p. 364, Elsevier/Kodansha, (1981).
22. Y. Nitta, F. Sekine, T. Imaida and S. Teranishi: *Bull. Chem. Soc. Jpn.* 54, 980 (1981).
23. Y. Nitta, F. Sekine, T. Imaida and S. Teranishi: *J. Catal.* 74, 382 (1982).
24. Y. Orito, S. Imai and S. Niwa: *Yuki Gosei Kagaku Kyokai Shi* 34, 236, 672 (1976); Y. Orito, S. Imai and S. Niwa: *ibid.* 35, 672, 753 (1977).
25. T. Harada, Y. Hiraki, Y. Izumi, J. Muraoka, H. Ozaki and A. Tai: *Proc. 6th Int. Congr. Catal. London (1976)*, p. 1024, The Chemical Society, London (1977).
26. G. V. Smith and M. Musow: *J. Catal.* 60, 184 (1979).
27. T. Harada, S. Onaka, A. Tai and Y. Izumi: *Chem. Lett.* 1131 (1977).
28. A. Hoek, H. W. Woerde, and W. M. H. Sachtler: *Proc. 7th Int. Congr. Catal. Tokyo (1980)*, p. 376. Elsevier/Kodansha, (1981).
29. L. H. Gross and R. Rys: *J. Org. Chem.* 39, 2429 (1974).
30. E. N. Lipyart, Y. L. Petrov and E. I. Klabunovskii: *Kinet. Katal.* 12, 1330 (1971).
31. Y. Nitta, F. Sekine, J. Sasaki, T. Imanaka and S. Teranishi: *Chem. Lett.* 541 (1981).
32. H. Ozaki: *Bull. Chem. Soc. Jpn.* 51, 257 (1978).
33. E. N. Lipgart, Y. L. Petrov and E. I. Klabunovskii: *Kinet. Katal.* 12, 1491 (1971).
34. Y. Orito, S. Niwa and S. Imai: *Yuki Gosei Kagaku Kyokai Shi* 34, 236 (1976).
35. T. Harada and Y. Izumi: *Chem. Lett.* 1195 (1978).
36. F. Higashi, T. Ninomiya and Y. Izumi: *Bull. Chem. Soc. Jpn.* 44, 1333 (1971).
37. Y. Izumi, H. Takizawa, K. Nakagawa, R. Imamura, M. Imaida, T. Ninomiya and S. Yajima: *Bull. Chem. Soc. Jpn.* 43, 1792 (1970).
38. A. Tai, T. Harada, Y. Hiraki and S. Murakami: *Bull. Chem. Soc. Jpn.* 56, 1414 (1983).
39. M. Nakahata, M. Imaida, H. Ozaki, T. Harada and A. Tai: *Bull. Chem. Soc. Jpn.* 55, 730 (1982).
40. A. Tai, M. Nakahata, T. Harada, Y. Izumi, S. Kusumoto, M. Inage and T. Shiba: *Chem. Lett.* 1125 (1980).
41. A. Tai, K. Ito and T. Harada: *Bull. Chem. Soc. Jpn.* 54, 223 (1981).
42. K. Ito, T. Harada, A. Tai and Y. Izumi: *Chem. Lett.* 1049 (1979).
43. K. Ito, T. Harada and A. Tai: *Bull; Chem. Soc. Jpn.* 53, 3367 (1980).
44. Y. Hiraki, K. Ito, T. Harada and A. Tai: *Chem. Lett.* 131 (1981).
45. T. Osawa and T. Harada: *Chem. Lett.* 315 (1982).

46. T. Osawa, T. Harada, H. Ozaki, A. Tai and Y. Izumi: *Prepr. 50th Meet. Catal. Soc. Jpn. (1982)*, 4G 204.
47. Y. Orito, S. Imai and S. Niwa: *Nippon Kagaku Kaishi* 1118 (1979); Y. Orito, S. Imai and S. Niwa: *ibid.* 670 (1980); S. Niwa, S. Imai and Y. Orito: *ibid.* 137 (1982).
48. Y. Orito, S. Imai, S. Niwa and Ngulyen-Gia-Hung: *Yuki Gosei Kagaku Kyokai Shi* **37**, 173 (1979).
49. H. Ozaki, A. Tai, S. Kobatake, H. Watanabe and Y. Izumi: *Bull. Chem. Soc. Jpn.* **51**, 3559 (1978).
50. I. Yasumori, Y. Inoue and K. Okabe: *Proc. Int. Symp. Relation between Heterogeneous and Homogeneous Catal. Phenom.*, p. 41, Elsevier, Amsterdam (1975).
51. I. Yasumori, Y. Inoue and K. Okabe: *Prepr. Japan–USSR Symp. Catal., Almata (1975)*.
52. I. Yasumori, M. Yokozeki and Y. Inoue: *Faraday Discuss. Chem. Soc.* 72, 385 (1981).
53. H. M. Woerde, L. J. Bostelaar, A. Hoek and W. M. H. Sachtler: *J. Catal.* 76, 316 (1982).
54. Y. Nitta, S. Imanaka and S. Teranishi: *J. Catal.* 80, 31 (1983).
55. A. Tai, H. Watanabe and T. Harada: *Bull. Chem. Soc. Jpn.* **52**, 1468 (1979).
56. T. Harada: *Bull. Chem. Soc. Jpn.* 48, 3236 (1975).
57. T. Harada: *Bull. Chem. Soc. Jpn.* **53**, 1019 (1980).
58. A. Hoek and W. M. H. Sachtler: *J. Catal.* 58, 276 (1979).
59. Y. Inoue, K. Okabe and I. Yasumori: *Bull. Chem. Soc. Jpn.* **54**, 613 (1981).
60. T. Tanabe: *Bull. Chem. Soc. Jpn.* 46, 1482 (1973); Y. Izumi: *Proc. Jpn. Acad.* **53**, 38 (1977).
61. A. Hatta and W. Suetaka: *Bull. Chem. Soc. Jpn.* 48, 2428 (1975).
62. M. J. Fish and D. F. Ollis: *J. Catal.* **50**, 353 (1977).
63. J. A. Groenewegen and W. M. H. Sachtler: *J. Catal.* 27, 369 (1972); *ibid.* **33**, 176 (1974).
64. A. Hatta, Y. Moriya and W. Suetaka: *Bull. Chem. Soc. Jpn.* **48**, 3441 (1975).
65. K. Tanabe and N. Uyeda: *Bull. Chem. Res. Kyoto Univ.* **52**, 616 (1974).
66. T. Harada, Y. Imachi, A. Tai and Y. Izumi: *Metal–Support and Metal–Additive Effects in Catalysis*, p. 377, Elsevier, Amsterdam (1982).
67. I. Yasumori: *Pure Appl. Chem.* **50**, 971 (1978).
68. E. I. Klabunovskii: *Proc. 7th Int. Congr. Catal. Tokyo (1980)*, discussion part of ref. 21.
69. J. A. Groenewegen and W. M. H. Sachtler: *J. Catal.* **38**, 501 (1975); J. A. Groenewegen and W. M. H. Sachtler: *Proc. 6th Int. Congr. Catal. London (1976)*, p. 1014, The Chemical Society, London (1977).
70. C. H. Bartholomew, R. B. Pannell and J. L. Butler: *J. Catal.* **65**, 335 (1980).
71. E. I. Klabunovskii, A. A. Vedenyapin, E. I. Krapeiskaya and V. A. Parrlon: *Proc. 7th Int. Congr. Catal. Tokyo (1980)*, p. 390, Elsevier/Kodansha, (1981).
72. S. Okamoto, T. Harada and A. Tai: *Bull. Chem. Soc. Jpn.* **52**, 2670 (1979).
73. A. Tai, K. Tsukioka, H. Ozaki, T. Harada and Y. Izumi: *Chem. Lett.* 2083 (1984).

INDEX